Synthesis and Application of Nanoparticles in Novel Composites

Synthesis and Application of Nanoparticles in Novel Composites

Guest Editors

Edgar O'Rear
Fernando Esteban Florez

MDPI

Basel • Beijing • Wuhan • Barcelona • Belgrade • Novi Sad • Cluj • Manchester

Guest Editors
Edgar O'Rear
School of Chemical, Biological and Materials Engineering
University of Oklahoma
Norman, OK
USA

Fernando Esteban Florez
Department of Restorative Sciences
University of Oklahoma Health Sciences Center
Oklahoma City, OK
USA

Editorial Office
MDPI AG
Grosspeteranlage 5
4052 Basel, Switzerland

This is a reprint of the Special Issue, published open access by the journal *Nanomaterials* (ISSN 2079-4991), freely accessible at: https://www.mdpi.com/journal/nanomaterials/special_issues/nanoparticles_composites.

For citation purposes, cite each article independently as indicated on the article page online and as indicated below:

Lastname, A.A.; Lastname, B.B. Article Title. *Journal Name* **Year**, *Volume Number*, Page Range.

ISBN 978-3-7258-5389-2 (Hbk)
ISBN 978-3-7258-5390-8 (PDF)
https://doi.org/10.3390/books978-3-7258-5390-8

Contents

About the Editors

Edgar O'Rear

Edgar Allen O'Rear is the Francis W. Winn Professor in the School of Sustainable Chemical, Biological and Materials Engineering at the University of Oklahoma. He received advanced degrees in Organic Chemistry (S.M.) from M.I.T. and in Chemical Engineering from Rice University (Ph.D.). His research centers on biomedical engineering with an emphasis on the injury to blood cells by non-physiologic forces and on materials such as those incorporating carbon nanotubes modified by admicellar polymerization. With his students, he has published 200 articles in the literature. His current projects include the production of extracellular vesicles with extensional flow and the composition of quinary high-entropy alloys for optimal performance. Dr. O'Rear has held several leadership positions during his career as an NSF Program Director, as the Associate Dean for Research in the College of Engineering, as the Director of the University of Oklahoma Bioengineering Center and as the Chair of the Faculty Senate. His honors include the Regents' Award for Superior Research and the Regents' Award for Superior Teaching. He is a Fellow of both AIChE and AIMBE.

Fernando Esteban Florez

Fernando L. Esteban Florez is a tenured Associate Professor and the Head of the Dental Biomaterials division, as well as the Director of Faculty Development. He earned his M.Sc. and Ph.D. in Restorative Dentistry from São Paulo State University. In March 2012, he joined the University of Oklahoma Health Sciences Center College of Dentistry as a postdoctoral fellow. He has received several awards, including the OUHSC Provost's Research Award for Junior Faculty in 2020, the OUCOD Outstanding Service Award in 2021 and the OUCOD Outstanding Professorial Achievement Award in 2023. Dr. Florez specializes in nanofabrication, the development of nanofilled polymers and advanced material characterization techniques such as AFM, Dual-FIB SEM, SASX, SANS, ToF-SIMS and He-Ion microscopy. He leads the Medical and Dental Applications team in CUPI2D, a $33.6 million project funded by the U.S. Department of Energy, aimed at developing a novel multi-role neutron beamline for material characterization. His research focuses on improving oral health by reducing secondary caries and extending the service life of polymer-based adhesive restorations. Dr. Florez has developed high throughput bioluminescence assays to evaluate the viability of *Streptococcus mutans* biofilms in real-time. Since joining the faculty in July 2015, he has served as the principal investigator on seven grants, with significant funding from various local, state, national and international sources. Dr. Florez has filed four international patents, published 50 peer-reviewed manuscripts and three book chapters and presented his research at numerous national and international conferences. He has also delivered 12 international presentations as a featured or keynote speaker.

Preface

The *Nanomaterials* Special Issue "Synthesis and Application of Nanoparticles in Novel Composites" attracted submissions across diverse areas of technology, confirming their ongoing value to societal needs and their potential for more in the future. The topics covered range from novel drug delivery systems to specialized dental materials to improved cementitious composites for the construction of roads and buildings. A common theme in the articles is the importance of the composition and structure of the composite nanoparticle or the nanoparticle fillers in a matrix. Two articles describe the incorporation of prophylactic and therapeutic agents in nanoparticles for controlled release based on acidity. Popova et al. present the synthesis of a calcium carbonate composite with doxorubicin for the pH-dependent release of the anticancer agent, while Zhu et al. employ chitosan nanoparticles to protect histatins from the proteolytic environment of the oral cavity for dental hygiene. Additionally, the use of doped composite nanoparticles in dental procedures is also reported in this issue. Gels with nitrogen and fluorine containing nano-titanium dioxide enhance radiative bleaching treatments with hydrogen peroxide while reducing adverse effects on the mineral ratio in the enamel (Kury). Doping allows for the use of visible light (violet) instead of UV light. Other studies in this Special Issue report results on the properties of dental adhesive resins containing doped titanium dioxide nanoparticles which offer an antibacterial feature. Hong et al. found that the incorporation of titanium dioxide in bonding agents at higher concentrations improved color stability while retaining effective shear bond strength. After examining the conversion, biaxial flexural strength, surface roughness, elastic modulus and action against *Streptococcus mutans* of modified dental adhesives, Hiers et al. concluded that a composite adhesive with doped titanium dioxide would increase the service life of tooth restorations.

Nanomaterials are known to improve the mechanical strength of cement and mortars in part by reducing the void size and space. This reflects the results reported by Kongsat et al., who have demonstrated that the size and size distribution of hematite nanoparticles affected the compressive strength of fly ash blended cement. With the high surface area of nanomaterials, the surface free energy can be a significant factor in processing (e.g. dispersibility) and in bulk properties dependent on the interphase region (e.g., mechanical strength). Li et al. describe the hydrothermal synthesis of a complex material comprised of graphene oxide nanoribbons, functionalized carbon nanotubes and graphene oxide and report its superior strength performance in a cement paste after hydration for 28 days. A critical review of the literature by O'Rear et al. confirms the need for the functionalization of carbon nanotubes for optimal increases in mechanical strength.

Lastly, two other contributions to the Special Issue illustrate the value of surface modifications in other applications. A strategy is presented for conjugation oligopeptides within porous silica nanoparticles for the purpose of selective peptide-based separations (Khan). Fajstavr prepared nanocomposite PEGylated gold nanoparticles with the use of N-acetyl-L-cyteine to immobilize them on polyethylene naphthalate.

The editors would like to thank all authors for their contributions to this Special Issue. We also express our appreciation of the staff at *Nanomaterials*, especially Greta Zhang, for their assistance in processing submissions.

Edgar O'Rear and Fernando Esteban Florez
Guest Editors

nanomaterials

MDPI

Review

Mechanical Strength and Conductivity of Cementitious Composites with Multiwalled Carbon Nanotubes: To Functionalize or Not?

Edgar A. O'Rear [1,2,*], Suthisa Onthong [1,3,4] and Thirawudh Pongprayoon [3,4]

1 School of Sustainable Chemical, Biological and Materials Engineering, University of Oklahoma, Norman, OK 73019, USA; suthisa.on@gmail.com
2 Institute for Applied Surfactant Research, University of Oklahoma, Norman, OK 73019, USA
3 Department of Chemical Engineering, Faculty of Engineering, King Mongkut's University of Technology North Bangkok, Bangkok 10800, Thailand; thirawudh.p@eng.kmutnb.ac.th
4 Center of Eco-Materials and Cleaner Technology, King Mongkut's University of Technology North Bangkok, Bangkok 10800, Thailand
* Correspondence: eorear@ou.edu

Abstract: The incorporation of carbon nanotubes into cementitious composites increases their compressive and flexural strength, as well as their electrical and thermal conductivity. Multiwalled carbon nanotubes (MWCNTs) covalently functionalized with hydroxyl and carboxyl moieties are thought to offer superior performance over bare nanotubes, based on the chemistry of cement binder and nanotubes. Anionic carboxylate can bind to cationic calcium in the hydration products, while hydroxyl groups participate in hydrogen bonding to anionic and nonionic oxygen atoms. Results in the literature for mechanical properties vary widely for both bare and modified filler, so any added benefits with functionalization are not clearly evident. This mini-review seeks to resolve the issue using an analysis of reports where direct comparisons of cementitious composites with plain and functionalized nanotubes were made at the same concentrations, with the same methods of preparation and under the same conditions of testing. A focus on observations related to the mechanisms underlying the enhancement of mechanical strength and conductivity helps to clarify the benefits of using functionalized MWCNTs.

Keywords: cement; composites; carbon nanotubes; functionalization; concrete; compressive strength; mortar

1. Introduction

Cementitious materials played a key role in the urbanization of modern economies, with utilization in the construction of commercial and residential buildings. Today, concrete remains an essential element in building construction and in the development of transportation infrastructure as an integral component to bridges, overpasses and pavement, such as in the interstate system of the United States. The importance of these materials has meant ongoing efforts to improve the properties of binder and concrete and to develop new uses with sustainability in mind. Many recent advances in the properties of cement paste, mortar and concrete have been through the incorporation of nanoparticles (NPs). Several reviews describe the benefits of composites with NPs of SiO_2, TiO_2, Al_2O_3 and other oxides which can increase strength under compression and tension [1–6].

Carbon nanotubes stand out as a subgroup of NPs because of their distinctive structure and remarkable features including high aspect ratio and electrical conductivity. Composed of an array of unsaturated hexagonal carbon rings, carbon nanotubes exist in zigzag, armchair and chiral forms (Figure 1). Multi-walled carbon nanotubes (MWCNTs) consist of a series of concentric, coaxial cylinders with diameters typically 20–50 nm, much larger than the 1 nm diameter of a single-walled carbon nanotube. These small diameters mean a high aspect ratio for lengths as high as 10 microns or more [7,8].

Due to lower cost and greater resistance to compression, MWCNTs are more widely used in composites than in single-wall forms. With tensile strengths on the order of 50–150 GPa [9,10], far greater than steel, carbon nanotubes offer the prospect of mitigating susceptibility to failure under tension, a well-known weakness of cementitious composites. Indeed, carbon nanotubes, with their high tensile strength, have shown the ability to reduce cracking and improve the flexural stress of mortars and cement pastes.

In addition to their physical strength, MWCNTs have high electrical and thermal conductivities of 2.2×10^4 S/cm [11] and 3×10^3 W/m K [12]. Their electrical conductivity supports features like electromagnetic shielding, dissipation of static charge and the piezocharacter that forms the basis for structural health monitoring. With high thermal conductivity, carbon nanotubes may help to reduce the degradation of structures at extreme temperatures, though the temperature dependence of MWCNTs thermal conductivity poses a challenge for this application [13]. These extraordinary properties make MWCNTs promising components for composites.

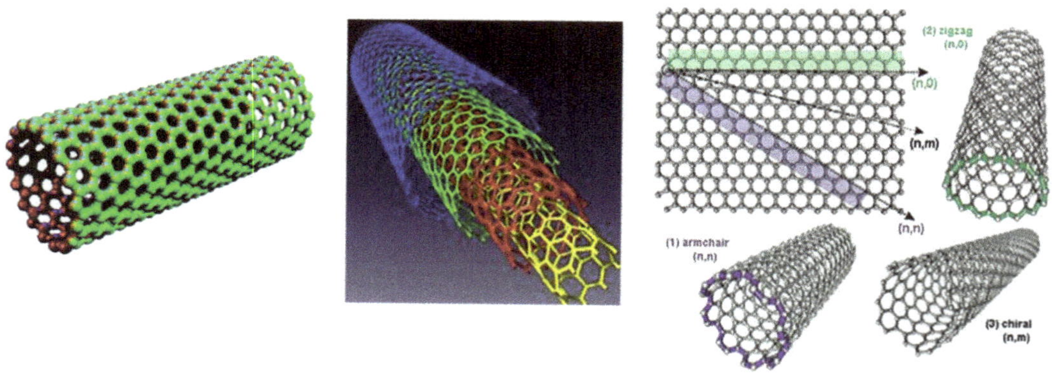

Figure 1. Multi-walled carbon nanotubes have been used more frequently than single-walled due to lower cost. Properties like electrical conductivity vary with the zig-zag, armchair and chiral chemical structures [14].

Interest in cementitious materials with MWCNTs has been high, with many articles appearing, and comprehensive reviews available [15–17]. Researchers have focused on the mechanical properties of compressive and flexural strength in cement paste and mortar composites with carbon nanotubes. However, findings in the literature have been inconsistent as to the extent of any benefit or even whether there is a benefit with functionalization. The addition of plain MWCNTs to cement paste [18–22] or mortar [23] has been shown to raise the compressive strength, while others have observed decreases or no change for cement paste and mortar composites [24–28]. Flexural strength with pristine MWCNTs at concentrations of 0.025–1.0 wt% of cement can increase as much as a factor of 2–3 [8] for cement paste, though increases are more typically 20–30% [18,20,29–31], and to a degree in mortar [23,32] as well. Yet, there are also reports with little or no improvement in flexural strength [33,34] for concentrations of 0.015–0.5 wt%.

Understanding the nature of carbon nanotubes led some researchers to propose functionalization of the nanotubes for better performance. The proposal considered the hydrophobicity of bare nanotubes, the interaction of MWCNTs with hydration products of cement, the aqueous processing conditions required and the mechanisms underlying the enhanced strength of cement composites. Chemical modification has shown promise, with the most common methods introducing hydroxyl and carboxylic acid groups to improve properties. Tests on composites with functionalized MWCNTs have yielded increases as high as 50% in the compressive strength for both cement paste [10,27,33,35] and mortar [22,36–39].

Similar levels of improvement for flexural strength are known for paste [10,33,40,41] and mortar [27,36,38,41]. While the modification of MWCNTs can certainly be rationalized, there are potential problems. The chemical reactions can damage the structure of the carbon nanotubes, weakening the properties that undergird the potential benefits. Some investigators found that functionalization yields comparable or worse performance of mechanical strength than bare MWCNTs [26,39,42,43]. Altogether, these reports create uncertainty about the merits of using functionalized carbon nanotubes.

It is clearly possible to improve mechanical strength with both nonfunctionalized and functionalized carbon nanotubes. There is, however, great variability in the results with pristine MWCNTs, and in the findings with functionalized MWCNTs as well, so reported values for compressive and flexural strength do not consistently and clearly favor functionalization. As such, the benefits of functionalization are not readily apparent. The failure in many investigations of functionalized nanotubes to include pristine nanotube composite samples for comparison has clouded the picture and not helped to resolve the question.

The purpose of this study is to examine the hypothesis that nanotube covalent functionalization improves mechanical and other properties of cementitious composites better than pristine nanotubes. The question is significant because the treatment of added carboxylic acid and alcohol substituents has been widely employed with the objective of improving dispersion and performance. For this review, an extensive search of the literature was conducted to find studies of cementitious composites incorporating functionalized carbon nanotubes. Of the many reports in the literature, articles were selected for inclusion and analysis based on a single, simple objective criterion of whether a pristine nanotube control was included. The approach yielded results for samples with pristine and functionalized carbon nanotubes from the same lab. These studies are more likely to have MWCNTs before and after functionalization of comparable structure and purity, consistent methods of dispersion, and similar techniques for sample preparation and testing. The scope is not limited to how functionalization affects mechanical strength. It includes the effect of functionalization on the mechanisms and properties underlying improved mechanical strength with MWCNTs with the same criterion applied in the selection of studies. Findings for properties of cementitious composites associated with strength (e.g., porosity and aggregation) are insightful. The results produce a clearer picture of the functionalization and mechanical strength of cementitious composites. As for electrical properties, fewer studies exist where direct comparisons have been conducted, but findings from direct comparisons of electrical conductivity are also presented. The addition of pristine or functionalized MWCNTs does reduce resistance or increase the conductivity of cement pastes [44–49] and mortars [50,51], though not always [52]. To affect this property, the MWCNTs must reach a minimum concentration for connectivity. The high aspect ratio of the nanotubes acts to lower the concentration of the percolation threshold over other fillers [53].

2. Functionalization of Multiwalled Carbon Nanotubes

The term functionalization has been used broadly to represent various methods to cause disaggregation of carbon nanotubes leading to their suspension in water with the objective of effective mixing to obtain distributed, better integrated nanotubes in the final composite. Functionalization changes the surface of the carbon nanotubes and helps to address the challenge that arises from the hydrophobic chemical nature of MWCNTs. Hydrophobicity impedes the distribution of the nanotubes in water, and ultimately in the composite, as the aqueous media is combined with cement and mixed. Not only are nanotubes not drawn into aqueous media, but their high surface energy with strong van der Waals attractive forces causes them to resist separation and to reaggregate during mixing. Distinct nanotubes in process water help ultimately to achieve separated nanotubes throughout the cementitious composite after processing. This is important because the disaggregated nanotubes, whether functionalized or not, contribute to mechanisms underlying strength improvement.

2.1. Methods of Functionalization

There is some ambiguity in what functionalization means to different researchers. Broadly applied, it includes the use of surfactants and polymers to improve the dispersion of MWCNTs in aqueous media to achieve better mechanical properties of pastes and mortar. Many, however, consider it limited to methods leading to the addition of covalently bonded polar substituents to the nanotubes. In this review, our use of functionalization focuses on nanotubes with covalently linked polar substituents. A common method to add alcohol and/or carboxylic acid groups employs a mixture of nitric and sulfuric acids. Other oxidizing conditions for functionalization have been treatment with $KMnO_4/H_2SO_4$ [34], exposure to a low-temperature oxygen plasma [34] and reaction with ozone [49]. Singer et al. compare methods of oxidation and recommend a milder treatment with hydrogen peroxide [54]. Harsh conditions can cause degradation, so purification procedures after oxidation can be important. Removal of carboxylated carbonaceous fragment byproducts from the oxidation reaction has been reported to have a great effect on performance [25]. In some studies, the method of functionalization is not specified, only a supplier is indicated. MWCNTs purchased from companies may be superior in quality, with more time invested by the company in optimizing preparation and purification after functionalization.

2.2. Surfactants

The challenges to the dispersion of nanotubes in water can be successfully overcome with surfactants and some polymers [55–57]. The choice and amount of surfactant are important [58]. For example, the block copolymer surfactant Pluronic F127 outperformed the anionic sodium dodecylbenzene sulfonate to yield improved flexural and compressive strengths of mortar loaded with single-walled carbon nanotubes by 7% and 19%, respectively [59]. The optimal concentration of Pluronic F127 was determined from an absence of aggregation in optical micrographs. Konsta-Gdoutos et al. [29] reported that the optimum surfactant-to-CNTs mass ratio for a uniform dispersion is four, while Zou et al. [30] suggested a mass ratio of eight for achieving a good dispersion. Use of surfactants generally requires sonication, which means significant energy, extra time and possible damage to the nanotubes [58,60]. Li et al. [61] presented the mechanisms of surfactant-modified MWCNTs treated by ultrasound in an aqueous phase for the comparison of three different types of surfactants: anionic surfactant (sodium dodecyl sulfate, SDS), cationic surfactant (cetyltrimethylammonium bromide, CTAB) and nonionic surfactant (octylphenol polyoxyethylene ether, OP-10), as shown in Figure 2. The surfactants were employed to disperse the MWCNTs in aqueous solution with the assistance of ultrasonic waves. In comparison to anionic and nonionic surfactants (SDS and OP-10), the cationic surfactant (CTAB) yielded better dispersity of MWCNTs, because of greater electrostatic repulsion and higher steric resistance [61].

While it is possible to achieve favorable results for mechanical properties with surfactant [58,60], poor results are often obtained [25,28,62]. Surfactants may introduce voids, and weaken cementitious materials [63]. Siddique et al. found that the use of the surfactant sodium dodecyl sulfate in the preparation of an MWCNT mortar composite led to foaming and voids with less compressive strength [62]. Others have attributed a decrease in performance to surfactants carrying excess water into the cement matrix.

Treatment of MWCNTs to form -OH and -COOH groups would seem to offer advantages over surfactants. Functionalization of this type enhances the ability of water to break up aggregates with little use of sonication. Moreover, surfactants do not provide a covalent bond to the CNTs, so they are more susceptible to pullout under tension and less able to transfer forces within the cementitious composite [41]. Functionalization provides a basis for bonding as detailed below.

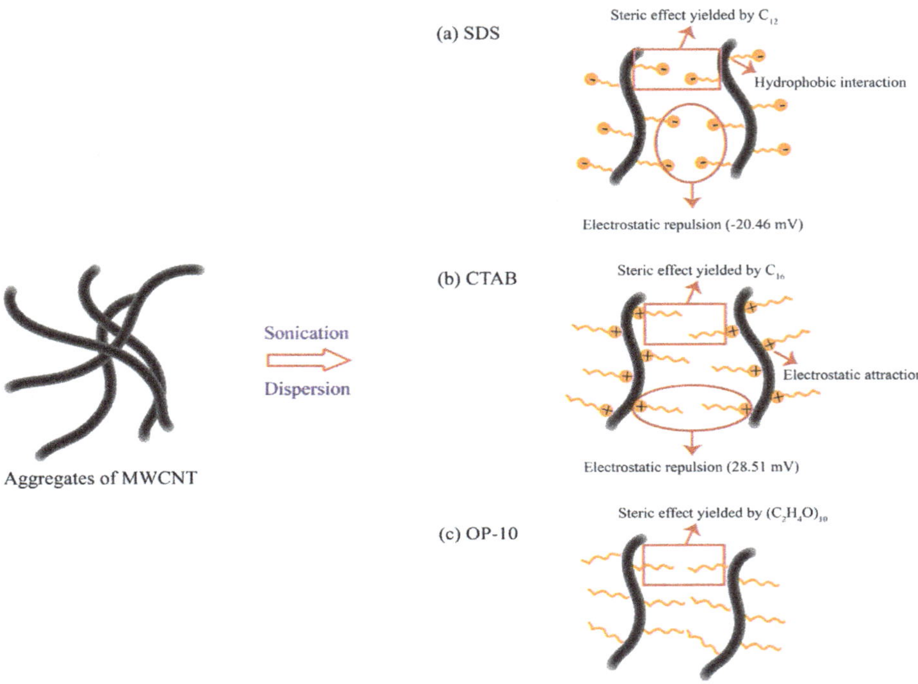

Figure 2. Interactions between MWCNT and surfactant with ultrasonic treatment [61]. Adsorbed surfactant on the surface of nanotubes acts to overcome van der Waals forces and facilitates disaggregation with ultrasound. Anionic surfactants like sodium dodecyl sulfate (SDS), and cationic surfactants like cetyltrimethylammonium bromide (CTAB), work by repulsion of like charges and steric effects. The nonionic octylphenol othoxylate (OP-10) functions by steric interactions only.

3. Mechanisms of Increased Mechanical Strength with Nanotubes and the Effects of Functionalization

If the hypothesis that functionalization of MWCNTs improves mechanical strength of composites better than pristine nanotubes is correct, then the difference should be evident in factors and mechanisms tied to mechanical behavior. In the sections below, properties known to affect compressive and flexural strength are examined. Results from head-to-head comparisons are emphasized for support or rejection of the hypothesis.

3.1. Dispersion and Functionalization

Dispersion is important for all NPs, but it is a particular challenge for carbon nanotubes. Functionalization alone will not achieve adequate dispersion. Mechanical energy input is needed with sonication to disaggregate the nanotubes in aqueous media and mix to form pastes and composites. In the absence of a means to overcome van der Waals forces, the addition of MWCNTs to cementitious materials can have a detrimental effect on mechanical properties. They must be spread throughout a cementitious composite to be effective, and that has proven to be difficult. For example, Malikov et al. observed a 27.8% drop in compressive strength with 0.01% pristine nanotubes that dispersed poorly in concrete [26].

By sonication of nonfunctionalized MWCNTs, Kumar et al. obtained a 15% increase in compressive strength and a 36% increase in tensile strength of a cement paste, that decreased at higher concentrations of nanotubes [20]. At higher concentrations, the drop in performance was attributed to poorer, nonuniform dispersion, which was evident in SEM imaging. Paste and mortar composites of both pristine and functionalized MWCNTs often exhibit decreasing compressive strength (Table 1A,B) above an intermediate weight percentage. Similar results have been reported for flexural strength (Table 2A,B). The reduction in mechanical strength occurring at higher concentrations is caused by reaggregation leading to agglomeration (Figure 3). In contrast, the effect of agglomeration on conductivity is not clear, due to the inconsistent trends of electrical conductivity (Table 3), though some data seem to show increasing conductivity over a wider concentration range of 0.05–2.0% [44,48,64]. Comparison is difficult, with concentrations in these studies based variously on the volume of cement paste, the weight percent of cement and the weight percent of the nanocomposite.

Functionalization improves dispersion. In comparative studies, functionalized MWCNTs by various methods displayed more effective dispersibility than untreated MWCNTs [10,25,34]. Superior dispersibility of modified MWCNTs in water has been demonstrated by UV-vis absorption, while SEM imaging of microstructure in the composite shows better dispersion stability in the cement matrix [34].

Whether functionalized or not, nanotube geometry and dimensions affect dispersion. Compared to short ones, longer, high aspect ratio nanotubes are more difficult to disperse [10,33,65], which can contribute to clumps of MWCNTs or agglomeration (Figure 3). Even so, long MWCNTs are more effective in improving strength, albeit short nanotubes at higher concentrations can achieve similar performance [8,33,66]. The diameter of the nanotubes is also a factor. Manzur et al. examined the compressive strength of cement composites for seven different sizes of nanotubes, with outer diameters (OD) ranging from less than 8 nm to greater than 50 nm and with lengths of 10–20 μm or 10–30 μm [39]. Both pristine and bare nanotube samples were tested. Composite compressive strength was greater for OD less than 20 nm with the smallest diameter functionalized sample yielding the highest value of 42 MPa for a 0.3% loading at day 28. The result was explained with smaller size nanotubes filling nanopore void space in the cement matrix more efficiently. At the same time, smaller diameters mean more surface area and a greater need to overcome surface energy requirements for dispersion.

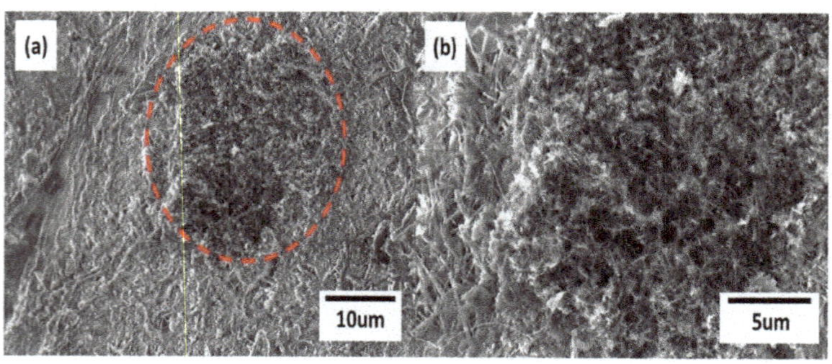

Figure 3. Agglomeration of bare MWCNTs in a cement paste composite [67]. MWCNTs reaggregate at higher concentrations during mixing to form agglomerates. The presence of these structures weakens the composite. An agglomerate in a composite is outlined in red (**a**) and shown at higher magnification (**b**).

Manzur et al. also concluded that the treatment of MWCNTs with a mixture of sulfuric and nitric acids introduced functional groups to reduce agglomeration and more uniformly disperse filler [39]. Similarly, Cui et al. found that functionalization led to better dispersion, significantly elevating the compressive and flexural strength of both long and short MWCNTs, relative to the improvement seen with pristine MWCNTs [10]. Hydroxyl-substituted nanotubes proved superior to carboxyl substituents in their study, though both produced excellent results.

Table 1. (**A**) Comparison of compressive strength for cement paste with functionalized, pristine or no MWCNTs *. (**B**) Comparison of compressive strength for cement mortar with functionalized, pristine or no MWCNTs *.

(A)						
Materials Cure Time	**Compression Control**	**Compression Pristine**	**Compression Functionalized**	**Concentration Range †**	**Functional Group or Method of Functionalization ‡**	**Ref.**
MWCNTs Cement Paste (silica fume)@14d	~45 MPa	~52 MPa	~75 MPa (no surf) ~72 MPa (with surf)	0.1%	Nitric/Sulfuric Acid	[35]
MWCNTs Cement Paste Cure time?	36 MPa	13 MPa	66 MPa 71 MPa 65 MPa 59 MPa	0.02% of paste 0.03% 0.05% 0.09%	Nitric and Sulfuric Acids	[25]
MWCNTs Cement Paste@28d	96.0 MPa	96.5 MPa 93.8 MPa	101.1 MPa 100.8 MPa	0.05% of paste 0.1%	-COOH	[63]
Long MWCNTs Cement Paste@28d	~95 MPa	~115 MPa ~110 MPa ~112 MPa	~125 MPa ~158 MPa ~120 MPa	0.1% 0.5% 0.8%	-OH	[10]
Long MWCNTs Cement Paste@28d	~95 MPa	~115 MPa ~110 MPa ~112 MPa	~130 MPa ~145 MPa ~145 MPa	0.1% 0.5% 0.8%	-COOH	[10]
Short MWCNTs Cement Paste@28d	~95 MPa	~100 MPa ~115 MPa ~75 MPa	~90 MPa ~155 MPa ~95 MPa	0.1% 0.5% 0.8%	-OH	[10]
Short MWCNTs Cement Paste@28d	~95 MPa	~100 MPa ~115 MPa ~75 MPa	~95 MPa ~135 MPa ~85 MPa	0.1% 0.5% 0.8%	-COOH	[10]
MWCNTs Cement Paste@28d	54.1 MPa	58.9 MPa 67.9 MPa 64.7 MPa 57.9 MPa 57.6 MPa	64.1 MPa 66.6 MPa 56.4 MPa 55.3 MPa 53.4 MPa	0.015% 0.05% 0.1% 0.25% 0.5%	-COOH	[33]
MWCNTs Cement Paste@28d	54.1 MPa	58.9 MPa 67.9 MPa 64.7 MPa 57.9 MPa 57.6 MPa	64.4 MPa 66.1 MPa 62.1 MPa 57.6 MPa 57.7 MPa	0.015% 0.05% 0.1% 0.25% 0.5%	-OH	[33]
MWCNTs Cement Paste Cure?	48.6 MPa	50.8 MPa	54.5 MPa	0.1%	Nitric/Sulfuric Acid	[68]

(B)						
Materials Cure Time	**Compression Control**	**Compression Pristine**	**Compression Functionalized**	**Concentration Range †**	**Functional Group or Method of Functionalization ‡**	**Ref.**
MWCNTs Cement Mortar@28d	~46 MPa	~51 MPa ~49 MPa ~49 MPa	~52 MPa ~54 MPa ~50 MPa	0.05% 0.1% 0.2%	-COOH	[34]
MWCNTs Cement Mortar@28d	~46 MPa	~51 MPa ~49 MPa ~49 MPa	~53 MPa ~54 MPa ~57 MPa	0.05% 0.1% 0.2%	Low temperature plasma	[34]

Table 1. *Cont.*

(**B**)						
Materials Cure Time	**Compression Control**	**Compression Pristine**	**Compression Functionalized**	**Concentration Range** †	**Functional Group or Method of Functionalization** ‡	**Ref.**
MWCNTs Cement Mortar@28d	~37 MPa	~38 MPa ~38 MPa ~39 MPa ~37 MPa	~40 MPa ~40 MPa ~41 MPa ~35 MPa	0.1% 0.2% 0.3% 0.5%	-COOH	[40]
MWCNTs Cement Mortar@28d	~37 MPa	~38 MPa ~38 MPa ~39 MPa ~37 MPa	~40 MPa ~40 MPa ~40 MPa ~36 MPa	0.1% 0.2% 0.3% 0.5%	-COOH	[40]
MWCNTs Cement Mortar@28d	~37 MPa	~38 MPa ~38 MPa ~39 MPa ~37 MPa	~40 MPa ~42 MPa ~42 MPa	0.1% 0.2% 0.3% 0.5%	-COOH	[40]
MWCNTs Cement Mortar@28d	~37 MPa	~38 MPa ~38 MPa ~39 MPa ~37 MPa	~40 MPa ~40 MPa ~36 MPa ~33 MPa	0.1% 0.2% 0.3% 0.5%	-COOH	[40]
MWCNTs Cement Mortar@28d	72.1 MPa	88.8 MPa	82.1 MPa	0.05%	-COOH	[69]
MWCNTs Cement Mortar@28d	72.1 MPa	88.8 MPa	85.3 MPa	0.05%	-OH	[69]
MWCNTs Cement Mortar@14d	~46 MPa	~50 MPa	~51 MPa ~50 MPa	0.15%	-COOH -OH	[22]
MWCNTs Mortar with 30% fumed sllica@14d	~59 MPa	~56 MPa	~71 MPa ~69 MPa	0.15%	-COOH -OH	[22]
MWCNTs Mortar with 30% fumed sllica@14d	~59 MPa	~56 MPa	~71 MPa ~69 MPa	0.15%	-COOH -OH	[22]

* The tilde symbol "~" indicates values estimated from graphical presentations of results. † Weight percent of cement. ‡ Functional group if specified or method of modification otherwise.

Table 2. (**A**) Comparison of flexural strength for cement pastes with functionalized, pristine or no MWCNTs *. (**B**) Comparison of flexural strength for cement mortar with functionalized, pristine or no MWCNTs *.

(**A**)						
Materials Cure Time	**Flexural Control**	**Flexural Pristine**	**Flexural Functionalized**	**Concentration Range** †	**Functional Group or Method of Functionalization** ‡	**Ref.**
Long MWCNTs Cement Paste@28d	~8.5 MPa	~13 MPa ~11.5 MPa ~12.5 MPa	~9 MPa ~11 MPa ~11 MPa	0.1% 0.5% 0.8%	-OH	[10]
Long MWCNTs Cement Paste@28d	~8.5 MPa	~13 MPa ~11.5 MPa ~12.5 MPa	~13 MPa ~12 MPa ~13.5 MPa	0.1% 0.5% 0.8%	-COOH	[10]

Table 2. *Cont.*

(A)						
Materials Cure Time	**Flexural Control**	**Flexural Pristine**	**Flexural Functionalized**	**Concentration Range †**	**Functional Group or Method of Functionalization ‡**	**Ref.**
Short MWCNTs Cement Paste@28d	~8.5 MPa	~9 MPa ~10.5 MPa ~9 MPa	~11 MPa ~15.5 MPa ~8 MPa	0.1% 0.5% 0.8%	-OH	[10]
Short MWCNTs Cement Paste@28d	~8.5 MPa	~9 MPa ~10.5 MPa ~9 MPa	~10.5 MPa ~13.5 MPa ~10 MPa	0.1% 0.5% 0.8%	-COOH	[10]
MWCNTs Cement Paste (silica fume)@14d	~4.3 MPa	~4.8 MPa (with surf)	~6.5 MPa (no surf) ~5.7 MPa (with surf)	0.1% of paste	Sulfuric and nitric acid	[35]
MWCNTs Cement Paste Cure?	1.5 MPa	1.6 MPa	4.7 MPa	0.1%	Sulfuric and nitric acid	[68]
MWCNTs Cement Paste@28d	7.7 MPa	9.0 MPa 10.3 MPa 9.4 MPa 8.8 MPa 8.3 MPa	8.7 MPa 9.3 MPa 8.4 MPa 7.1 MPa 6.4 MPa	0.015% 0.05% 0.1% 0.25% 0.5%	-COOH	[33]
MWCNTs Cement Paste@28d	7.7 MPa	9.0 MPa 10.3 MPa 9.4 MPa 8.8 MPa 8.3 MPa	8.7 MPa 9.3 MPa 8.8 MPa 8.1 MPa 6.7 MPa	0.015% 0.05% 0.1% 0.25% 0.5%	-OH	[33]

(B)						
Materials Cure Time	**Flexural Control**	**Flexural Pristine**	**Flexural Functionalized**	**Concentration Range †**	**Functional Group or Method of Functionalization ‡**	**Ref.**
MWCNTs Cement Mortar@28d	~7 MPa	~7.9 MPa ~8.2 MPa ~7.2 MPa	~8.5 MPa ~9.6 MPa ~9.0 MPa	0.05% 0.1% 0.2%	-COOH	[34]
MWCNTs Cement Mortar@28d	~7 MPa	~7.9 MPa ~8.2 MPa ~7.2 MPa	~8.9 MPa ~8.1 MPa ~8.3 MPa	0.05% 0.1% 0.2%	Low temperature plasma	[34]
MWCNTs Cement Mortar@28d	10.3 MPa	13.3 MPa	12.1 MPa	0.05%	-COOH	[69]
MWCNTs Cement Mortar@28d	10.3 MPa	13.3 MPa	11.6 MPa	0.05%	-OH	[69]
MWCNTs Mortar with 30% fumed sllica@14d	~4.4 MPa	5.9 MPa	~6.6 MPa ~6.4 MPa	0.15%	-COOH -OH	[22]

* The tilde symbol "~" indicates values estimated from graphical presentations of results. † Weight percent of cement. ‡ Functional group if specified or method of modification otherwise.

Table 3. Comparison of electrical conductivities for cement pastes with functionalized, pristine or no MWCNTs *.

Materials Cure Time	Electrical Control	Electrical Pristine	Electrical Functionalized	Concentration Range †	Functional Group or Method of Functionalization ‡	Ref.
MWCNTs Cement Paste@28d	~2.03 × 10^{-7} S/cm	~2.54 × 10^{-7} S/cm ~3.09 × 10^{-7} S/cm ~3.97 × 10^{-5} S/cm ~9.02 × 10^{-4} S/cm	~4.46 × 10^{-7} S/cm ~8.15 × 10^{-7} S/cm ~8.56 × 10^{-4} S/cm ~5.20 × 10^{-3} S/cm	0.05% 0.1% 0.3% 0.5%	-C=O from PVAc -NH from PIn (Admicellar polymerization)	[45]
MWCNTs Cement Paste@28d	~1.32 × 10^{-7} S/cm	~3.97 × 10^{-7} S/cm ~3.82 × 10^{-5} S/cm ~6.94 × 10^{-4} S/cm	~8.46 × 10^{-7} S/cm ~5.96 × 10^{-4} S/cm ~3.12 × 10^{-3} S/cm	0.1% 0.3% 0.5%	-C=O from PVAc -NH from PIn (Grafting polymerization)	[46]
Short MWCNTs Cement Paste@28d	~200 Ω·m	~155 Ω·m ~130 Ω·m ~179 Ω·m	~162 Ω·m ~117 Ω·m ~190 Ω·m	0.1% 0.5% 0.8%	-COOH	[48]
Short MWCNTs Cement Paste@28d	~200 Ω·m	~155 Ω·m ~130 Ω·m ~179 Ω·m	~151 Ω·m ~178 Ω·m ~159 Ω·m	0.1% 0.5% 0.8%	-OH	[48]
Long MWCNTs Cement Paste@28d	~200 Ω·m	~150 Ω·m ~145 Ω·m ~149 Ω·m	~140 Ω·m ~100 Ω·m ~90 Ω·m	0.1% 0.5% 0.8%	-COOH	[48]
Long MWCNTs Cement Paste@28d	~200 Ω·m	~150 Ω·m ~145 Ω·m ~149 Ω·m	~210 Ω·m ~120 Ω·m ~130 Ω·m	0.1% 0.5% 0.8%	-OH	[48]
MWCNTs Cement Paste@30d	-	~610 ΔR/Ω ~600 ΔR/Ω ~575 ΔR/Ω	~596 ΔR/Ω - -	0.1% 0.5% 2.0%	-COOH	[70]
MWCNTs Cement Paste@28d& 90d	17.16 Ω·m 401.07 Ω·m	15.13 Ω·m 291.03 Ω·m	12.91 Ω·m 207.47 Ω·m	1%	H_2SO_4-HNO_3	[49]
MWCNTs Cement Paste@28d & 90d	17.16 Ω·m 401.07 Ω·m	15.13 Ω·m 291.03 Ω·m	14.14 Ω·m 126.96 Ω·m	1%	Ozone (O_3)	[49]
MWCNTs Cement Paste@28d & 90d	17.16 Ω·m 401.07 Ω·m	15.13 Ω·m 291.03 Ω·m	14.73 Ω·m 50.35 Ω·m	1%	O_3-NaOH	[49]
MWCNTs Cement Paste@1d	3.7 × 10^6 Ω·m	3.6 × 10^6 Ω·m	3.0 × 10^6 Ω·m, 15 min 3.3 × 10^6 Ω·m, 30 min 1.9 × 10^6 Ω·m, 45 min 1.1 × 10^6 Ω·m, 60 min	0.1%	-COOH, Oxidation time	[68]

* The tilde symbol "~" indicates values estimated from graphical presentations of results. † Weight percent of cement. ‡ Functional group if specified or method of modification otherwise.

Findings for aspect ratio were consistent with the work of Mazur et al. Long MWCNTs (10–30 µm, outer diameter less than 8 nm) generally outperformed short nanotubes (0.5–2 µm, outer diameter less than 8 nm) in mechanical strength (Tables 1A and 2A). Ahmed et al. noted the strong influence of aspect ratio on dispersion and reinforcement and calculated theoretical nanotube spacing for ideal dispersion and uniform distribution in cement paste. This analysis showed that spacing decreased with aspect ratio as well as concentration [33]. While nanotubes with smaller aspect ratios are easier to disperse, the greater theoretical spacing means fewer are available at crack surfaces and there is less strength from MWCNT-matrix bonding. It was concluded that high aspect ratio MWCNTs are desirable if well dispersed [33].

Other methods, less common than covalent functionalization by oxidation, have been reported to enhance dispersion. Isopropanol can be used to separate nanotubes with sonication before suspending them in water [71]. A recent report describes the use of sodium hydroxide to disperse MWCNTs in composites [72]. Li et al. found synergism with the use of polyvinyl alcohol latex to help disperse hydroxy-modified MWCNTs [73]. These recent papers indicate ongoing interest in addressing dispersion.

In summary, functionalized MWCNTs enhance dispersity better than bare MWCNTs. This is important because improved mechanical properties with nanotubes depend on effective dispersion. Direct comparative studies by several groups have shown that MWCNTs with covalent functionalization by oxidation improved dispersion with concomitant superior strength. Consistency for long and short MWCNTs with theoretical analysis for the effect of aspect ratio helps to validate the results for dispersity.

3.2. Porosity and Functionalization

The void volume of a cementitious material is described by its porosity, a major determinant of compressive strength. An important property, porosity, as well as pore size distribution, can be determined by Mercury Intrusion Porosimetry (MIP) and by the digital analysis of Environmental SEM (ESEM) images [74]. In general, the addition of MWCNTs to composites causes a reduction in porosity [35,63,74,75]. This reduction in porosity with MWCNTs can be understood as their filling the small voids between hydration products of cement (Figure 4), as well as contributing to nucleation [10,18,29,40,76].

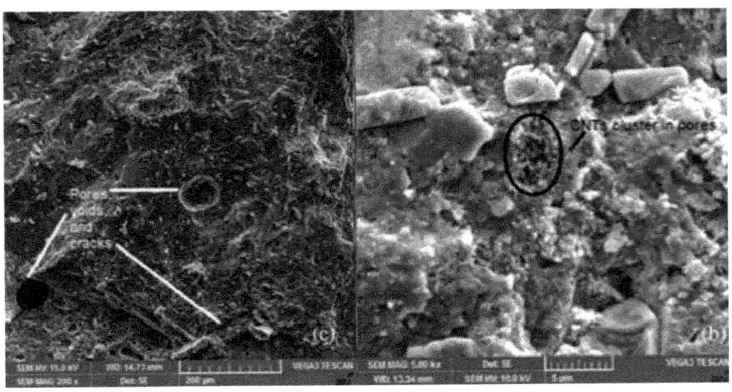

Figure 4. SEM images of mortar at 200× show pores (**c**), and at 5000× magnification (**b**) show a pore filled with plain MWCNTs [22]. MWCNTs can fill pores and reduce porosity to increase the compressive strength of cement pastes and mortars.

Mechanical strength is closely related to the pore structure of the composites [75]. Pores create sites for crack propagation and failure, so enhancing the effect of MWCNTs on porosity is a mechanism substantiating the benefits of functionalization. Kang et al. attributed the observed greater increase in compressive and tensile strength with acid-treated nanotubes to lower porosity compared to pristine nanotubes. They examined the porosity of cement paste composites containing either 0.1 wt% plain nanotubes or 0.1 wt% acid-treated nanotubes for porosity [35]. In a head-to-head comparison, the cement composite containing acid-treated, functionalized nanotubes was found to have lower porosity and greater strength than the one with plain nanotubes [35]. Hu measured the porosities of cement pastes with 0.1 wt% nanotubes and found that pristine MWCNTs led to a porosity of 24.84%, compared to 15.70% for 0.1 wt% functionalized MWCNTs [63]. The reduction in porosity corresponded with greater strength for functionalized MWCNTs [38,63].

Composite strength depends more on the total size distribution rather than just the total pore volume [63]; macropores with diameters greater than 50 nm affect strength more than smaller pores [43,63]. As such, it is notable that the addition of nanotubes causes a greater percentage reduction in the presence of larger pores than smaller pores. Li et al. used MIP to demonstrate that 0.5% acid-treated carbon nanotubes produced a Portland cement mortar with a total porosity 64% lower than the control mortar sample, yielding a value of 10.8% [38]. This is compared to the important macropore porosity at 1.47%, representing a reduction of 82%.

Porosity and strength are very sensitive to the concentration of nanotubes, whether functionalized or not [34,35,63]. As the amount of MWCNTs modified with an oxygen plasma [34] increased for concentrations of 0, 0.05, 0.1 and 0.2%, the cumulative pore volume decline indicated a continuing decrease in micropores, corresponding to a continual increase in compressive strength. In other studies, with varying concentrations of MWCNTs, the compressive strength of the composites tends to exhibit a maximum (Table 1A,B), then decreases due to the formation of agglomerates (Figure 3). Bare hydrophobic MWCNTs may reaggregate at higher concentrations to form agglomerates, leading to defects, and acting as voids [17,34,65] to lessen the improvement in mechanical properties. The formation of agglomerates to create voids has the same effect as higher porosity.

A different mechanism for the reaggregation of functionalized MWCNTs must exist in place of the van der Waals forces. Li et al. proposed that high levels of Ca^{2+} and other multivalent cations may bind with carboxylic acids on functionalized MWCNTs to provide a chemical basis for agglomeration [38]. The work by Ahmed et al. illustrates the effect of agglomeration as observed by SEM [33]. They found optimal compression strength values of 66–67 MPa for cement paste modified with either pristine, hydroxyl functionalized or carboxyl functionalized MWCNTs at levels of 0.05% [33]. Smaller increases at higher concentrations for each were attributed to agglomerates with samples approaching the unmodified paste compressive strength of 54.1 MPa.

With its effect on porosity and agglomeration, the concentration of nanotubes has also been shown to affect the properties of concrete, with decreases in mechanical strength reported for concretes with functionalized [43,77] and bare MWCNTs [65]. In working with UHPC containing bare CNTs, Jung et al. found that the mechanisms for decreasing mechanical performance observed above the critical incorporation concentration (CIC) were CNT agglomeration and formation of air voids [65].

Functionalized MWCNTs have been shown to be superior to pristine MWCNTs in reducing porosity in cement pastes. Studies finding lowered porosity also observed improved compressive strength.

3.3. Fracture Resistance, Bridging and Interfacial Bonding with Nanotubes

Another important mechanism by which MWCNTs increase the strength of cementitious composites involves load transfer [10] by distributing stresses in an extensive network. Nanotubes can bridge micro-cracks (Figure 5a) and voids, as first shown by Makar and Beaudoin [8] and observed subsequently by others [8,20,31,77]. Smaller size MWCNTs allow for finer dispersion and the ability to stop crack propagation faster than reinforcement with larger fibers [31], though long, well-dispersed nanotubes are more likely to span cracks, according to a theoretical model [33]. The presence of MWCNTs acts to increase the fracture resistance of the composite [17].

The mechanism requires strong interfacial bonding for effective bridging, though failure may still occur if well-anchored nanotubes break (Figure 5b) [41]. Nonfunctionalized MWCNTs lack chemical bonds with the hydration products of the composite, so the interface is weak [22]. As a result, the pullout of nanotubes can happen, undermining the reinforcement of the composite (Figure 5c) [41]. Pullout and weak bridging compromise fracture resistance. Poor interfacial bonding can be addressed by treatment to create functional groups, commonly -OH and -COOH moieties [41]. Hu found that increases in fracture toughness for 0.1 wt% pristine MWCNTs were only 11.4% over plain cement, compared to 19.4% for MWCNTs-COOH [63]. Computer simulations support the benefit of the interactions afforded by functionalization. Theoretical investigations using molecular dynamics of CNTs have shown enhanced adhesion with tobermorite that grows with the number of acid functionalities to promote hydrogen bonding and ionic bonding with divalent calcium [78–80]. Acid-treated MWCNTs have covalently integrated carboxyl and hydroxyl groups that can react with the C-S-H and $Ca(OH)_2$ hydration products of cement to produce strong bonds [38]. Using SEM and FTIR, Li et al. obtained evidence showing that reactions occur between the carboxylic acid and the cement matrix [38]. A greater increase in fracture resistance with functionalized MWCNTs results from the bonding of carboxylate and alcohol groups with C-S-H hydration products [22]. Ionic bonds can form with calcium and carboxylate or with hydrogen bonding (Figure 6) [81]. The mechanism of stronger bonding along with bridging and pull-out, evident in the microstructure, has been used to explain the greater impact resistance of UHPC with -OH and -COOH modified MWCNTs [82].

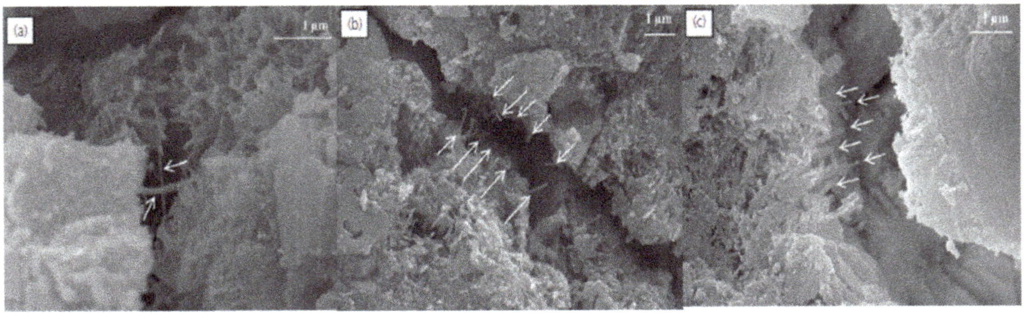

Figure 5. SEM images showing (**a**) carboxyl functionalized nanotubes bridging a crack, (**b**) broken nanotubes and (**c**) pullout of pristine MWCNTs [33]. Nanotubes well anchored to the hydration products can inhibit crack propagation but may fail due to weak bonding or breakage.

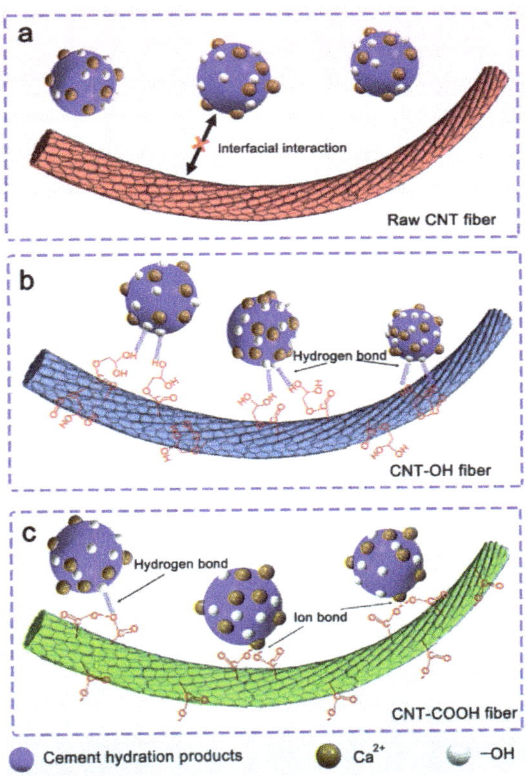

Figure 6. Comparison of interfacial interaction mechanism between cement hydration products with (**a**) bare MWCNT, (**b**) MWCNT-OH and (**c**) MWCNT-COOH [81]. Alcohol functional groups can hydrogen bond to C-S-H while carboxylic acid functionality can form both hydrogen and ionic bonds.

Covalent links to the nanotube structure, like those for carboxyl and hydroxyl moieties, seem to be a key element. In the absence of good interfacial bonding between the MWCNTs and the cement matrix, slippage and pullout occur, undermining force transmission. This is the issue with surfactant. Cwirzen et al. described how MWCNTs tend to pull out under tension due to slippage with nanotubes lacking the covalently linked polar groups that come with functionalization [83]. The group noted, as an example, that while noncovalently linked polyacrylic acid could help with dispersion, it still had weak bond strength and was subject to pullout [83]. In the absence of polar groups on the nanotubes, they concluded in their study that bare MWCNTs did not increase the compressive or bending strength of cement paste.

Many factors can influence the interface bonding contributing to pullout. For example, increasing the site density of carboxylate groups on the MWCNTs can mean a greater number of bonds to the matrix to avoid slippage. Processing conditions resulting in insufficient hydration and poor wetting can weaken bonding, while other species can interfere with bonding. Carboxylate will not compete effectively against residual oxide (O^{2-}) for bonding to Ca^{2+}. Bonding can apparently be blocked, as Nasibulina et al. found that using surfactants in combination with functionalized nanotubes resulted in compromised compressive strength. They hypothesized that the surfactant blocked the interaction of the functional group with cement [25].

3.4. Nucleation and Hydration

According to some researchers, the high surface area of pristine MWCNTs creates more sites for nucleation and accelerates the hydration process [21,32,69,77,84,85]. Enhanced nucleation and hydration lead to the formation of more portlandite of higher crystallinity [33] and integration of MWCNTs with calcium silicate hydrate (C-S-H) (Figure 6). Moreover, the promotion of the hydration reactions to form C-S-H in small interstitial spaces leads to densification and a shift in pore size distribution that contributes to greater strength [10,21,35]. This densification occurs with both pristine [21,32,74] and functionalized MWCNTs [35], but, in a direct comparison, Kang et al. found the hydration products to be denser with the acid-treated nanotubes [35]. That result is consistent with reported strength enhancement due to reduced crack formation with accelerated hydration [84].

Others observing a decrease in mechanical strength with functionalized MWCNTs have attributed the loss to lower hydration and a shift in reaction products. Cui et al. found that the degree of hydration decreased with functionalized MWCNTs, though to a lesser degree than with pristine nanotubes [10]. This can be explained by functionalized nanotubes being so hydrophilic that they absorb water to impede hydration.

The presence of functionalized MWCNTs can alter the reaction products of hydration. Ahmed et al. suggested more formation of ettringite [33] with functionalized MWCNTs, due to the presence of sulfate from acid treatment [34], thereby causing a less dense structure of the hydration products and poorer mechanical properties. Similarly, Musso et al. explained a reduction in the compressive and flexural strength of a cement paste upon the addition of carboxyl MWCNTs with poorer hydration and less formation of tobermorite [42]. Pristine CNTs, on the other hand, have been reported not to accelerate or change hydration reactions [86].

4. Electrical and Thermal Conductivity of Cementitious Composites with Functionalized and Pristine MWCNTs

Fewer studies of composites were found for electrical conductivity, and none were found for thermal conductivity, using the criterion of a pristine MWCNT control sample. It is known that the thermal conductivity of cementitious composites improves with the addition of nanotubes [12,81,87] for both pristine and functionalized MWCNTs.

Dispersion of MWCNTs during the hydration reaction of cement is important to developing a network structure inside the cement matrix to enhance the electrical and thermal conductivity [64]. Better electrical conductivity is found with higher MWCNT loading (0.05 to 0.8 wt% of cement) in accordance with a denser network of MWCNTs inside the cement (Figure 7) [45,48,64,70]. When mixing the MWCNT suspensions with the cement, one portion of the dispersed MWCNTs will be incorporated into the hydrating cementitious matrix, particularly during the formation of C-S-H and ettringite. This helps the cured composite to attain the critical distance of 2–3 nm for electron tunnelling between individual MWCNTs in a denser MWCNT network [64]. Ruan et al. found no clear difference in the resistance of cement paste between short MWCNTs with either short hydroxyl or short carboxyl functionalized nanotubes [48]. However, cement pastes with functionalized long MWCNTs did outperform the composite with pristine carbon nanotubes (Table 3). The lower resistivity of the long MWCNTs compared to the short MWCNTs is consistent with smaller theoretical spacing for a higher aspect ratio [33]. Onthong et al. modified the surface of MWCNT by coating it with a conducting polymer, polyindole and polyvinyl acetate using concurrent admicellar polymerization [45]. The coated MWCNTs obtained provided good water dispersion and yielded a cement paste (0.3 wt% MWCNTs) with an electrical conductivity of 8.56×10^{-4} S/cm, more than 20 times greater than the paste with bare MWCNTs. With the exception of the inconclusive results for the short MWCNTs, the findings overall in Table 3 indicate increased conductivity or reduced resistivity with covalent functionalization.

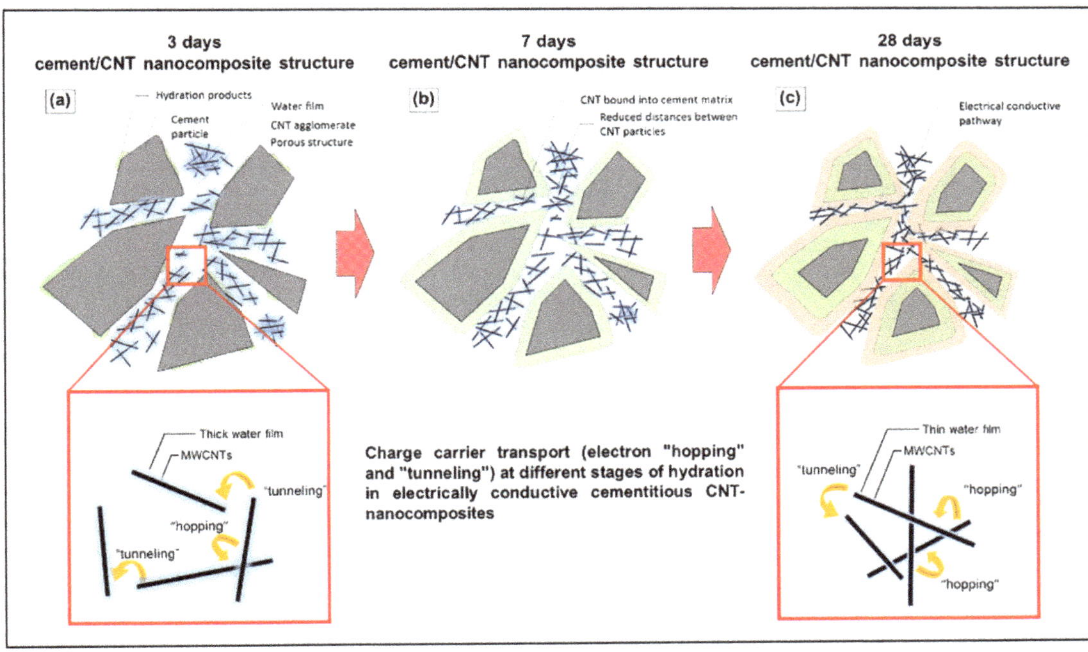

Figure 7. Scheme of the MWCNT interconnected network within a cementitious matrix and the occurring charge-carrier mechanisms (hopping–tunnelling) as a function of time [64]. Liebscher et al. describe how increasing density of carbon nanotubes in a composite changes their proximity and alters the contribution of hopping and tunneling to charge transport.

Thermal conductivity of cementitious composites improves on the addition of nanotubes [12,81,87]. To the authors' knowledge, there are no studies where the thermal conductivity of cementitious composites was compared for bare and functionalized MWCNTs. The thermal conductivity of grouting material rose steadily with concentration, from 0.39 W/m-K with no nanotubes to 0.57 W/m-K at the concentration tested of 2 wt% [81]. Batiston et al. investigated thermal conductivity of cement pastes with pristine MWCNTs as a function of concentration and aspect ratio. At 0.05%, experimental values increased from 0.73 W/m-K for the control to an optimum of 0.84 W/m-K at an aspect ratio of 250. Values were lower for a loading of 0.10%.

5. Conclusions

Addressing the value of covalent functionalization is significant because it has been a common technique to enhance performance of cementitious composites. The question of the value of functionalization is understandable, with complex processes for cementitious composites with many options for components, composition and methods of preparation. Methods of dispersion vary with whether to use surfactant, type and concentration of surfactant and power and duration of sonication. Aside from the question of functionalization or not, carbon nanotubes come in different sizes and purity which can have a significant effect. With so many factors and seemingly conflicting reports, uncertainty related to functionalization is not surprising. The selection of studies for analysis based on the inclusion of a pristine MWCNT control reduced ambiguity and led to the following observations:

(1) Both covalently functionalized and bare nanotubes improve the compressive and flexural strength of cement paste and mortar.
(2) Covalent functionalization by oxidation of MWCNTs leads to a greater improvement in mechanical strength. Analyzing the tabulated values supports the superiority of functionalized MWCNTs. The average increase in compressive strength among different research groups for cement paste (Table 1A) is about 14% for pristine and 34% for functionalized. For mortar (Table 1B), the numbers are 11% and 16%. For flexural strength of the paste, pristine MWCNTs yield an average 23% increase compared to 43% for functionalized (Table 2A); for mortar (Table 2B), the results are 28% and 32%.
(3) Greater reduction in porosity, greater increase in dispersity and greater fracture resistance occur with functionalized MWCNTs. These changes align with the improved mechanical properties in cementitious composites.
(4) Data for cement pastes show that composites with functionalized MWCNTs have higher electrical conductivity than those with pristine MWCNTs. The aspect ratio appears to be a critical factor for conductivity, though additional work is needed. The average reduction of tabulated values of resistivity for cement paste is about 35% for pristine MWCNTs and 50% for functionalized.

Overall, these suggest that functionalized nanotubes are superior to pristine nanotubes. Trends in properties that affect mechanical strength reinforce the direct measurements of improved compressive and flexural strength. The electrical conductivity of cementitious composites is also enhanced with covalent functionalization. Thus, we conclude that a direct comparison of mechanical strength, as well as electrical conductivity, makes a strong case for using functionalized nanotubes where performance alone is considered.

Author Contributions: Conceptualization, E.A.O.; formal analysis, E.A.O., S.O. and T.P.; writing—original draft preparation, E.A.O.; writing—review and editing, S.O. and T.P. All authors have read and agreed to the published version of the manuscript.

Funding: This research was supported in part with funds from the Francis W. Winn Professorship at the University of Oklahoma Foundation and the Institute for Applied Surfactant Research.

Data Availability Statement: Data are contained within the article.

Conflicts of Interest: The authors declare no conflict of interest.

References

1. Shi, T.; Li, Z.X.; Guo, J.; Gong, H.; Gu, C.P. Research progress on CNTs/CNFs-modified cement-based composites. *Constr. Build. Mater.* **2019**, *202*, 290–307. [CrossRef]
2. Metaxa, Z.S.; Tolkou, A.K.; Efstathiou, S.; Rahdar, A.; Favvas, E.P.; Mitropoulos, A.C.; Kyzas, G.Z. Nanomaterials in Cementitious Composites: An Update. *Molecules* **2021**, *26*, 1430. [CrossRef] [PubMed]

3. Zhao, Z.F.; Qi, T.Q.; Zhou, W.; Hui, D.; Xiao, C.; Qi, J.Y.; Zheng, Z.H.; Zhao, Z.G. A review on the properties, reinforcing effects, and commercialization of nanomaterials for cement-based materials. *Nanotechnol. Rev.* **2020**, *9*, 303–322. [CrossRef]
4. Bautista-Gutierrez, K.P.; Herrera-May, A.L.; Santamaria-Lopez, J.M.; Honorato-Moreno, A.; Zamora-Castro, S.A. Recent Progress in Nanomaterials for Modern Concrete Infrastructure: Advantages and Challenges. *Materials* **2019**, *12*, 3548. [CrossRef] [PubMed]
5. Paul, S.C.; van Rooyen, A.S.; van Zijl, G.P.A.G.; Petrik, L.F. Properties of cement-based composites using nanoparticles: A comprehensive review. *Constr. Build. Mater.* **2018**, *189*, 1019–1034. [CrossRef]
6. Reches, Y. Nanoparticles as concrete additives: Review and perspectives. *Constr. Build. Mater.* **2018**, *175*, 483–495. [CrossRef]
7. Li, Q.H.; Liu, J.T.; Xu, S.L. Progress in Research on Carbon Nanotubes Reinforced Cementitious Composites. *Adv. Mater. Sci. Eng.* **2015**, *2015*, 307435. [CrossRef]
8. Abu Al-Rub, R.K.; Ashour, A.I.; Tyson, B.M. On the aspect ratio effect of multi-walled carbon nanotube reinforcements on the mechanical properties of cementitious nanocomposites. *Constr. Build. Mater.* **2012**, *35*, 647–655. [CrossRef]
9. Peng, B.; Locascio, M.; Zapol, P.; Li, S.Y.; Mielke, S.L.; Schatz, G.C.; Espinosa, H.D. Measurements of near-ultimate strength for multiwalled carbon nanotubes and irradiation-induced crosslinking improvements. *Nat. Nanotechnol.* **2008**, *3*, 626–631. [CrossRef]
10. Cui, X.; Han, B.G.; Zheng, Q.F.; Yu, X.; Dong, S.F.; Zhang, L.Q.; Ou, J.P. Mechanical properties and reinforcing mechanisms of cementitious composites with different types of multiwalled carbon nanotubes. *Compos. Part A-Appl. Sci. Manuf.* **2017**, *103*, 131–147. [CrossRef]
11. Abdulhameed, A.; Abd Wahab, N.Z.; Mohtar, M.N.; Hamidon, M.N.; Shafie, S.; Halin, I.A. Methods and Applications of Electrical Conductivity Enhancement of Materials Using Carbon Nanotubes. *J. Electron. Mater.* **2021**, *50*, 3207–3221. [CrossRef]
12. Zhang, S.Q.; Zhang, H.; Jia, Z.J.; Cao, R.L.; Xiong, Y.L.; Chen, C.; Zhang, Y.M. Thermal properties of amino-functionalized multi-walled carbon nanotubes reinforced epoxy-based transducers embedded in concrete. *Cem. Concr. Comp.* **2022**, *127*, 104411. [CrossRef]
13. Kumanek, B.; Janas, D. Thermal conductivity of carbon nanotube networks: A review. *J. Mater. Sci.* **2019**, *54*, 7397–7427. [CrossRef]
14. Monea, B.F.; Ionete, E.I.; Spiridon, S.I.; Ion-Ebrasu, D.; Petre, E. Carbon Nanotubes and Carbon Nanotube Structures Used for Temperature Measurement. *Sensors* **2019**, *19*, 2464. [CrossRef] [PubMed]
15. Rashad, A.M. Effect of carbon nanotubes (CNTs) on the properties of traditional cementitious materials. *Constr. Build. Mater.* **2017**, *153*, 81–101. [CrossRef]
16. Zhang, P.; Su, J.; Guo, J.J.; Hu, S.W. Influence of carbon nanotube on properties of concrete: A review. *Constr. Build. Mater.* **2023**, *369*, 130388. [CrossRef]
17. Ramezani, M.; Dehghani, A.; Sherif, M.M. Carbon nanotube reinforced cementitious composites: A comprehensive review. *Constr. Build. Mater.* **2022**, *315*, 125100. [CrossRef]
18. Xu, S.L.; Liu, J.T.; Li, Q.H. Mechanical properties and microstructure of multi-walled carbon nanotube-reinforced cement paste. *Constr. Build. Mater.* **2015**, *76*, 16–23. [CrossRef]
19. Camacho, M.D.; Galao, O.; Baeza, F.J.; Zornoza, E.; Garces, P. Mechanical Properties and Durability of CNT Cement Composites. *Materials* **2014**, *7*, 1640–1651. [CrossRef]
20. Kumar, S.; Kolay, P.; Malla, S.; Mishra, S. Effect of Multiwalled Carbon Nanotubes on Mechanical Strength of Cement Paste. *J. Mater. Civil. Eng.* **2012**, *24*, 84–91. [CrossRef]
21. Naqi, A.; Abbas, N.; Zahra, N.; Hussain, A.; Shabbir, S.Q. Effect of multi-walled carbon nanotubes (MWCNTs) on the strength development of cementitious materials. *J. Mater. Res. Technol* **2019**, *8*, 1203–1211. [CrossRef]
22. Tamimi, A.; Hassan, N.M.; Fattah, K.; Talachi, A. Performance of cementitious materials produced by incorporating surface treated multiwall carbon nanotubes and silica fume. *Constr. Build. Mater.* **2016**, *114*, 934–945. [CrossRef]
23. Danoglidis, P.A.; Konsta-Gdoutos, M.S.; Gdoutos, E.E.; Shah, S.P. Strength, energy absorption capability and self-sensing properties of multifunctional carbon nanotube reinforced mortars. *Constr. Build. Mater.* **2016**, *120*, 265–274. [CrossRef]
24. Kim, H.K.; Nam, I.W.; Lee, H.K. Enhanced effect of carbon nanotube on mechanical and electrical properties of cement composites by incorporation of silica fume. *Compos. Struct.* **2014**, *107*, 60–69. [CrossRef]
25. Nasibulina, L.I.; Anoshkin, I.V.; Nasibulin, A.G.; Cwirzen, A.; Penttala, V.; Kauppinen, E.I. Effect of Carbon Nanotube Aqueous Dispersion Quality on Mechanical Properties of Cement Composite. *J. Nanomater.* **2012**, *2012*, 169262. [CrossRef]
26. Malikov, E.Y. The effect of polyvinyl alcohol functionalized multiwall carbon nanotubes on the improvement of the compressive strength of concrete. *Fuller. Nanotub. Carbon Nanostructures* **2020**, *28*, 781–785. [CrossRef]
27. Yan, X.T.; Cui, H.Z.; Qin, Q.H.; Tang, W.C.; Zhou, X.M. Study on Utilization of Carboxyl Group Decorated Carbon Nanotubes and Carbonation Reaction for Improving Strengths and Microstructures of Cement Paste. *Nanomaterials* **2016**, *6*, 153. [CrossRef]
28. Cwirzen, A.; Habermehl-Cwirzen, K.; Penttala, V. Surface decoration of carbon nanotubes and mechanical properties of cement/carbon nanotube composites. *Adv. Cem. Res.* **2008**, *20*, 65–73. [CrossRef]
29. Konsta-Gdoutos, M.S.; Metaxa, Z.S.; Shah, S.P. Highly dispersed carbon nanotube reinforced cement based materials. *Cem. Concr. Res.* **2010**, *40*, 1052–1059. [CrossRef]
30. Zou, B.; Chen, S.J.; Korayem, A.H.; Collins, F.; Wang, C.M.; Duan, W.H. Effect of ultrasonication energy on engineering properties of carbon nanotube reinforced cement pastes. *Carbon* **2015**, *85*, 212–220. [CrossRef]
31. Tyson, B.M.; Abu Al-Rub, R.K.; Yazdanbakhsh, A.; Grasley, Z. Carbon Nanotubes and Carbon Nanofibers for Enhancing the Mechanical Properties of Nanocomposite Cementitious Materials. *J. Mater. Civil. Eng.* **2011**, *23*, 1028–1035. [CrossRef]

32. Fakhim, B.; Hassani, A.; Rashidi, A.; Ghodousi, P. Preparation and microstructural properties study on cement composites reinforced with multi-walled carbon nanotubes. *J. Compos. Mater.* **2015**, *49*, 85–98. [CrossRef]
33. Ahmed, H.; Bogas, J.A.; Guedes, M.; Pereira, M.F.C. Dispersion and reinforcement efficiency of carbon nanotubes in cementitious composites. *Mag. Concr. Res.* **2019**, *71*, 408–423. [CrossRef]
34. Li, S.J.; Zhang, Y.L.; Cheng, C.; Wei, H.; Du, S.G.; Yan, J. Surface-treated carbon nanotubes in cement composites: Dispersion, mechanical properties and microstructure. *Constr. Build. Mater.* **2021**, *310*, 125262. [CrossRef]
35. Kang, S.T.; Seo, J.Y.; Park, S.H. The Characteristics of CNT/Cement Composites with Acid-Treated MWCNTs. *Adv. Mater. Sci. Eng.* **2015**, *2015*, 308725. [CrossRef]
36. Sarvandani, M.M.; Mahdikhani, M.; Aghabarati, H.; Fatmehsari, M.H. Effect of functionalized multi-walled carbon nanotubes on mechanical properties and durability of cement mortars. *J. Build. Eng.* **2021**, *41*, 102407. [CrossRef]
37. Elkashef, M.; Wang, K.; Abou-Zeid, M.N. Acid-treated carbon nanotubes and their effects on mortar strength. *Front. Struct. Civ. Eng.* **2016**, *10*, 180–188. [CrossRef]
38. Li, G.Y.; Wang, P.M.; Zhao, X.H. Mechanical behavior and microstructure of cement composites incorporating surface-treated multi-walled carbon nanotubes. *Carbon* **2005**, *43*, 1239–1245. [CrossRef]
39. Manzur, T.; Yazdani, N.; Emon, M.A.B. Effect of Carbon Nanotube Size on Compressive Strengths of Nanotube Reinforced Cementitious Composites. *J. Mater. Civil. Eng.* **2014**, *2014*, 960984. [CrossRef]
40. Manzur, T.; Yazdani, N.; Emon, M.A. Potential of Carbon Nanotube Reinforced Cement Composites as Concrete Repair Material. *J. Nanomater.* **2016**, *2016*, 1421959. [CrossRef]
41. Sun, G.X.; Liang, R.; Lu, Z.Y.; Zhang, J.R.; Li, Z.J. Mechanism of cement/carbon nanotube composites with enhanced mechanical properties achieved by interfacial strengthening. *Constr. Build. Mater.* **2016**, *115*, 87–92. [CrossRef]
42. Musso, S.; Tulliani, J.M.; Ferro, G.; Tagliaferro, A. Influence of carbon nanotubes structure on the mechanical behavior of cement composites. *Compos. Sci. Technol.* **2009**, *69*, 1985–1990. [CrossRef]
43. Hawreen, A.; Bogas, J.A. Creep, shrinkage and mechanical properties of concrete reinforced with different types of carbon nanotubes. *Constr. Build. Mater.* **2019**, *198*, 70–81. [CrossRef]
44. Yoo, D.Y.; You, I.; Lee, S.J. Electrical Properties of Cement-Based Composites with Carbon Nanotubes, Graphene, and Graphite Nanofibers. *Sensors* **2017**, *17*, 1064. [CrossRef] [PubMed]
45. Onthong, S.; O'Rear, E.A.; Pongprayoon, T. Enhancement of electrically conductive network structure in cementitious composites by polymer hybrid-coated multiwalled carbon nanotube. *Mater. Struct.* **2022**, *55*, 232. [CrossRef]
46. Onthong, S.; O'Rear, E.A.; Pongprayoon, T. Composite nanoarchitectonics by interfacial bonding for conductivity and strength development of grafted multiwall carbon nanotube/cement. *Constr. Build. Mater.* **2023**, *392*, 131940. [CrossRef]
47. Konsta-Gdoutos, M.S.; Aza, C.A. Self sensing carbon nanotube (CNT) and nanofiber (CNF) cementitious composites for real time damage assessment in smart structures. *Cem. Concr. Comp.* **2014**, *53*, 162–169. [CrossRef]
48. Ruan, Y.F.; Han, B.G.; Wang, D.N.; Zhang, W.; Yu, X. Electrical properties of carbon nanotubes filled cementitious composites. *Mater. Res. Express* **2018**, *5*, 105704. [CrossRef]
49. Del Moral, B.; Gullón, I.M.; Navarro, R.; Galao, O.; Baeza, F.J.; Zornoza, E.; Calderón, B.; Rodríguez, I.; Arnaiz, N.; Sánchez, M.D.R.; et al. The Effect of Different Oxygen Surface Functionalization of Carbon Nanotubes on the Electrical Resistivity and Strain Sensing Function of Cement Pastes. *Nanomaterials* **2020**, *10*, 807. [CrossRef]
50. Dalla, P.T.; Dassios, K.G.; Tragazikis, I.K.; Exarchos, D.A.; Matikas, T.E. Carbon nanotubes and nanofibers as strain and damage sensors for smart cement. *Mater. Today Commun.* **2016**, *8*, 196–204. [CrossRef]
51. Han, B.G.; Zhang, L.Q.; Sun, S.W.; Yu, X.; Dong, X.F.; Wu, T.J.; Ou, J.P. Electrostatic self-assembled carbon nanotube/nano carbon black composite fillers reinforced cement-based materials with multifunctionality. *Compos. Part A-Appl. Sci. Manuf.* **2015**, *79*, 103–115. [CrossRef]
52. Yu, X.; Kwon, E. A carbon nanotube/cement composite with piezoresistive properties. *Smart Mater. Struct.* **2009**, *18*, 055010. [CrossRef]
53. Nilsson, F.; Krückel, J.; Schubert, D.W.; Chen, F.; Unge, M.; Gedde, U.W.; Hedenqvist, M.S. Simulating the effective electric conductivity of polymer composites with high aspect ratio fillers. *Compos. Sci. Technol.* **2016**, *132*, 16–23. [CrossRef]
54. Singer, G.; Siedlaczek, P.; Sinn, G.; Rennhofer, H.; Micusík, M.; Omastová, M.; Unterlass, M.M.; Wendrinsky, J.; Milotti, V.; Fedi, F.; et al. Acid Free Oxidation and Simple Dispersion Method of MWCNT for High-Performance CFRP. *Nanomaterials* **2018**, *8*, 912. [CrossRef] [PubMed]
55. Seneewong-Na-Ayutthaya, M.; Pongprayoon, T.; O'Rear, E.A. Colloidal Stability in Water of Modified Carbon Nanotube: Comparison of Different Modification Techniques. *Macromol. Chem. Phys.* **2016**, *217*, 2635–2646. [CrossRef]
56. Seneewong-Na-Ayutthaya, M.; Pongprayoon, T. Water-dispersible carbon nanotube prepared by non-destructive functionalization technique of admicellar polymerization. *Diam. Relat. Mater.* **2015**, *60*, 111–116. [CrossRef]
57. D'Alessandro, A.; Tiecco, M.; Meoni, A.; Ubertini, F. Improved strain sensing properties of cement-based sensors through enhanced carbon nanotube dispersion. *Cem. Concr. Comp.* **2021**, *115*, 103842. [CrossRef]
58. Luo, J.L.; Duan, Z.D.; Li, H. The influence of surfactants on the processing of multi-walled carbon nanotubes in reinforced cement matrix composites. *Phys. Status Solidi A-Appl. Mater. Sci.* **2009**, *206*, 2783–2790. [CrossRef]
59. Parveen, S.; Rana, S.; Fangueiro, R.; Paiva, M.C. Microstructure and mechanical properties of carbon nanotube reinforced cementitious composites developed using a novel dispersion technique. *Cem. Concr. Res.* **2015**, *73*, 215–227. [CrossRef]

60. Sobolkina, A.; Mechtcherine, V.; Khavrus, V.; Maier, D.; Mende, M.; Ritschel, M.; Leonhardt, A. Dispersion of carbon nanotubes and its influence on the mechanical properties of the cement matrix. *Cem. Concr. Comp.* **2012**, *34*, 1104–1113. [CrossRef]
61. Li, X.; Pu, C.S.; Bai, Y.; Huang, F.F. Effect of surfactant types on the foam stability of multiwalled carbon nanotube stabilized foam. *Colloids Surf. A-Physicochem. Eng. Asp.* **2022**, *648*, 129389. [CrossRef]
62. Siddique, R.; Mehta, A. Effect of carbon nanotubes on properties of cement mortars. *Constr. Build. Mater.* **2014**, *50*, 116–129. [CrossRef]
63. Hu, Y.; Luo, D.N.; Li, P.H.; Li, Q.B.; Sun, G.Q. Fracture toughness enhancement of cement paste with multi-walled carbon nanotubes. *Constr. Build. Mater.* **2014**, *70*, 332–338. [CrossRef]
64. Liebscher, M.; Tzounis, L.; Junger, D.; Dinh, T.T.; Mechtcherine, V. Electrical Joule heating of cementitious nanocomposites filled with multi-walled carbon nanotubes: Role of filler concentration, water content, and cement age. *Smart Mater. Struct.* **2020**, *29*, 125019. [CrossRef]
65. Jung, M.; Lee, Y.S.; Hong, S.G.; Moon, J. Carbon nanotubes (CNTs) in ultra-high performance concrete (UHPC): Dispersion, mechanical properties, and electromagnetic interference (EMI) shielding effectiveness (SE). *Cem. Concr. Res.* **2020**, *131*, 106017. [CrossRef]
66. Shah, S.P.; Konsta-Gdoutos, M.S.; Metaxa, Z.S.; Mondal, P. Nanoscale Modification of Cementitious Materials. In *Nanotechnology in Construction (NICOM3)*; Bittnar, Z., Bartos, P.J.M., Němeček, J., Šmilauer, V., Zeman, J., Eds.; Springer: Berlin/Heidelberg, Germany, 2009; pp. 125–130.
67. Jang, S.H.; Hochstein, D.P.; Kawashima, S.; Yin, H.M. Experiments and micromechanical modeling of electrical conductivity of carbon nanotube/cement composites with moisture. *Cem. Concr. Comp.* **2017**, *77*, 49–59. [CrossRef]
68. Lavagna, L.; Bartoli, M.; Suarez-Riera, D.; Cagliero, D.; Musso, S.; Pavese, M. Oxidation of Carbon Nanotubes for Improving the Mechanical and Electrical Properties of Oil-Well Cement-Based Composites. *Acs Appl. Nano Mater.* **2022**, *5*, 6671–6678. [CrossRef]
69. Hawreen, A.; Bogas, J.A.; Dias, A.P.S. On the mechanical and shrinkage behavior of cement mortars reinforced with carbon nanotubes. *Constr. Build. Mater.* **2018**, *168*, 459–470. [CrossRef]
70. Han, B.G.; Zhang, K.; Yu, X.; Kwon, E.; Ou, J.P. Fabrication of Piezoresistive CNT/CNF Cementitious Composites with Superplasticizer as Dispersant. *J. Mater. Civil. Eng.* **2012**, *24*, 658–665. [CrossRef]
71. Rocha, V.V.; Ludvig, P. Nanocomposites Prepared by a Dispersion of Cnts on Cement Particles. *Archit. Civ. Eng. Environ.* **2018**, *11*, 73–77.
72. Mesquita, E.; Sousa, I.; Vieira, M.; Matos, A.M.; Santos, L.P.M.; Silvestro, L.; Salvador, R.; D'Alessandro, A.; Ubertini, F. Investigation of the electrical sensing properties of cementitious composites produced with multi-wall carbon nanotubes dispersed in NaOH. *J. Build. Eng.* **2023**, *77*, 107496. [CrossRef]
73. Li, G.Y.; Wang, L.B.; Yu, J.; Yi, B.L.; He, C.B.; Wang, Z.K.; Leung, C.K.Y. Mechanical properties and material characterization of cement mortar incorporating CNT-engineered polyvinyl alcohol latex. *Constr. Build. Mater.* **2022**, *345*, 128320. [CrossRef]
74. Chen, J.X.; Akono, A.T. Influence of multi-walled carbon nanotubes on the hydration products of ordinary Portland cement paste. *Cem. Concr. Res.* **2020**, *137*, 106197. [CrossRef]
75. Wang, J.L.; Dong, S.F.; Pang, S.D.; Zhou, C.S.; Han, B.G. Pore structure characteristics of concrete composites with surface-modified carbon nanotubes. *Cem. Concr. Comp.* **2022**, *128*, 104453. [CrossRef]
76. Nochaiya, T.; Chaipanich, A. Behavior of multi-walled carbon nanotubes on the porosity and microstructure of cement-based materials. *Appl. Surf. Sci.* **2011**, *257*, 1941–1945. [CrossRef]
77. Carrico, A.; Bogas, J.A.; Hawreen, A.; Guedes, M. Durability of multi-walled carbon nanotube reinforced concrete. *Constr. Build. Mater.* **2018**, *164*, 121–133. [CrossRef]
78. Merodio-Perea, R.G.; Lado-Tourino, I.; Paez-Pavon, A.; Talayero, C.; Galan-Salazar, A.; Ait-Salem, O. Mechanical Properties of Cement Reinforced with Pristine and Functionalized Carbon Nanotubes: Simulation Studies. *Materials* **2022**, *15*, 7734. [CrossRef]
79. Sanchez, F.; Zhang, L. Molecular dynamics modeling of the interface between surface functionalized graphitic structures and calcium-silicate-hydrate: Interaction energies, structure, and dynamics. *J. Colloid Interf. Sci.* **2008**, *323*, 349–358. [CrossRef]
80. Wang, P.; Qiao, G.; Hou, D.S.; Jin, Z.Q.; Wang, M.H.; Zhang, J.R.; Sun, G.X. Functionalization enhancement interfacial bonding strength between graphene sheets and calcium silicate hydrate: Insights from molecular dynamics simulation. *Constr. Build. Mater.* **2020**, *261*, 120500. [CrossRef]
81. Yang, Z.P.; Yang, J.F.; Shuai, B.; Niu, Y.T.; Yong, Z.Z.; Wu, K.J.; Zhang, C.J.; Qiao, X.Y.; Zhang, Y.Y. Superflexible yet robust functionalized carbon nanotube fiber reinforced sulphoaluminate cement-based grouting materials with excellent mechanical, electrical and thermal properties. *Constr. Build. Mater.* **2022**, *328*, 126999. [CrossRef]
82. Wang, J.L.; Dong, S.F.; Pang, S.D.; Yu, X.; Han, B.G.; Ou, J.P. Tailoring Anti-Impact Properties of Ultra-High Performance Concrete by Incorporating Functionalized Carbon Nanotubes. *Engineering* **2022**, *18*, 232–245. [CrossRef]
83. Cwirzen, A.; Habermehl-Cwirzen, K.; Nasibulin, A.G.; Kaupinen, E.I.; Mudimela, P.R.; Penttala, V. SEW/AFM studies of cementitious binder modified by MWCNT and nano-sized Fe needles. *Mater. Charact.* **2009**, *60*, 735–740. [CrossRef]
84. Han, B.G.; Zhang, L.Q.; Zeng, S.Z.; Dong, S.F.; Yu, X.; Yang, R.W.; Ou, J.P. Nano-core effect in nano-engineered cementitious composites. *Compos. Part A-Appl. Sci. Manuf.* **2017**, *95*, 100–109. [CrossRef]
85. Makar, J.M.; Chan, G.W. Growth of Cement Hydration Products on Single-Walled Carbon Nanotubes. *J. Am. Ceram. Soc.* **2009**, *92*, 1303–1310. [CrossRef]

86. Amin, M.S.; El-Gamal, S.M.A.; Hashem, F.S. Fire resistance and mechanical properties of carbon nanotubes—Clay bricks wastes (Homra) composites cement. *Constr. Build. Mater.* **2015**, *98*, 237–249. [CrossRef]
87. Batiston E, G.P.; Mezzomo, P.; Pelisser, F.; Matos, P.R. Effect of Carbon Nanotubes (CNTs) aspect ratio on the rheology, thermal conductivity and mechanical performance of Portland cement paste. *Rev. IBRACON De Estrut. E Mater.* **2021**, *14*. [CrossRef]

nanomaterials

MDPI

Article

Effect of Morphologically Controlled Hematite Nanoparticles on the Properties of Fly Ash Blended Cement

Pantharee Kongsat [1,2], Sakprayut Sinthupinyo [3], Edgar A. O'Rear [4] and Thirawudh Pongprayoon [1,2,*]

[1] Department of Chemical Engineering, Faculty of Engineering, King Mongkut's University of Technology North Bangkok, Bangkok 10800, Thailand; pantharee99@gmail.com

[2] Center of Eco-Materials and Cleaner Technology, King Mongkut's University of Technology North Bangkok, Bangkok 10800, Thailand

[3] Siam Research and Innovation Co., Ltd., Saraburi 18260, Thailand; sakprays@scg.com

[4] School of Chemical, Biological and Materials Engineering and Institute for Applied Surfactant Research, University of Oklahoma, Norman, OK 73019, USA; eorear@ou.edu

* Correspondence: thirawudh.p@eng.kmutnb.ac.th; Tel.: +66-8-9182-6168

Abstract: Several types of hematite nanoparticles (α-Fe_2O_3) have been investigated for their effects on the structure and properties of fly ash (FA) blended cement. All synthesized nanoparticles were found to be of spherical shape, but of different particle sizes ranging from 10 to 195 nm depending on the surfactant used in their preparation. The cement hydration with time showed 1.0% α-Fe_2O_3 nanoparticles are effective accelerators for FA blended cement. Moreover, adding α-Fe_2O_3 nanoparticles in FA blended cement enhanced the compressive strength and workability of cement. Nanoparticle size and size distribution were important for optimal filling of various size of pores within the cement structure.

Keywords: hematite nanoparticles; fly ash blended cement; cement hydration; compressive strength; workability

1. Introduction

Nanomaterials have been widely studied for their beneficial effects on the properties of cement and concrete. Nanoparticles of silica, titanium dioxide, hematite, alumina, clay and other substances increase the mechanical strength and durability of concrete [1,2]. The mechanisms underlying these improvements rely on structural characteristics of the nano-modifier. Increased surface area accelerates the pozzolonic reactions while the smaller size facilitates densification on filling voids. This means the method of synthesis and control of structure of the nanoparticles is important.

Cement consists mainly of Ordinary Portland Cement (OPC) as calcium oxide (CaO) [3,4]. Partial replacement of OPC with fly ash (FA) can reduce CO_2 emissions and help address an environmental waste problem of the FA byproduct from coal combustion in electrical power production [5,6]. Moreover, FA blended cement has been found in actual use to improve workability, strength, durability and hardened cement composites [7–9].

The hardening of cement materials is the result of hydration reactions of OPC with water. Basically, hydration can be separated into five stages (Supplementary Figure S1) [4,10–13]. The first stage, very soon after mixing, involves C_3A reacting with water to form an aluminate-rich gel. A few minutes later, one observes low heat evolution in the induction period, when C_3S and C_2S in the cement start to react and form a C-S-H gel. For several hours, early formed ettringite converts to CAS^-, allowing continuation of the C_3A hydration process. As C_3A is hydrating within the cement paste, reaction species create a shoulder in the decelerating rate region of the peak. Lastly, the C_4AF reacts in a similar manner as C_3A but more slowly [13–16].

Many researchers have demonstrated the enhanced properties of OPC with added iron oxides. Hematite (α-Fe_2O_3) enhances cement hydration, increases compressive strength and lowers drying shrinkage [17,18]. The added hematite nanoparticles act in part by filling pores of the cement mortar [19,20]. The α-Fe_2O_3 also acts as a foreign nucleation site that accelerates the formation of C-S-H gel as the hydrated product of crystalline $Ca(OH)_2$, especially at stage 3 hydration, leading to greater cement strength [13, 15]. As noted above, the size of a nucleation particle can change the microstructure of hydrated cement and much improve its mechanical strength. This work is aimed at understanding how properties of nano-hematite affect the hydration, workability and strength of FA blended cement.

Researchers have described several approaches to the synthesis of α-Fe_2O_3 nanoparticles including coprecipitation [21], hydrothermal [22] and surfactant-assisted hydrothermal methods [23]. Of these techniques, synthesis of nanoparticles using a surfactant template offers some advantages. Especially, the property of a surfactant to form micellar aggregates in solution at its critical micelle concentration (CMC) provides a means to control size and structure of the nanoparticles [23,24]. The structure, size and other properties of α-Fe_2O_3 nanoparticles are affected by the shape and size of the micelles, which are determined by the chemical structure of the surfactant molecule.

The function and mechanisms of α-Fe_2O_3 on FA blended cement have yet to be established in the literature. Research is needed to develop the fundamental mechanisms relating hydration and porosity to macroscopic observations of mechanical strength of the composite upon hematite nanoparticle addition. This study focused on the effects of α-Fe_2O_3 of various sizes on the hydration reaction and workability of FA blended cement. Three structure-controlled hematite nanoparticles have been synthesized by the surfactant-assisted method using anionic, cationic and nonionic surfactants and each characterized by X-ray diffraction (XRD), Transmission electron microscope (TEM) and Field-emission-scanning-electron-microscopy (FE-SEM). FA blended cements were prepared with each of the three synthesized α-Fe_2O_3 nanoparticles as well as with a commercial α-Fe_2O_3 nanoparticle. Heat flow calorimetry was used to follow cement hydration, while SEM was used for microstructural analysis with the aim of understanding morphological structure for optimal performance. The workability during hydration, an important behavior for utilization of cement composites, was also evaluated. Strength and workability were determined following standard tests for fresh cement.

2. Materials and Methods

2.1. Materials

Ordinary Portland Cement (OPC) type 1 and fly ash were provided by cement plant (SCG Cement Co., Ltd., Saraburi, Thailand). Commercial α-Fe_2O_3 came from a vendor for cement factory and its properties were analyzed by XRD and SEM. The synthesized α-Fe_2O_3 nanoparticles were prepared by surfactant-assisted method as described previously [23]. Iron (II) sulfate heptahydrate ($FeSO_4{\cdot}7H_2O$, 99%) from Quality Reagent Chemical (QREC, Auckland, New Zealand) served as the precursor for the nanoparticle synthesis. Ammonium hydroxide solution (NH_4OH, 30 wt%) in water was purchased from Sigma–Aldrich (Sigma-Aldrich In., Saint Louis, MO, USA). Sodium dodecyl sulfate (SDS) was obtained from Ajax Finechem (Ajex finechem Inc., Taren Point, Australia). Cetyltrimethylammonium bromide (CTAB) provided from Amresco (Amresco Inc., Solon, OH, USA). Polyoxyethlyene tert-octylphenyl ether (TX100) was supplied from Applichem Panreac (Applichem Gmbh Inc., Darmstadt, Germany). All chemicals for α-Fe_2O_3 synthesis were used as-received.

2.2. Synthesis and Characterization of α-Fe_2O_3

In this case, α-Fe_2O_3 nanoparticles were synthesized by the surfactant-assisted method to provide morphological control. Three surfactants (SDS, CTAB and TX100 as anionic, cationic and nonionic surfactants, respectively) were used in this study. For each surfactant, an aqueous solution (200 mL distilled, deionized water) at twice the CMC (8.40 mM for SDS, 0.98 mM for CTAB and 0.24 mM for TX100) was prepared. The surfactant solution was stirred at 60 °C for 1 h until it was clear. Stirring continued for 2 h after addition of precursor iron (II) sulfate heptahydrate (20 g). After that 30 wt% ammonia solution (5 mL) was carefully added into the solution with continuous stirring for another 2 h. The mixture color changed from light yellow to dark blue as the hematite nanoparticles were formed in the solution. Lastly the nanoparticles were filtered and washed with distilled water for removing the excess surfactants and other impurities. The filtrate was dried at 80 °C in an oven for 24 h. Dried nanoparticles were further calcined at 600 °C for 4 h under air for crystallization.

X-ray diffraction (X-Ray Diffractometer; Bruker AXS Model D8 Discover) was used for crystal structure and phase analysis by CuK_{α} radiation (λ = 1.5406 Angstrom) with a scanning speed of 0.02 step/s in 2θ ranging from 10 to 70 degree. A transmission electron microscope (TEM), Philips-TECNAI 20, was used to observe the nanoparticle morphology as 2D. Field-emission-scanning-electron-microscopy (FE-SEM), JSM-6480LV, was also used for morphological observation of the synthesized nanoparticles and FE-SEM micrographs were used to observe the nanoparticle morphology as 3D and evaluate the particle size by ImageJ software.

2.3. FA Blended Cement Preparation

Seven samples were prepared by varying the compositions of binders (Table 1). The basic binders of this study included OPC and FA. In this case, α-Fe_2O_3 were varied in both quantity and quality. The percentage of FA was fixed at 40% by weight due to its high performance in cement mixtures designed for both high strength and high durability [5,9,25,26]. An increasing mass percentage of α-Fe_2O_3 nanoparticles was replaced with a decreasing percentage OPC.

Table 1. The fraction of OPC, fly ash and α-Fe_2O_3 nanoparticles as a binder and their codes.

Sample Name	Proportions		
	OPC	Fly Ash	Amount and Type of α-Fe_2O_3
FA40	60%	40%	No α -Fe_2O_3 used as reference
C1	59%	40%	1.0% commercial α-Fe_2O_3
C3	57%	40%	3.0% commercial α-Fe_2O_3
C5	55%	40%	5.0% commercial α-Fe_2O_3
SDS	59%	40%	1.0% α-Fe_2O_3 synthesized with SDS
CTAB	59%	40%	1.0% α-Fe_2O_3 synthesized with CTAB
TX100	59%	40%	1.0% α-Fe_2O_3 synthesized with Triton X 100

For hydration and setting time investigation, samples were prepared as cement pastes with the same compositions shown in Table 1. For slump test, flow table and compressive strength measurements, the samples were constituted as the mortar by adding fine aggregate sand in the size range of 2.0–2.8 mm, with the mass ratio specified for the slump test experiment. Sample codes are indicated in Table 1.

2.4. Hydration Analysis

The heat evolution emitted from cement hydration was observed by isothermal calorimetry (TA instrument, TAM Air 8 channel). The calorimeter gauges heat flow associated with physical processes and chemical reactions for observation of the hydration stages and heat reaction rate, respectively. In this work, a twin-type calorimeter was used with separate sample and reference chambers.

The behavior of cement hydration can generally be divided into five stages (Supplementary Figure S1). In line with previous work from Yuenyongsuwan et al. [12] and Kim et al. [27], this study concentrated on the acceleration period (stage 3) and the deceleration period (stage 4) of the hydration behavior curve. The effect of the nanofiller on hydration can be observed at the maximum heat flow values which occur in these stages [14,27]. Binders tested included mixtures of OPC, fly ash and α-Fe_2O_3 nanoparticles (Table 1). The total binder (40.0 g) was mixed and stirred in the cup for 40 s, then 5.0 g of binder was weighed and carefully filled into a testing glass bottle for the calorimetry test. Water (3.8 g) in the syringe that was placed in the calorimeter was slowly combined with the binder for hydration observation (Supplementary Figure S2). All measurements were conducted over 40 h to cover the hydration period.

2.5. Workability and Compressive Strength

2.5.1. Setting Time

Setting time, the required time for stiffening of cement paste, was investigated by the Vicat needle method, ASTM C 191 standard. The cement pastes for all samples of this test were prepared to a total mass of 650 g. Water and binder (w/b) at a mass ratio of 0.3:1 were put in the mix pot while continuously stirring at low speed (140 ± 5 rpm/min) for 30 s and followed by continuous stirring at medium speed (285 ± 10 rpm/min) for 60 s. A mold (an internal diameter of 70 mm at the top and 80 mm at the bottom and a height of 40 mm) was filled with the cement paste within 5 min after mixing. The bearing surface of the needle was brought into contact with the cement paste, the scale zeroed, and the plunger immediately released with the needle allowed to settle for 30 s when depth of penetration was recorded. The measurement was repeated every 15 min until a penetration of 25 mm was reached as indicated by the scribe mark. The time at which cement starts to harden and completely loses its plasticity is called initial setting time, while the time of the change from the plastic state to solid state is called the final setting time.

2.5.2. Mini-Slump Test and Flow Table Test

The mini-slump test, according to ASTM C-143 (ASTM 2004), was used to measure the mortar behavior under the action of gravity in a compacted inverted cone or mold for testing. The mini-slump cone has top and bottom diameters of 60 mm and 70 mm, respectively, with a cone height of 100 mm. The mortar for each sample (Table 1) was prepared to a total mass of 500 g. Water–binder (w/b) and binder-sand mass ratios were fixed at 0.45:1 and 1:3, respectively. Firstly, binder, sand and water were put into the mix pot and then continuously mixed at 140 rpm/min for 3 min. Next, the mixture was filled into the mini slump cone that had been placed on a table. The inverted cone mold was slowly lifted up and the height of the mortar placed on the table was measured and recorded (Supplementary Figure S3).

The flow table (according to ASTM C1437) was also used to evaluate workability. The mortar sample and cone mold were similar to that in the slump test. After the cone mold was raised, the mortar sample was dropped 25 times from the height of the cone mold, approximately 100 mm, within 15 s. Then the diameter of the spread mortar on the table was measured and recorded.

2.5.3. Compressive Strength

The compressive strength of mortar samples was tested according to the standard ASTM C109/C109M. The mortar samples were prepared by mixing binder and fine aggregate in a 1:3 mass ratio. The w/b ratio of mass was fixed at 0.45:1. A total 200 g binder and 600 g fine aggregate were mixed together under dry condition for 1 min and then water was added. The mortar was cast in a mold with the dimension of 50 × 50 × 50 mm. When casting was completed, the mortar was left in the mold for 24 h while curing at 25 °C. After that, the mortar specimen was removed from the mold and placed in water at 25 °C. The specimen was used to test the compressive strength at 7-, 28- and 90-days curing. The test at 90 days curing was of interest for further application in self-compacting concrete.

2.6. Morphology of Cement Paste

A scanning-electron-microscope (SEM), Quanta-450, was used for morphological observation of the FA blended cement paste specimens at 7-, 28- and 90-days curing. The specimens were prepared by mixing water and binder as shown in Table 1 with w/b ratio at 0.3:1. After mixing, the cement pastes were put into plastic tubes. After 24 h, the plastic tubes were peeled off and the cement specimens were soaked in water for curing for 7, 28 and 90 days at 25 °C.

3. Results and Discussion

The purpose of this study was to evaluate the performance of fly ash cement modified with a series of synthetic α-Fe_2O_3 with particle sizes on the order of 10–195 nm. Knowledge gained with different α-Fe_2O_3 nanoparticles can be used to improve fly ash blended concrete.

3.1. Synthesized α-Fe_2O_3 Characterization

In this case, α-Fe_2O_3 nanoparticles were characterized by XRD, TEM and FE-SEM to determine phase, size and shape of the synthesized nanoparticles.

XRD patterns of α-Fe_2O_3 are depicted in Figure 1. The recorded and indexed diffraction patterns of the sharp peaks are as expected for highly crystalline samples. The results show peak positions of 2θ values at 23.88°, 33.47°, 36.25°, 41.54°, 49.58°, 54.82°, 57.55°, 63.44° and 64.67°, indexed as (012), (104), (110), (113), (024), (116), (018), (214) and (300) planes that fit the standard pattern of α-Fe_2O_3 [28]. No diffraction peaks corresponding to other phases were present, indicating a high purity of α-Fe_2O_3 for all synthesized samples. These clearly indicate the formation of fully crystalline iron oxide of α-Fe_2O_3 structure without other iron oxide phases.

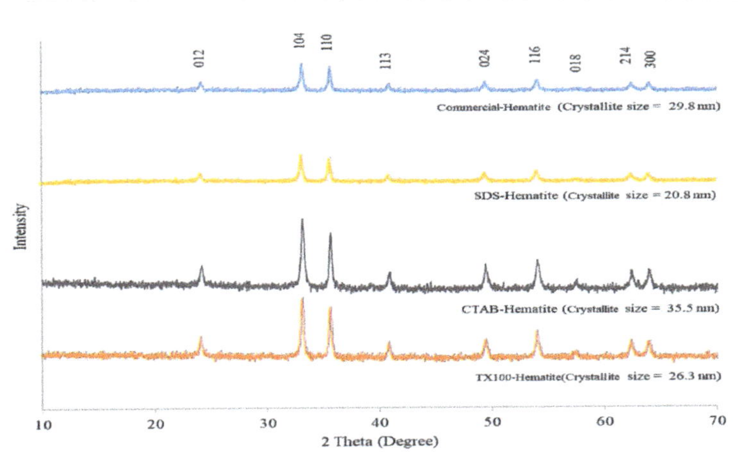

Figure 1. X-ray diffraction (XRD) patterns of commercial hematite (α-Fe_2O_3) nanoparticles and α-Fe_2O_3 nanoparticles synthesized using sodium dodecyl sulfate (SDS), cetyltrimethylammonium bromide (CTAB) and polyoxyethlyene tert-octylphenyl ether (TX100) as templates.

The crystallite size was calculated from XRD patterns using Scherrer equation and applying full-width half-maximum (FWHM) of characteristic peak (at 104 or 2θ = 33.3°) of α-Fe_2O_3 with the following equation:

$$Crystallite\ size = \frac{0.9\lambda}{FWHMcos\theta} \quad (1)$$

where λ is the X-ray wavelength (1.5406 Å in this study) and θ is the diffraction angle for the (104) plane. The crystallite sizes of the synthesized α-Fe_2O_3 using SDS, CTAB and TX100 at 2 CMC were approximately 20.8, 35.5 and 26.3 nm diameters, respectively, whereas the crystallite size of commercial α-Fe_2O_3 was approximately 29.8 nm. Basically, crystallization occurs in two major steps, the first step is nucleation from the reaction and then crystal growth to increase the size of particles, leading to a stable crystalline state. For the second step, an important feature may occur due to the surfactant micelle with crystal defects themselves appearing as open inconsistencies such as pores and cracks [29]. These results for the crystallite size show some correspondence to the micelle size of SDS (3.5–4.0 nm) [30], CTAB (118.0–192.0 nm) [31] and TX100 (10.2 nm) [32].

Both TEM and FE-SEM were used to observe the agglomerate particles in this study. Grain size was measured by TEM and the particle morphology was determined by FE-SEM.

The 2D images of TEM in Figure 2 clearly show that the individual α-Fe_2O_3 were rounded as the sphere-like shape with various curvatures in all samples. The grain size of 2D images in Figure 2 were analyzed by ImageJ analysis software. A single grain within a nanoparticle corresponds to a crystallite, while multiple aggregated grains exist within most nanoparticles. 3D images of FE-SEM micrographs and ImageJ analysis software were used for morphological observation and size estimation of all samples (Figure 3). Results confirmed the spheroidal shape of all samples and relative sizes depending on surfactant used. The smallest ones occurred when using SDS for the synthesis, whereas CTAB yielded the largest nanoparticles. The average grain size and average particle size of the α-Fe_2O_3 synthesized via SDS were 54.1 nm and 65.2 nm. Grain sizes ranged between 10 and 110 nm while particle sizes varied from 10 to 130 nm. The grain size is smaller than particles size because one particle can have several grains. The average grain sizes of synthesized α-Fe_2O_3 using CTAB and TX100 were different at 100.6 nm and 79.8 nm, even though particle sizes were similar. Particle sizes were 131.0 nm and 122.0 nm average diameters, respectively, in a similar particle size range of 35 to 195 nm. These compared to an average particle size for the commercial α-Fe_2O_3 nanoparticles of approximately 174.1 nm with particle sizes ranging from 100 to 240 nm.

The relative dimensions of the surfactant-synthesized particles from TEM and SEM confirmed the XRD crystal size estimation. Others have found that the size of nanoparticles could be controlled by micelles in water, often with increasing particle size for larger micelles [29–31]. Summarized results of crystal size and particle size are shown for comparison in Figure 4. Results confirmed that surfactants significantly influence the shape and size of the nanoparticles, in accord with the findings of Jing et al. [33] and Colombo et al. [34].

3.2. *Hydration of FA Blended Cement: Effect of Amount and Types of* α-Fe_2O_3 *Addition*

The hydration mechanisms of FA blended cement with added α-Fe_2O_3 nanoparticles were studied by calorimetry, commonly used to assess acceleration (stage 3) of the hydration reactions [16]. Normally, FA acts to reduce the heat of hydration as indicated by a lowering of the highest peak for OPC or pure cement. In contrast, it has been reported that α-Fe_2O_3 improves the rate of heat generation depending on both the particle size distribution and quantity of α-Fe_2O_3 [13,16]. Thus, the effect of added hematite to an FA blend is of interest.

Results for the FA blended cements without and with α-Fe_2O_3 up to an age of 40 h are shown in Figures 5 and 6. The general shape of the calorimetry curves is typical for cement hydration. A quick heat flow release in stage 1 (0 to 1 h) occurs due to the neutralization of electrostatic charge on the particle surfaces, dissolution of calcium sulfate and alkali sulfates. Stage 2 (1 to 3 h) starts with the onset C-S-H crystallization, precipitating a layer around the cement particles, which slows down the rate of hydration reactions with formation of a barrier around the particles. Stage 3 (3 to 10 h) proceeds with the main C-S-H gel creating an outer shell around the particles with high heat flow release. This is the acceleration period of the reaction. After peak heat flow release, the reaction rate of C-S-H gel formation decreases during the deceleration period from 10 to 40 h (stage 4).

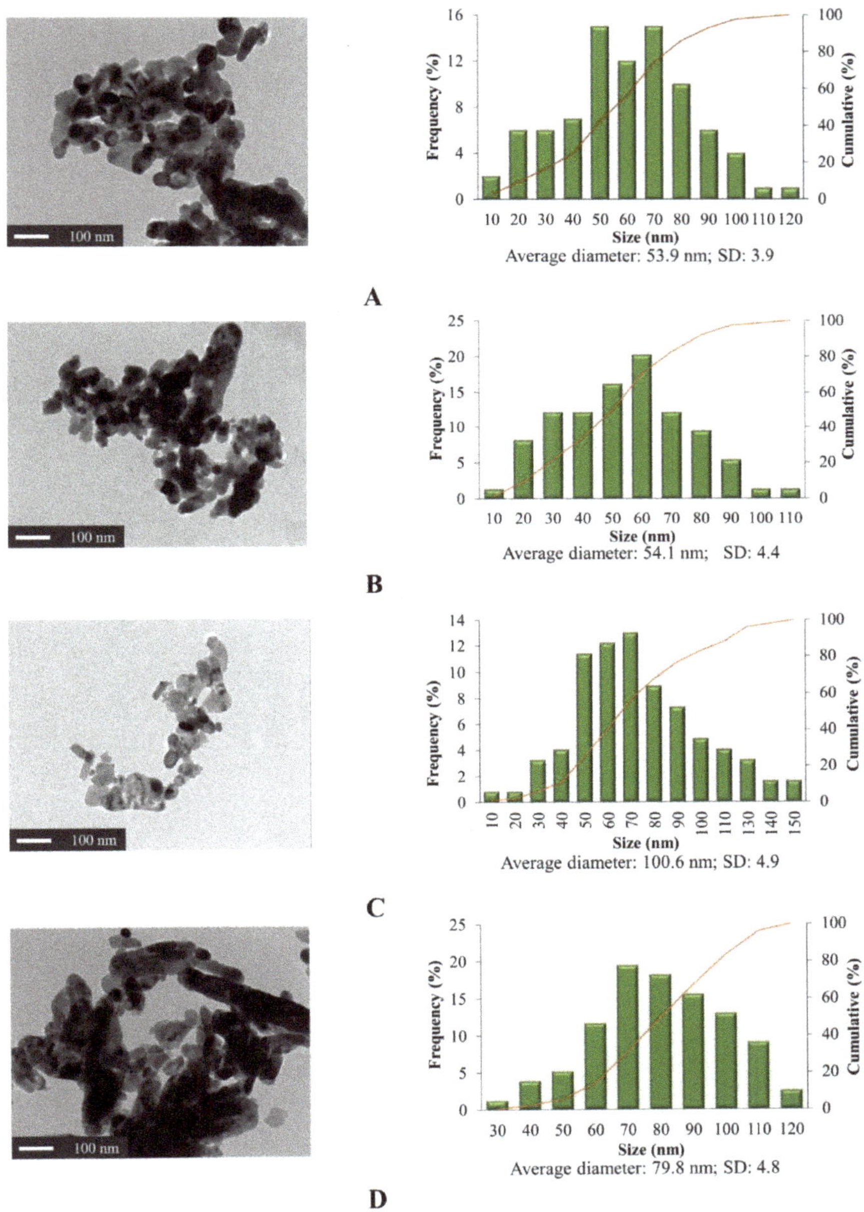

Figure 2. Transmission electron microscope (TEM) micrographs and grain size analysis of (**A**) commercial α-Fe_2O_3, (**B**) synthesized α-Fe_2O_3 at 2× critical micelle concentration (CMC) SDS, (**C**) synthesized α-Fe_2O_3 at 2× CMC CTAB and (**D**) synthesized α-Fe_2O_3 at 2× CMC TX100.

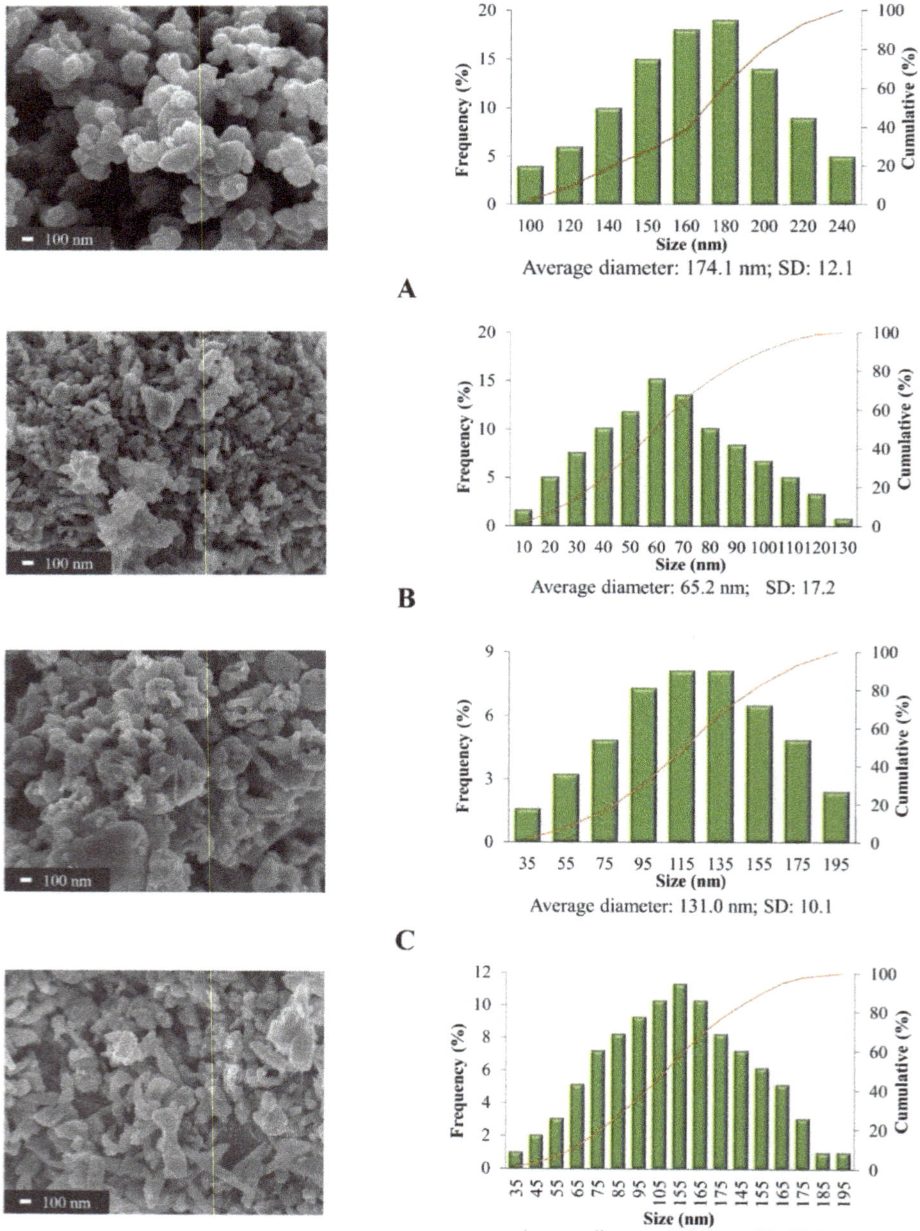

Figure 3. Morphology by field-emission-scanning-electron-microscopy (FE-SEM) imaging by secondary electron detector and size analysis of (**A**) commercial α-Fe_2O_3, (**B**) synthesized α-Fe_2O_3 at 2× CMC SDS, (**C**) synthesized α-Fe_2O_3 at 2× CMC CTAB and (**D**) synthesized α-Fe_2O_3 at 2× CMC TX100.

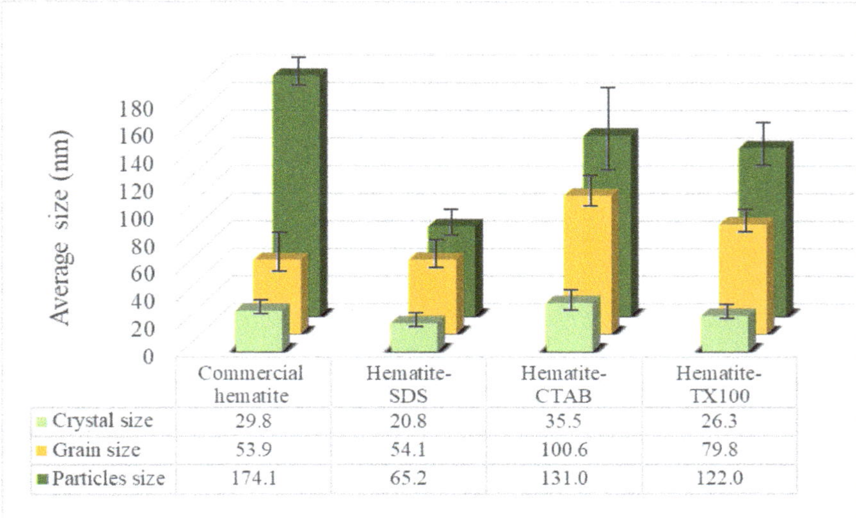

Figure 4. Crystal size obtained by XRD, grain size from ImageJ analysis of TEM images and particle size from ImageJ analysis of FE-SEM images of α-Fe_2O_3 nanoparticles.

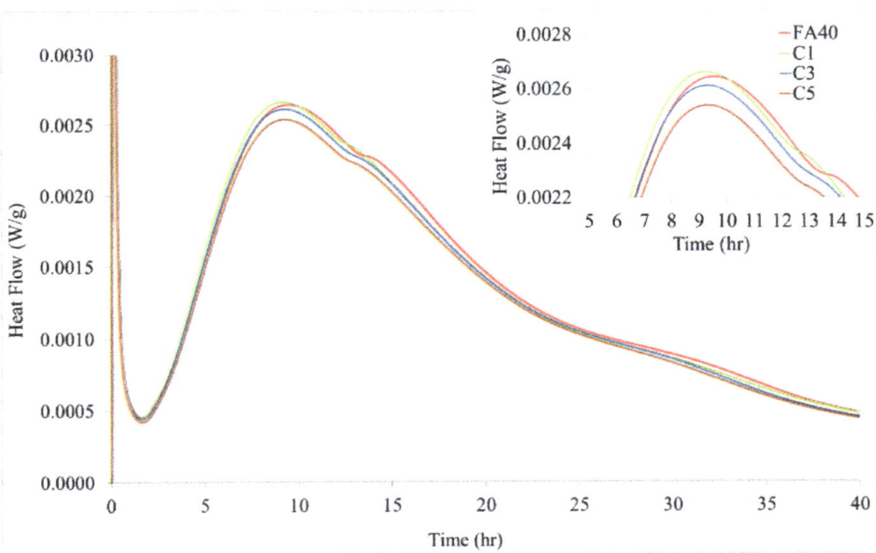

Figure 5. Effect of amount of commercial α-Fe_2O_3 at 0.0%, 1.0%, 3.0% and 5.0% on the time dependent heat of hydration for fly ash (FA) blended cement. Inset: Enlarged image for heat flow during stages 3 and 4 to show peak maxima for samples C1-C3 and control FA40.

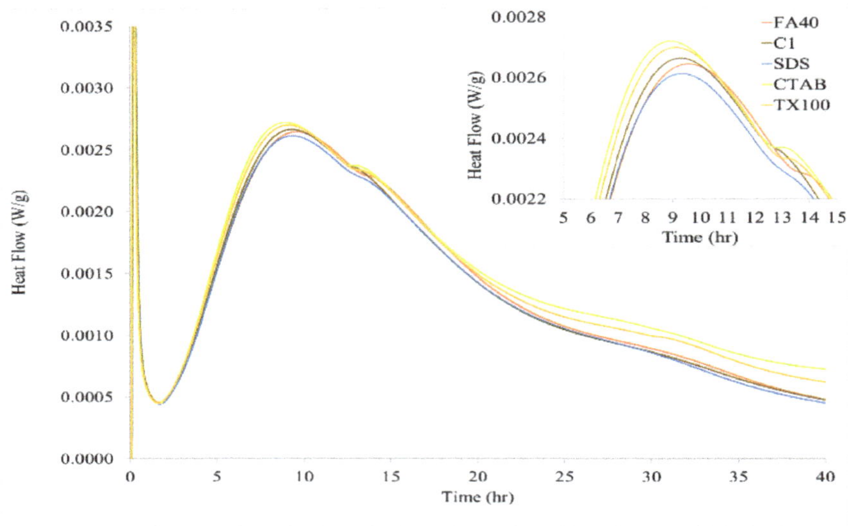

Figure 6. Effect of various synthesized α-Fe_2O_3 nanoparticles (1.0%) on the heat of hydration of FA blended cement. Inset: Enlarged image for heat flow during stages 3 and 4 showing peak maxima for control FA40 and nanoparticle samples.

With added commercial α-Fe_2O_3, the maximum rate of heat release shifts to earlier times while the maximum value decreases as the level of nanoparticles goes from 1.0 to 5.0% (Figure 5). At 1.0% α-Fe_2O_3 nanoparticles (C1), the time period of C_3S production from hydration was altered with the peak maximum appearing at 9.3 h, faster than without added α-Fe_2O_3 by approximately 0.5 h. This showed that the addition of α-Fe_2O_3 nanoparticles advanced the initiation and end of the acceleration period of cement hydration, resulting in the increase of the C–S–H crystal growth as reported by Kiamahalleh et al. [35]. After this enhanced reactivity, cement hydration slowed down in the deceleration period of stage 4 [14,36]. Maximum heat release of the α-Fe_2O_3 added FA cement decreased with the amount of the nanoparticles, becoming lower than that of FA cement alone. These findings imply that α-Fe_2O_3 nanoparticles acted as an accelerator to form C-S-H gel [37,38] with less exothermic heat overall being released as a result of reduced formation of C-S-H gel and ettringite [11,18]. Kishar et al. [39] similarly found that 1.0% α-Fe_2O_3 nanoparticles added in modified cementitious materials enhanced the cement hydration rate.

How the various surfactant-synthesized α-Fe_2O_3 nanoparticles affected the hydration of FA blended cement is shown in Figure 6. Peak heat flow generation observed for the 1.0% α-Fe_2O_3 added FA cement samples was approximately 0.0026 W/g at 9.3 h for the SDS synthesized nanoparticles, 0.0027 W/g at 9.1 h for the CTAB synthesized nanoparticles, and 0.0027 W/g at 9.2 h for TX100 synthesized nanoparticles. These results reflect the effect of size and size distribution since the type of surfactant influenced these characteristics of the α-Fe_2O_3 nanoparticles. For comparison, peak heat flow generation for the FA control sample was 0.0025 W/g at 9.8 h while the 1.0% commercial α-Fe_2O_3 value (C1) of 0.0026 W/g occurred at 9.3 h. A relationship can be seen between the heat of hydration and particle size for the FA blended cement with added hematite. The lowest heat of hydration occurred with the smallest nanoparticles synthesized (SDS, 10–130 nm) while the larger particles and broader size distribution of the nanoparticles synthesized with CTAB and TX100 (35–195 nm) had higher maximum heat flow values.

When α-Fe_2O_3 was added to FA blended cement, the maxima in the acceleration period occurred sooner at 9.1–9.4 h compared to 9.8 h in the absence of added α-Fe_2O_3. This indicated that the α-Fe_2O_3 nanoparticles function as hydration catalyst or accelerator, depending on the percentage added and particle diameter. The FA blended cement with added α-Fe_2O_3 nanoparticles, synthesized using CTAB and TX100, had the earliest maxima heat flow values along with their bigger particles and broader particle size distribution (35–195 nm), whereas the FA blended cement with added α-Fe_2O_3 nanoparticles, synthesized using SDS, had the latest maximum according to the smallest particles (10–130 nm). Kayali et al. [13,14] and Kocaba et al. [15] reported similar observations.

How the various sizes of the α-Fe_2O_3 nanoparticles synthesized using CTAB and TX100 advanced the acceleration period of cement hydration (higher heat of hydration) can be explained by greater nucleation to yield more C-S-H gel [40]. Several nucleation sites of C-S-H gel can produce a branch-like nanostructure, interconnected via electrostatic and van der Waals forces. This structural aspect of the C-S-H gel represents a scaffolding component enhancing mechanical properties of hardened cement. The CTAB and TX100 synthesized hematite nanoparticles had a broad range of sizes causing a different morphology of C-S-H and granular agglomeration of C-S-H gel as suggested by the slightly higher peak of the heat evolution (0.0027 W/g) compared to that of the commercial hematite-added FA blended cement (0.0026 W/g). The comparison of the highest peak hydration in the relation of time and heat flow of all studied samples (Figures 5 and 6) was concluded in Supplementary Figure S4.

Particles the size of α-Fe_2O_3 nanoparticles can also affect packing density of cementitious materials [41,42]. At the nanoscale (on the order of 10–100 nm), α-Fe_2O_3 nanoparticles can contribute to higher packing density and lower water content in the scaffolding component so the hydration reaction proceeds more slowly [41–43]. At the mesoscale with bigger particles and a broader particle size distribution (on the order of 100 nm), high water content in the scaffolding component leads to an increased hydration reaction [41–43]. This knowledge implies that cement-based materials could be further upgraded by fine-tuning the nano and microstructure [41].

3.3. *Workability and Compressive Strength of FA Blended Cement: Effect of α-Fe_2O_3 Addition*

Workability, a property of freshly mixed FA blended cement directly tied to strength and appearance, is typically gauged by setting time, mini-slump test and flow table test, though attempts have been made to relate workability directly to the hydration reaction. Increased hydration products or C-S-H gel correspond to decreased excess water and improved workability. In practice, it is necessary to place and consolidate the cement product before initial setting starts and then not to disturb the sample until the final setting of mortar or concrete. If delayed, the cement will lose strength.

3.3.1. Setting Time

Setting times of cement paste, related to workability, are defined as of the moment cement changes from liquid state to plastic state and then to solid state. Two important properties, initial and final setting times, indicated the stiffness of the cementitious matrix and the rate of solidification of mortar or concrete with added nanoparticles.

Setting times were observed for all samples as shown in Figure 7. The initial and final setting times of FA blended cement control were 4.5 and 6.0 h, respectively. Normally, the initial and final setting times of cement are specified as a requirement of related standards (EN-197, 2011; ASTM C150, 2015). For the initial setting time, the value should not be less than 0.5 h while the final setting time should not be greater than 10.0 h. This means that the FA blended cement must be in place within 4.5 h after mixing with water before it will consolidate and then must be left undisturbed until 6.0 h.

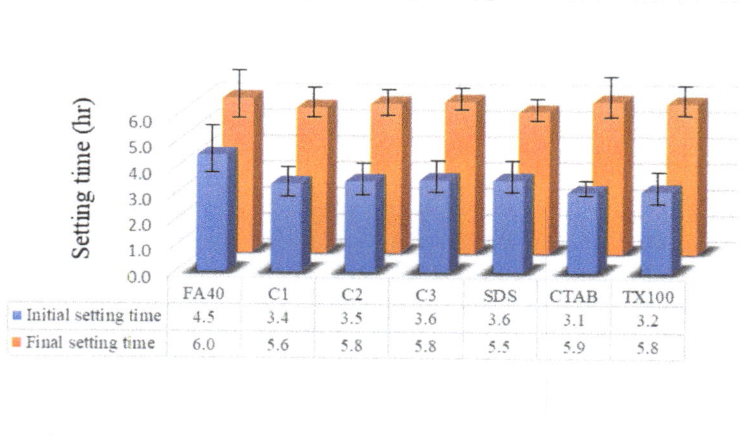

Figure 7. Effect of concentration and synthesized α-Fe_2O_3 nanoparticles on setting time of FA blended cement mortar.

When commercial α-Fe_2O_3 nanoparticles were added at 1.0% in the FA blended cement, the initial setting time and the final setting time decreased to 3.4 h and 5.6 h. FA blended cement with the hematite nanoparticles synthesized using CTAB gave the lowest initial setting time at 3.1 h in agreement with its highest rate of hydration reaction (shown in Figures 5 and 6). The slightly shorter initial setting time of the FA blended cement with the added hematite nanoparticles led to higher strength gain from the effect of nucleation on high degree of hydration, low porosity and permeability [42]. For the final setting times of α-Fe_2O_3 modified samples, the FA blended cement with CTAB nanoparticles gave the highest final setting time at 5.9 h, while the FA blended cement with the SDS nanoparticles exhibited the lowest final setting time at 5.5 h. Final setting times for all samples met the applicable standards of 10.0 h or less.

3.3.2. Mini-Slump Test and Flow Table

The mini-slump test measures the workability of fresh mortar from its behavior under the action of gravity while the flow table test reflects the consistency or wetness of OPC, giving another measure of the workability. Results for these tests are presented in Figure 8. The mini-slump test of control sample FA blended cement exhibited the greatest height at 6.7 cm, while the FA blended cement containing α-Fe_2O_3 synthesized with CTAB showed the least height at 5.1 cm. The trend of the data from the flow table test was the same as that for the mini-slump test results. The flow table test showed that the FA blended cement with hematite nanoparticles synthesized via CTAB had the lowest spread at 22.0 cm, while the FA blended cement was the highest at 34.5 cm.

Results of the mini-slump test and flow table of the FA blended cement control were higher than those of all α-Fe_2O_3 added FA blended cements. As the amount of α-Fe_2O_3 was increased, values obtained for flow test also increased. These findings with the heat flow results show workability depends upon the water inside the cement during hydration. Water exists in four forms including capillary water, adsorbed water, interlayer water and chemically combined water. In general, capillary water, or generally called "free water", affects the flow of cement with the attractive forces exerted by the solid surface [43,44]. When the nanoparticles of α-Fe_2O_3 were added, values of the mini-slump test and flow table for the cement composites slightly decreased due to the displacement of free water caused by the reduction of interstitial volume inside the cement during reaction [45]. Currently the α-Fe_2O_3 nanoparticles are used in concrete work to reduce the amount of free water by accelerating the hydration reaction [46].

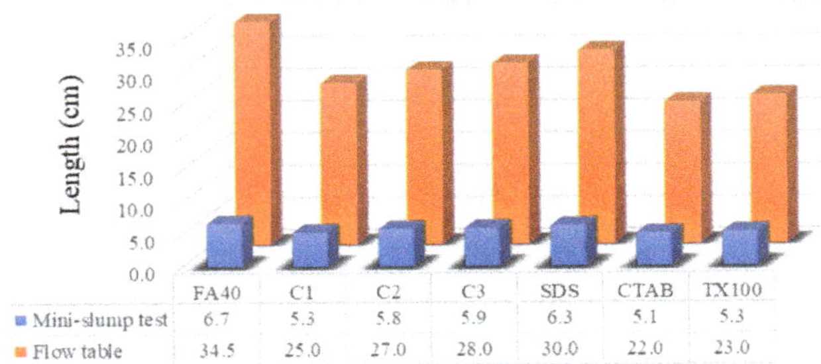

Figure 8. Effect of α-Fe_2O_3 nanoparticles on mini-slump test and flow table of FA blended cement mortar.

3.3.3. Compressive Strength

The compressive strength of a cured cement composite is commonly measured at 7, 28 and 90 days. As expected, the compressive strength increased with time (Figure 9). Compressive strength of the FA blended mortar control rose with the value at 7 days being 62% of that at 90 days. This can be compared to the work of Abd elaty et al., who found that a cement composite reaches as much as 60% of the equilibrium value for all mechanical properties at 7 days [47]. At 7 days, the compressive strength of FA blended mortar with commercial hematite 3.0% was the lowest at 24.0 MPa, while FA blended mortar with CTAB-synthesized α-Fe_2O_3 or TX100-synthesized α-Fe_2O_3 of various sizes was the highest at 26.0 MPa. At 90 days, the same CTAB mortar still had the highest compressive strength, now at 47.1 MPa, with a similar value for the TX100 synthesized nanoparticle modified mortar.

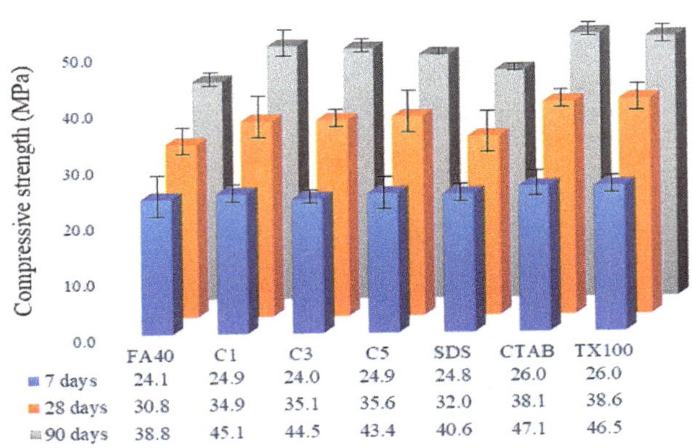

Figure 9. Compressive strength of FA blended cement mortar at 7, 28 and 90 days via varying α-Fe_2O_3.

Higher compressive strength among the α-Fe_2O_3 modified samples corresponded to faster rates of the hydration reactions. The highest compressive strength occurred with the variable nanosize of the CTAB-assisted synthesized α-Fe_2O_3 in line with the hydration behavior within the various voids in the mortar. The results imply that the α-Fe_2O_3 nanoparticles function as an accelerator of cement hydration and filled in the voids during reaction.

3.4. Microstructural Analysis of FA Blended Cement Added α-Fe_2O_3 by SEM Images

SEM micrographs of FA blended cement without and with added α-Fe_2O_3 were used to observe the microstructure of the cement paste at 7, 28 and 90 curing days (Figure 10). With partial replacement of OPC by FA, the cement paste shows looser and more inhomogeneous microstructure with presence of less C-S-H gel (Figure 10A, point 1) and more pores (point 2). The morphology of FA blended cement with hematite nanoparticles of various curing times shows that when the curing time was longer, the C-S-H gel products formed a denser structure around the embedded FA particles. The α-Fe_2O_3 added mortar enhanced the compressive strength by reducing the porosity of cementitious composites, a result in accord with previous studies where iron oxides increased density [37,48,49]. The size and size distribution had the effect of fitting into available pore space in the cement composites during hydration. Moreover, α-Fe_2O_3 nanoparticles help the hydration reaction by increasing the nucleation of hydration products to form more C-S-H gel [40].

At 7 curing days (in the first column) all cement samples show incomplete integration of the FA particles in the overall structure of the cement reaction products. The images present many voids and incompletely covered C-S-H on FA particles. In contrast, the microstructure of the FA blended cement with added α-Fe_2O_3 nanoparticles consists of dense crystal hydrated products around FA according to the nucleation of α-Fe_2O_3 nanoparticles (Figure 10B–E). The cement paste with commercial α-Fe_2O_3 has C-S-H gel covered FA particles but lower than with the synthesized α-Fe_2O_3. Images of the synthesized α-Fe_2O_3 using CTAB (Figure 10D) or TX100 (Figure 10E) indicated C-S-H gel covered on FA particles at a higher level than that of synthesized α-Fe_2O_3 using SDS (Figure 10C).

At 28 curing days (in the second column), FA blended cement pastes were found to show two different forms including an incompletely covered C-S-H layer on FA in cement without α-Fe_2O_3 nanoparticles (Figure 10A) and a dense layered C-S-H on FA in cement with α-Fe_2O_3 nanoparticles (Figure 10B–E). These results implied that α-Fe_2O_3 nanoparticles acted not only as an inert material with the ability to increase the packing effect, but also led to hydration products forming around the FA. During cement hydration, the microstructure of FA blended cement with α-Fe_2O_3 nanoparticles becomes denser due to formation of more C-S-H gel. Especially, the microstructures of FA blended cement containing α-Fe_2O_3 synthesized by CTAB (D) or TX100 (E) showed a relatively densified network gel of hydration products layered on FA particles.

At 90 days (in the last column), all cement pastes were completely hydrated. The gel filling out the spaces around particles and the covering layer on FA surface can be observed. Almost all FA particles were substituted by hydrated products when FA was activated by $Ca(OH)_2$. The pozzolanic reaction of FA speeded up at the later stages and the consumed content of C-S-H gel increased. The surface of FA covered by C-S-H gel were caused by pozzolanic reaction between FA and $Ca(OH)_2$ and other hydration products [50]. As the process continued, the layer on the FA and cement grain thickened. The hydration products growing from the cement grains and FA are seen to be connected, although some particles still remain unreacted and acted solely as filler (Figure 10A). For FA blended cement adding α-Fe_2O_3 nanoparticles, the micrograph shows the surface of FA covered by C-S-H gel (Figure 10B–E). From the SEM images, more pores of the solidate cement were filled by hydration products and hematite nanoparticles. This microstructure is consistent with the enhanced compressive strength and workability.

Therefore, the influence of α-Fe_2O_3 nanoparticles was established as a promotor of the hydration reaction, and also as a filler, leading to increased strength of cement by filling voids between FA and the hydration products.

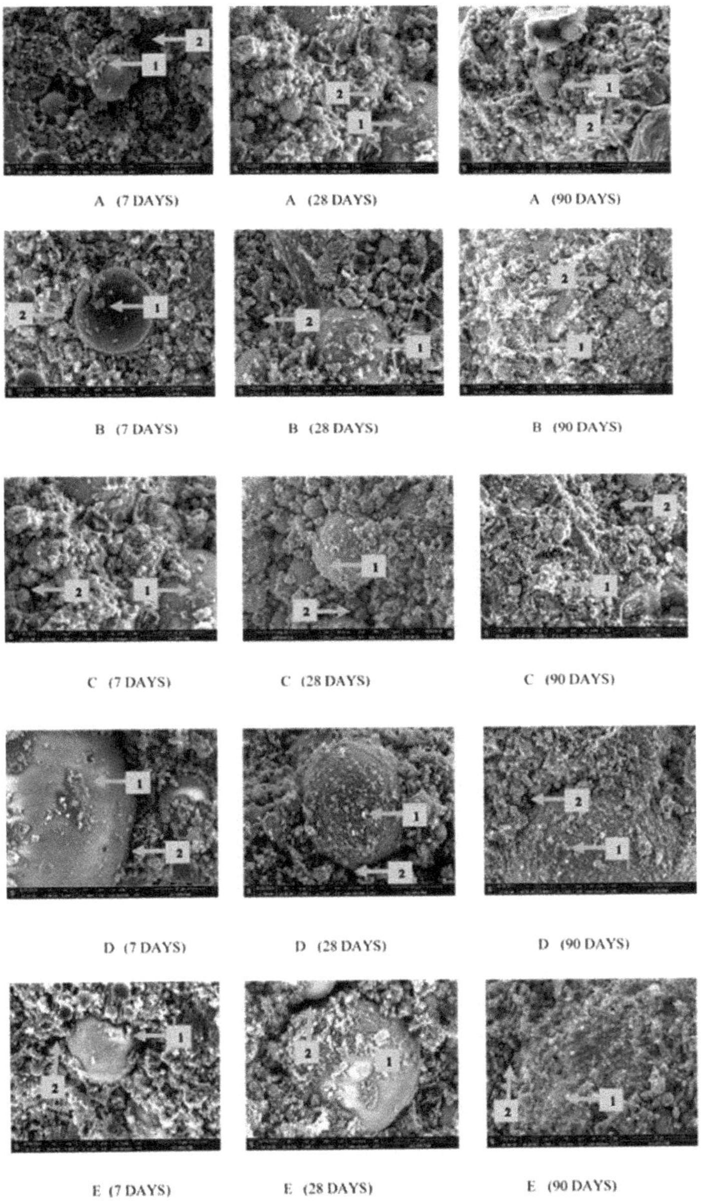

Figure 10. SEM micrographs of hydrated FA blended cement at 7, 28 and 90 days showing layered $Ca(OH)_2$ around FA (point 1) and the air voids (point 2): varying α-Fe_2O_3 of (**A**) FA40, (**B**) C1, (**C**) SDS, (**D**) CTAB and (**E**) TX100.

3.5. Structural Model of α-Fe_2O_3 Nanoparticle Modified FA Cement

The hydration of cement blended with FA in this work can be illustrated as three cases: (I) without nanoparticles added, (II) with smaller and uniform particle size nanoparticles present and (III) with bigger, broadly distributed particle size nanoparticles (Figure 11). Case I (Figure 11a), the cementitious composites produce the C-S-H gel and capillary pores during hydration [51]. The added FA particles fill in the capillary pore of C-S-H gel structure to reduce the porosity in cement [52,53]. Meanwhile, as hydration progresses, the capillary pores create even more spacing that may decrease the strength of cement composites. To reduce capillary voids, α-Fe_2O_3 nanoparticles function as both filler and hydration accelerator that can improve the microstructure and strength of cementitious composites as Cases II and III. Case II, by adding nanosize fillers smaller than the capillary pore size, the uniform or narrow particle size distribution of nanofillers still leaves small voids from the surface-to-surface distance between the particles and consolidated cement. When the SDS synthesized α-Fe_2O_3 filler of smaller size and narrower particle size distribution (10–130 nm) was used, the capillary pore volume inside the cement was higher and the hydration rate was lower (Figure 11b). For Case III, large and small capillary pores are effectively filled with the bigger size and broader size distribution of nanofillers to lower the porosity in the cementitious composites. The results from hydration and cement properties of this work confirmed this hypothesis. α-Fe_2O_3 nanoparticles synthesized by CTAB and TX100 (range size distribution 35–195 nm) facilitated cement hydration by accelerating the formation of C-S-H gel and filling up the capillary pores, resulting in a higher density and compact microstructure (Figure 11c). With the various sizes of α-Fe_2O_3, the capillary pore size in cement and agglomerated particles around the cement grains are reduced according to the size of the additives that play an effective role in acceleration of hydration reaction and densification of the cement matrix [48,53,54].

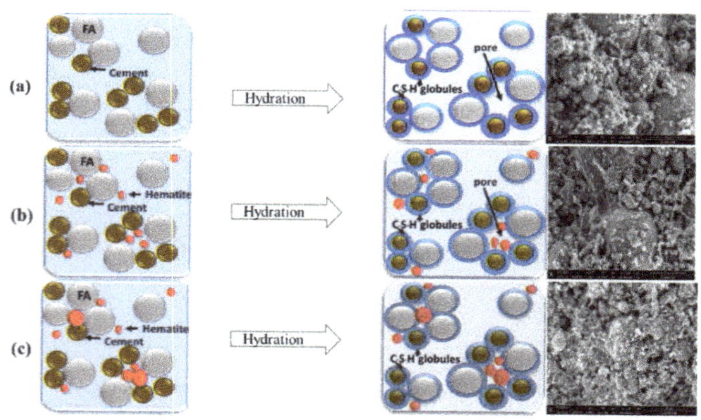

Figure 11. Schematic of cement hydration of (**a**) no α-Fe_2O_3, (**b**) α-Fe_2O_3 in the narrow range particle size distribution and (**c**) α-Fe_2O_3 in the broad particle size distribution.

4. Conclusions

Based on the morphologically controlled synthesis of α-Fe_2O_3, the spherical α-Fe_2O_3 of different sizes were produced using surfactant templates for nanoparticle synthesis. The smallest nanoparticles were produced when using SDS for the synthesis, whereas production using CTAB yielded the largest nanoparticles. The average particle sizes of the α-Fe_2O_3 synthesized via SDS and CTAB were 65.2 nm and 131.0 nm, relating to the size of their micelles. The effect of α-Fe_2O_3 nanoparticles on the hydration and properties of a FA cement blend was investigated as a function of concentration and size of the nanoparticles. The different nanoparticles in this study on FA blended cements illustrated how the particle size and particle size distribution of nanoparticles affect structure and properties of cementitious materials. The various particle size fractions of α-Fe_2O_3 nanoparticles increased the cement hydration rate and affected cementitious material pore structure development, leading to increased compressive strength. The optimum of the amount of α-Fe_2O_3 nanoparticles was 1.0% by weight replacement of OPC that led to the fastest hydration rate and the highest degree of hydration of cement samples. With the various sizes of synthesized α-Fe_2O_3 nanoparticles via CTAB and TX100, the capillary pore size in cement and agglomerated particles around the cement grains are reduced according to the size of the additives that play an effective role in acceleration of hydration reaction and densification of the cement matrix. Moreover, the α-Fe_2O_3 nanoparticles in binder content may help improve the overall stability and property of self-compacting concrete.

Supplementary Materials: The following are available online at https://www.mdpi.com/2079-4991/11/4/1003/s1, Supplementary Figure S1: Heat of hydration curve, Supplementary Figure S2. Syringe to fill water to the binder for hydration observation by TAM air calorimeter, Supplementary Figure S3. Mini-slump test operation, Supplementary Figure S4. Comparisons the heat flow and time of the hydration peak maximum for FA blended cement with commercial α-Fe_2O_3 and synthesized α-Fe_2O_3.

Author Contributions: Conception and design of study: P.K., S.S., E.A.O. and T.P.; Analysis and/or interpretation of data: P.K., S.S., E.A.O. and T.P.; Drafting the manuscript: P.K., T.P.; Revising the manuscript critically for important intellectual content: E.A.O., S.S.; Approval of the version of the manuscript to be published: P.K., S.S., E.A.O. and T.P. All authors have read and agreed to the published version of the manuscript.

Funding: This research was funded by The Thailand Science Research and Innovation (TSRI) cooperated with Siam Research and In-novation Co., Ltd., Thailand, for the "Research and Researcher for Industry (RRI) Fund", contract number PHD 59I0071, which provides for Pantharee Kongsat for her Ph.D. study and research. The publication was also supported by the University Libraries Open Access Fund of the University of Oklahoma.

Data Availability Statement: The data that support the findings of this study are available from the corresponding author upon reasonable request.

Acknowledgments: The authors wish to thank The Thailand Science Research and Innovation (TSRI) cooperated with Siam Research and Innovation Co., Ltd., Thailand, for the "Research and Researcher for Industry (RRI) Fund", contract number PHD 59I0071, which provides for Pantharee Kongsat for her study and research. The cement test was supported by Siam Research and Innovation Co., Ltd., Thailand. Special thanks to Edgar A. O'Rear for his critical reading of the manuscript. The authors would also like to thank the University Libraries Open Access Fund of the University of Oklahoma.

Conflicts of Interest: The authors declare that there is no conflict of interests.

Abbreviations

Abbreviations

α-Fe_2O_3	Hematite nanoparticles
OPC	Ordinary Portland Cement
FA	Fly ash
C_3A	Tricalcium aluminate or $3CaO.Al_2O_3$
C_3S	Alite or $3CaO.SiO_2$
C_2S	Belite or $2CaO.SiO_2$
C-S-H gel	Calcium-silicate-hydrate gel
CAS^-	Calcium sulfoaluminate or $Ca_4Al_2(SO_4)(OH)_{12}.6H_2O$
C_4AF	Ferrite phase or $4CaO.Al_2O_3.Fe_2O_3$

References

1. Paul, S.C.; Van Rooyen, A.S.; Van Zijl, G.P.; Petrik, L.F. Properties of cement-based composites using nanoparticles: A comprehensive review. *Constr. Build. Mater.* **2018**, *189*, 1019–1034. [CrossRef]

2. Balapour, M.; Joshaghani, A.; Althoey, F. Nano-SiO_2 contribution to mechanical, durability, fresh and microstructural characteristics of concrete: A review. *Constr. Build. Mater.* **2018**, *181*, 27–41. [CrossRef]
3. Demirboğa, R. Influence of mineral admixtures on thermal conductivity and compressive strength of mortar. *Energy Build.* **2003**, *35*, 4. [CrossRef]
4. Subramanian, N. Introduction to Reinforced Concrete. In *Design of Reinforced Concrete Structures*; Harper & Row: New York, NY, USA, 2013.
5. Chindaprasirt, P.; Jaturapitakkul, C.; Sinsiri, T. Effect of fly ash fineness on compressive strength and pore size of blended cement paste. *Cem. Concr. Compos.* **2005**, *27*, 425–428. [CrossRef]
6. Zeng, Q.; Li, K.; Fen-Chong, T.; Dangla, P. Pore structure characterization of cement pastes blended with high-volume fly-ash. *Cem. Concr. Res.* **2012**, *42*, 194–204. [CrossRef]
7. Feng, J.; Sun, J.; Yan, P. The Influence of Ground Fly Ash on Cement Hydration and Mechanical Property of Mortar. *Adv. Civ. Eng.* **2018**, *2018*, 4023178. [CrossRef]
8. Siddique, R. Effect of fine aggregate replacement with Class F fly ash on the abrasion resistance of concrete. *Cem. Concr. Res.* **2003**, *33*, 1877–1881. [CrossRef]
9. Wang, H.; Li, H.; Liang, X.; Zhou, H.; Xie, N.; Dai, Z. Investigation on the mechanical properties and environmental impacts of pervious concrete containing fly ash based on the cement-aggregate ratio. *Constr. Build. Mater.* **2019**, *202*, 387–395. [CrossRef]
10. Artioli, G.; Bullard, J.W. Cement hydration: The role of adsorption and crystal growth. *Cryst. Res. Technol.* **2013**, *48*, 903–918. [CrossRef]
11. Gani, M.S.J. *Cement and Concrete*; CRC Press: Boca Raton, FL, USA, 1997.
12. Yuenyongsuwan, J.; Sinthupinyo, S.; O'Rear, E.A.; Pongprayoon, T. Hydration accelerator and photocatalyst of nanotitanium dioxide synthesized via surfactant-assisted method in cement mortar. *Cem. Concr. Compos.* **2019**, *96*, 182–193. [CrossRef]
13. Scrivener, K.; Ouzia, A.; Juilland, P.; Mohamed, A.K. Advances in understanding cement hydration mechanisms. *Cem. Concr. Res.* **2019**, *124*. [CrossRef]
14. Scrivener, K.L.; Juilland, P.; Monteiro, P.J. Advances in understanding hydration of Portland cement. *Cem. Concr. Res.* **2015**, *78*, 38–56. [CrossRef]
15. Kocaba, V.; Gallucci, E.; Scrivener, K.L. Methods for determination of degree of reaction of slag in blended cement pastes. *Cem. Concr. Res.* **2012**, *42*, 511–525. [CrossRef]
16. Monteagudo, S.; Moragues, A.; Gálvez, J.; Casati, M.; Reyes, E. The degree of hydration assessment of blended cement pastes by differential thermal and thermogravimetric analysis. Morphological evolution of the solid phases. *Thermochim. Acta* **2014**, *592*, 37–51. [CrossRef]
17. Vidhya, R.D.K. An Experimental Investigation on Strength Characteristic of High Density Concrete Incorporating Hematite. *Int. J. Innov. Res. Sci. Technol.* **2015**, *2*, 6.
18. Tobbala, D. Effect of Nano-ferrite addition on mechanical properties and gamma ray attenuation coefficient of steel fiber reinforced heavy weight concrete. *Constr. Build. Mater.* **2019**, *207*, 48–58. [CrossRef]
19. Li, H.; Xiao, H.-G.; Yuan, J.; Ou, J. Microstructure of cement mortar with nano-particles. *Compos. Part B Eng.* **2004**, *35*, 185–189. [CrossRef]
20. Jennings, H.M.; Kumar, A.; Sant, G. Quantitative discrimination of the nano-pore-structure of cement paste during drying: New insights from water sorption isotherms. *Cem. Concr. Res.* **2015**, *76*, 27–36. [CrossRef]
21. Babay, S.; Mhiri, T.; Toumi, M. Synthesis, structural and spectroscopic characterizations of maghemite γ-Fe_2O_3 prepared by one-step coprecipitation route. *J. Mol. Struct.* **2015**, *1085*, 286–293. [CrossRef]
22. Tadic, M.; Panjan, M.; Damnjanovic, V.; Milosevic, I. Magnetic properties of hematite (α-Fe_2O_3) nanoparticles prepared by hydrothermal synthesis method. *Appl. Surf. Sci.* **2014**, *320*, 183–187. [CrossRef]
23. Kongsat, P.; Kudkaew, K.; Tangjai, J.; O'Rear, E.A.; Pongprayoon, T. Synthesis of structure-controlled hematite nanoparticles by a surfactant-assisted hydrothermal method and property analysis. *J. Phys. Chem. Solids* **2021**, *148*, 109685. [CrossRef]
24. Vaidya, S.; Ganguli, A.K. Microemulsion Methods for Synthesis of Nanostructured Materials. *Compr. Nanosci. Nanotechnol.* **2019**, 1–12. [CrossRef]
25. Shaikuthali, S.A.; Mannan, M.A.; Dawood, E.T.; Teo, D.C.L.; Ahmadi, R.; Ismail, I. Workability and compressive strength properties of normal weight concrete using high dosage of fly ash as cement replacement. *J. Build. Pathol. Rehabil.* **2019**, *4*, 26. [CrossRef]
26. Gilson, R.; Lomboy, K.W.; Sundararajan, S. Nanoscale Characterization of Cementitious Materials. In *Nanotechnology in Construction Proceedings of NICOM5*; Sobolev, S.P.S.K., Ed.; Springer International Publishing: Heidelberg, Germany, 2015; p. 45.
27. Kim, S.-J.; Yang, K.-H.; Moon, G.-D. Hydration Characteristics of Low-Heat Cement Substituted by Fly Ash and Limestone Powder. *Materials* **2015**, *8*, 5847–5861. [CrossRef]
28. Elahi, A.; Basheer, P.; Nanukuttan, S.; Khan, Q. Mechanical and durability properties of high performance concretes containing supplementary cementitious materials. *Constr. Build. Mater.* **2010**, *24*, 292–299. [CrossRef]
29. Atkin, R.; Craig, V.; Wanless, E.; Biggs, S. Mechanism of cationic surfactant adsorption at the solid–aqueous interface. *Adv. Colloid Interface Sci.* **2003**, *103*, 219–304. [CrossRef]
30. Lu, Y.; Yue, Z.; Xie, J.; Wang, W.; Zhu, H.; Zhang, E.; Cao, Z. Micelles with ultralow critical micelle concentration as carriers for drug delivery. *Nat. Biomed. Eng.* **2018**, *2*, 318–325. [CrossRef]

31. Das, N.C.; Cao, H.; Kaiser, H.; Warren, G.T.; Gladden, J.R.; Sokol, P.E. Shape and Size of Highly Concentrated Micelles in CTAB/NaSal Solutions by Small Angle Neutron Scattering (SANS). *Langmuir* **2012**, *28*, 11962–11968. [CrossRef] [PubMed]
32. Paradies, H.H. Shape and size of a nonionic surfactant micelle. Triton X-100 in aqueous solution. *J. Phys. Chem.* **1980**, *84*, 599–607. [CrossRef]
33. Jing, Z.; Han, D.; Wu, S. Morphological evolution of hematite nanoparticles with and without surfactant by hydrothermal method. *Mater. Lett.* **2005**, *59*, 804–807. [CrossRef]
34. Colombo, C.; Palumbo, G.; Di Iorio, E.; Song, X.; Jiang, Z.; Liu, Q.; Angelico, R. Influence of hydrothermal synthesis conditions on size, morphology and colloidal properties of Hematite nanoparticles. *Nano-Struct. Nano-Objects* **2015**, *2*, 19–27. [CrossRef]
35. Kiamahalleh, M.V.; Alishah, A.; Yousefi, F.; Astani, S.H.; Gholampour, A.; Kaiamahalleh, M.V. Iron oxide nanoparticle incorporated cement mortar composite: Correlation between physico-chemical and physico-mechanical properties. *Mater. Adv.* **2020**, *1*, 1835–1840. [CrossRef]
36. Pustovgar, E.; Sangodkar, R.P.; Andreev, A.S.; Palacios, M.; Chmelka, B.F.; Flatt, R.J.; De Lacaillerie, J.-B.D. Understanding silicate hydration from quantitative analyses of hydrating tricalcium silicates. *Nat. Commun.* **2016**, *7*, 10952. [CrossRef]
37. Cao, B.; Fan, S.; Tan, X.; Li, M.; Hu, Y. Cementitious materials modified with hematite nanoparticles for enhanced cement hydration and uranium immobilization. *Environ. Sci. Nano* **2017**, *4*, 1670–1681. [CrossRef]
38. Abo-El-Enein, S.; El-Hosiny, F.; El-Gamal, S.; Amin, M.; Ramadan, M. Gamma radiation shielding, fire resistance and physico-chemical characteristics of Portland cement pastes modified with synthesized Fe_2O_3 and ZnO nanoparticles. *Constr. Build. Mater.* **2018**, *173*, 687–706. [CrossRef]
39. Kishar, E.A.; Alasqalani, M.Y.; Sarraj, Y.R.; Ahmed, D.A. The Effect of Using Commercial Red and Black Iron Oxides as a Concrete Admixtures on its Physiochemical and Mechanical Properties. *Int. J. Sci. Res.* **2015**, *4*, 5.
40. Berodier, E.; Scrivener, K. Understanding the Filler Effect on the Nucleation and Growth of C-S-H. *J. Am. Ceram. Soc.* **2014**, *97*, 3764–3773. [CrossRef]
41. Lin, K.L.; Chang, W.C.; Lin, D.F.; Luo, H.L.; Tsai, M.C. Effects of nano-SiO(2) and different ash particle sizes on sludge ash-cement mortar. *J. Environ. Manag.* **2008**, *88*, 708–714. [CrossRef]
42. Li, H.; Ding, S.; Zhang, L.; Ouyang, J.; Hna, B. Effects of particle size, crystal phase and surface treatment of nano-TiO_2 on the rheological parameters of cement paste. *Constr. Build. Mater.* **2020**, *239*, 117897. [CrossRef]
43. Wang, Y.; Jin, Z.; Liu, S.; Yang, L.; Luo, S. Physical filling effect of aggregate micro fines in cement concrete. *Constr. Build. Mater.* **2013**, *41*, 812–814. [CrossRef]
44. Zhao, M.; Zhang, X.; Zhang, Y. Effect of free water on the flowability of cement paste with chemical or mineral admixtures. *Constr. Build. Mater.* **2016**, *111*, 571–579. [CrossRef]
45. Mehdipour, I.; Khayat, K.H. Effect of particle-size distribution and specific surface area of different binder systems on packing density and flow characteristics of cement paste. *Cem. Concr. Compos.* **2017**, *78*, 120–131. [CrossRef]
46. Ali Nazari, S.R.; Riahi, S.; Shamekhi, S.F.; Khademno, A. Benefits of Fe_2O_3 nanoparticles in concrete mixing matrix. *J. Am. Sci.* **2010**, *6*, 5.
47. Abd Elaty, M.A.A. Compressive strength prediction of Portland cement concrete with age using a new model. *HBRC J.* **2019**, *10*, 145–155. [CrossRef]
48. Ghazanlou, S.I.; Jalaly, M.; Sadeghzadeh, S.; Korayem, A.H. A comparative study on the mechanical, physical and morphological properties of cement-micro/nanoFe_3O_4 composite. *Sci. Rep.* **2020**, *10*, 2859. [CrossRef]
49. Chung, C.-W.; Chun, J.; Wang, G.; Um, W. Effects of iron oxides on the rheological properties of cementitious slurry. *Colloids Surf. A Physicochem. Eng. Asp.* **2014**, *453*, 94–100. [CrossRef]
50. Shi, C.; Day, R.L. Pozzolanic reaction in the presence of chemical activators Part II. Reaction products and mechanism. *Cem. Concr. Res.* **2000**, *30*, 607–613.
51. Manfroi, E.P.; Cheriaf, M.; Rocha, J.C. Microstructure, mineralogy and environmental evaluation of cementitious composites produced with red mud waste. *Constr. Build. Mater.* **2014**, *67*, 29–36. [CrossRef]
52. Zarzuela, R.; Luna, M.; Carrascosa, L.M.; Yeste, M.P.; Garcia-Lodeiro, I.; Blanco-Varela, M.T.; Cauqui, M.A.; Rodríguez-Izquierdo, J.M.; Mosquera, M.J. Producing C-S-H gel by reaction between silica oligomers and portlandite: A promising approach to repair cementitious materials. *Cem. Concr. Res.* **2020**, *130*, 106008. [CrossRef]
53. Yazdi, N.A.; Arefi, M.R.; Mollaahmadi, E.; Nejand, B.A. To study the effect of adding Fe_2O_3 nanoparticles on the morphology properties and microstructure of cement mortar. *Life Sci. J.* **2011**, *8*, 5.
54. Ng, D.S.; Paul, S.C.; Anggraini, V.; Kong, S.Y.; Qureshi, T.S.; Rodriguez, C.R.; Liu, Q.-F.; Šavija, B. Influence of SiO_2, TiO_2 and Fe_2O_3 nanoparticles on the properties of fly ash blended cement mortars. *Constr. Build. Mater.* **2020**, *258*, 119627.

nanomaterials

MDPI

Article

Synthesis of Highly-Dispersed Graphene Oxide Nanoribbons–Functionalized Carbon Nanotubes–Graphene Oxide (GNFG) Complex and Its Application in Enhancing the Mechanical Properties of Cementitious Composites

Peiqi Li [1], Junxing Liu [1], Sungwun Her [1], Erfan Zal Nezhad [2], Seungmin Lim [3] and Sungchul Bae [1,*]

1 Department of Architectural Engineering, Hanyang University, Seoul 04763, Korea; lipeiqi1995@hanyang.ac.kr (P.L.); liujx128119@hanyang.ac.kr (J.L.); sung0@hanyang.ac.kr (S.H.)
2 Department of Biomedical Engineering, University of Texas, San Antonio, TX 78249, USA; erfan.zalnezhad@utsa.edu
3 Department of Architecture, Kangwon National University, Chuncheon 24341, Korea; smlim@kangwon.ac.kr
* Correspondence: sbae@hanyang.ac.kr

Abstract: In this study, a graphene oxide nanoribbons–functionalized carbon nanotubes–graphene oxide (GNFG) complex was hydrothermally synthesized as a nanomaterial for reinforcing cementitious composites, using a modified Hummers' method. Three types of components existed in the GNFG: Type I, the functionalized carbon nanotubes–graphene oxide nanoribbons (FCNTs–GNR); and types II and III are graphene oxide (GO) and functionalized carbon nanotubes (FCNTs), respectively, which exist independently. The dispersivity of GNFG and its effects on the mechanical properties, hydration process, and microstructures of cement pastes were evaluated, and the results were compared with those using cement pastes incorporating other typical carbon nanomaterials. The results demonstrated that dispersion of GNFG in aqueous solutions was superior to that of the CNTs, FCNTs, and GO/FCNTs mixture. Furthermore, the highly-dispersed GNFG (0.05 wt.%) improved the mechanical properties of the cement paste after 28 days of hydration and promoted the hydration of cement compared to CNTs, GO, and GO/FCNTs mixture (0.05 wt.%). The results in this study validated the feasibility of using GNFG with enhanced dispersion as a new nano-reinforcing agent for various cementitious systems.

Keywords: graphene oxide nanoribbons; graphene oxide; functionalized carbon nanotubes; cement paste; mechanical properties

1. Introduction

Since the development of Portland cement in 1824, it has gradually become the most extensively used building material worldwide due to its excellent performance and low cost [1]. Portland cement is usually applied as a binder to form cementitious materials, including concrete, along with other aggregates (sand and gravel). However, owing to the brittleness and lack of flexural/tensile strength of cementitious materials, many studies have focused on improving its mechanical strength. The typical method for improving the flexural/tensile strength of cementitious materials is to employ fibers such as polypropylene fiber, carbon fiber, steel fiber, and glass fiber as reinforcing agents [2–6]. Xue et al. [7] and Cao et al. [8] reported that the fiber exhibited various reinforcement effects on cementitious materials due to their different types and lengths. Although these fibers can improve the tensile strength and toughness of cementitious materials, delaying the transformation of microcracks into microforms, they cannot constrain the development of microcracks [9]. Nanomaterials, such as nano-silica, nano-titanium dioxide (TiO_2), carbon nanotubes (CNTs), and graphene oxide (GO), have also been found to improve the mechanical properties of cementitious materials and hinder the spread of microcracks [10]. Among these materials, carbon nanomaterials, represented by CNTs and GO, are favored owing to their excellent mechanical properties [11,12].

CNTs, developed by Iijima in 1991 [13], are an allotropic form of carbon with unique characteristics that make them suitable reinforcing agents in various fields [14]. The carbon atoms of the tubes are sp^2 hybridized and combine through carbon-carbon bonds (σ bond) to form a hexagonal honeycomb structure [15]. This structure provides the unique mechanical properties of CNTs. The tensile strength and Young's modulus of CNTs exceed 60 GPa [16] and 1.2 TPa [17], respectively. Therefore, CNTs are considered to be an ideal nanomaterial and one of the most promising nanomaterials in the 21st century [18]. However, because of the strong van der Waals forces between the CNTs and the absence of hydrophilic functional groups on the surface, it is difficult to uniformly disperse CNTs in aqueous solutions. The precondition of using CNTs as a reinforcing material in cement-based materials is uniform dispersion. Three methods are currently used to disperse CNTs: ultrasonication treatment, use of surfactants, and functionalization of the CNTs [19–21]. Xu et al. [22] reported that CNTs could reduce the porosity of cement by filling the pores in the cement matrix, and found that the flexural strength of the cement paste increased by the addition of CNTs (0.05 wt.%) that sonication and surfactants were used.

In recent years, functionalized CNTs (FCNTs) have gradually attracted attention. The purpose of functionalizing the CNTs is to address the poor dispersion of pure CNTs [23]. CNTs can be functionalized using chemical (covalent) and physical (non-covalent) methods [24]. Generally, CNTs have several inherent structural defects. During chemical functionalization, oxygen-containing functional groups are inserted at these defect sites by damaging inherent defects using strong acids [25]. This method is easy to achieve in the laboratory; thus, in this study, we focus on the chemical method. Furthermore, Mousavi et al. [26] demonstrated that the ultrasonic treatment can increase the dispersion of FCNTs in aqueous solutions. When the FCNTs content was 0.05 wt.%, the compressive and flexural strengths of cement paste increased by 26.6% and 3.2%, respectively.

As a two-dimensional (2D) carbon nanomaterial, GO has also received widespread attention from researchers. It is a hydrophilic material with a variety of oxygen-containing functional groups (carboxyl and hydroxyl); thus, GO disperses better in aqueous solutions and is more uniformly distributed in the cement matrix than CNTs and FCNTs [27]. The 2D structure of GO confers excellent mechanical properties, which are superior to nano-fibrillar cellulose, nano calcium carbonate, and nano-clay [28]. Specifically, the elastic modulus and fracture toughness of GO are 32 GPa and 120 MPa, respectively [29]. Peng et al. [30] demonstrated that GO can improve the microscopic morphology of cement hydration products, thereby increasing the strength of hardened cement paste. When the water-to-cement ratio (w/c) was 0.35, the incorporation of 0.03 wt.% GO increased the flexural strength of the cement mortar by 21.86%. Pan et al. [31] also reported that GO interacted strongly with the cement matrix due to the wrinkled surface structure that provided more nucleation sites for cement hydration, enhancing the mechanical properties of the cement paste. An et al. [32] reported similar results: the large specific surface area of GO enhanced the nucleation effect and promoted hydration. After 28 days of curing, cement paste containing 0.05 wt.% GO exhibited increases in the compressive strength and flexural strength by 25.5% and 37.3%, respectively.

As research on the enhancing effect of single carbon nanomaterial in cement-based materials has proceeded, the synergistic effect that may exist between multiple nanomaterials has entered the vision of researchers [33]. Many studies have demonstrated the enhancing effects of CNTs or GO on cementitious materials; however, only a few studies have demonstrated the co-effect of CNTs and GO on the mechanical properties of cement paste. Lu et al. [34] found that the electrostatic repulsion between CNTs and GO in a solution can overcome the van der Waals force between CNTs, thus improve the dispersion of CNTs. Moreover, when the w/c ratio was 0.4, the flexural and compressive strengths of the cement paste containing 0.025 wt.% CNTs and 0.025 wt.% GO (total content of 0.05 wt.%) were increased by 21.13% and 24.21%, respectively. These increases were greater than those obtained from the addition of 0.05 wt.% CNTs (flexural: 10.14%, compressive: 6.40%) or 0.05 wt.% GO (flexural: 16.20%, compressive: 11.05%). Zhou et al. [35] also reported that GO/CNTs could form a relatively stable dispersion system; because of the filling effect of GO/CNTs, the porosity was reduced and the mechanical strength of the cement paste increased. Furthermore, Kaury et al. [36] reported that the dispersion of FCNTs and GO was improved after mixing owing to the co-effect, and the strength of mortars can be significantly improved when using this mixture compared to adding either FCNTs or GO. Although this synergistic effect can appropriately improve the poor dispersion of CNTs, the results are still not satisfying.

Herein, as a new type of nano-reinforcing agent in cement-based materials, highly dispersed graphene oxide nanoribbons (GNR)-functionalized carbon nanotubes (FCNTs)-graphene oxide (GO), GNFG complex was synthesized, which has good dispersion and also improves the mechanical properties of cementitious materials. The structure, functional groups, and dispersibility of GNFG were investigated and compared to those of CNTs, FCNTs, GO, and GO/FCNTs mixture. The morphologies and chemical bonds of the nanomaterials were characterized by transmission electron microscopy (TEM) and Fourier transform infrared spectroscopy (FTIR), respectively. X-ray diffraction (XRD) was employed to characterize the crystal structure of the nanomaterials and to qualitatively analyze the cement hydration products. The dispersion of these nanomaterials was tested by ultraviolet–visible (UV–vis) spectroscopy. To confirm the effect of GNFG in the cement-based materials, the hydration products, microstructure, and hydration kinetics of the cement paste with GNFG were investigated via thermogravimetric (TG) and differential thermogravimetric (DTG) analysis, scanning electron microscopy (SEM), and isothermal conduction calorimetry, respectively.

2. Experimental Procedure

2.1. Materials

Ordinary Portland cement (OPC) CEM I 42.5R provided by Sungshin Co. Ltd. (Seoul, Korea) was used as the binder material to prepare the cement paste. Multiwalled CNTs were purchased from Hengqiu Technology (Suzhou, China). Graphite flake powders were procured from Alfa Aesar (Haverhill, MA, USA). The chemical composition of OPC and the physical properties of the CNTs and graphite are shown in Tables 1–3, respectively. To disperse the nanomaterials in the cement matrix, a polycarboxylate superplasticizer (SP, ExCon SP20, Buildex Co., Ltd., Cheonan-si, Chungcheongnam-do, Korea.) was used, and its properties are listed in Table 4.

Table 1. Chemical composition of OPC.

Chemical Composition	SiO_2	Al_2O_3	Fe_2O_3	CaO	MgO	K_2O	SO_3	TiO_2	LOI	Total
(wt.%)	18.43	2.83	2.17	68.17	2.37	1.11	3.03	0.15	1.72	100

Table 2. Properties of the CNTs.

Outside Diameter (nm)	Inside Dimeter (nm)	Length (μm)	Ash (%)	Purity (%)	SSA (m^2/g)	Color
10–20	5–10	10–30	<1.5	95	>200	Black

SSA: specific surface area.

Table 3. Properties of graphite flakes.

Mesh	Purity (%)	Density (g/mL)	Boiling Point (°C)	Particle Size (μm)	Color
325	99	2.2	4830	<50	Black

Table 4. Properties of the polycarboxylate superplasticizer.

Water Content (%)	pH Value (10% Solution)	Active Component (%)	Bulk Density (kg/m^3)
<3	6.0–8.0	>90	450

2.2. Preparation of GNFG, GO, FCNTs, and GO/FCNTs Mixture

2.2.1. Synthesis of GNFG and GO

In 1958, Hummers et al. described a method to synthesize GO using a mixture of sodium nitrate ($NaNO_3$), potassium permanganate ($KMnO_4$), and concentrated sulfuric acid (H_2SO_4), which is also known as Hummers' method [37]. Although Hummers' method has been used widely to synthesize GO [38], it still suffers from several drawbacks, including the generation of toxic gases (NO_2 and N_2O_4), nitrate residues, and low yields [27]. The Hummers' method has been modified in the past 20 years to address these problems [39]. Currently, most researchers utilize modified Hummers' method to synthesize GO, wherein the proportions of the experimental materials and process are changed [40].

GNFG was also synthesized using the modified Hummers' method. First, a combined mass of 1 g of graphite and CNTs (mass ratio of graphite: CNTs = 2:1) and 0.65 g of $NaNO_3$ were added to a beaker and mixed. The beaker was kept in an ice-water bath below 5 °C. Subsequently, 30 mL of H_2SO_4 (95%) was transferred to the beaker and the mixture was stirred. After 3 min of stirring, $KMnO_4$ (3.9 g) was gradually added to the beaker and the mixture was stirred with a magnetic stirrer for 2 h. Then, the beaker was transferred to a water bath (35 °C) and kept in the water bath for 30 min. Next, 50 mL of distilled water was added dropwise to the mixture and the mixture was stirred for 30 min. The temperature of the water bath was increased to 98 °C and maintained for 20 min. Finally, a 30% H_2O_2 solution (5 mL) was added dropwise to terminate the oxidation reaction. The mixture was filtered, and the remaining metal ions were removed by adding a 10% HCl aqueous solution dropwise. Then, the solution was washed thoroughly with distilled water until the pH was approximately 7. Finally, the filtered solid was dissolved in distilled water, sonicated for 30 min, filtered again, and dried. A similar procedure was followed to synthesize GO using 1 g of graphite powder as the raw material. The GNFG and GO powders were obtained by drying in an oven at 60 °C for 24 h. In the preliminary experiment, the GNFG with the different mass ratio of graphite to CNTs (1:1, 2:1, and 4:1) were synthesized and their dispersity in aqueous solutions tested, as presented in Figure 1. GNFG with a mass ratio of 2:1 (graphite: CNTs) exhibited the best dispersion in the aqueous solution. Therefore, we used the GNFG with graphite and CNTs mass ratio of 2:1 for all measurements in this study.

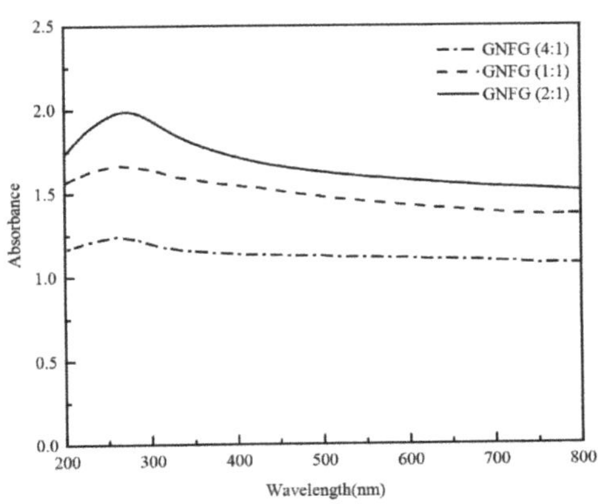

Figure 1. UV–vis spectra of GNFG synthesized using different ratios of graphite to CNTs.

2.2.2. Functionalization of the CNTs

Many methods of functionalizing CNTs have been introduced in previous studies [41–43]. In our experiments, a relatively safe and simple method was chosen, as follows: H_2O_2 (50 mL) was slowly added to 150 mL of H_2SO_4 (95%) to obtain a piranha solution; the mixture was left to cool and then used. Next, 0.3 g of CNTs were added to the piranha solution and the mixture was stirred for 30 min. The mixed solution was then sonicated for 1 h. After the ultrasonic treatment was completed, the solution was diluted with distilled water, filtered, and repeatedly washed to reach a pH of approximately 7. Finally, The FCNTs powder was obtained by drying the filtered solid in an oven at 60 °C for 24 h.

2.2.3. Mixing of GO/FCNTs

In order to compare with GNFG, the GO/FCNTs mixture was also prepared. We added 0.6 g of GO and 0.3 g of FCNTs (mass ratio of 2:1) to an aqueous solution and sonicated it for 15 min to ensure uniform mixing of GO and FCNTs. Then the mixture was filtered and the residue dried. Finally, a mixture of GO and FCNTs was obtained.

2.3. Characterization of the GRAPHITE, GO, CNTs, FCNTs, GO/FCNTs Mixture, and GNFG

The morphologies of graphite, GO, CNTs, FCNTs, GO/FCNTs mixture, and GNFG were determined using TEM (JEOL, JEM-2100F, Tokyo, Japan) with an electron source operating at an accelerating voltage of 220 kV. To prepare the samples, first, appropriate amounts of the graphite, GO, CNTs, FCNTs, GO/FCNTs mixture, and GNFG were dissolved in ethanol and ultrasonically dispersed for 15 min (Qsonica, Q700 Sonicator, Newtown, CT, USA, 20 kHz, Amplitude: 50%) [34]. Next, the dispersions of graphite, GO, CNTs, FCNTs, GO/FCNTs mixture, and GNFG were placed on a holey carbon film on 400 mesh Cu grids. Finally, the Cu grids were dried in an oven at 65 °C for 24 h.

XRD analysis was performed to study the crystalline structure of the graphite, GO, CNTs, FCNTs, GO/FCNTs mixture, and GNFG using a Bruker D2 X-ray diffractometer (λ = 1.54 Å) in the scattering (2θ) range of 5–60° with a step time of 1.5 s and step size of 0.01°. Phase identification was conducted using the DIFFRAC.EVA software. To evaluate the chemical bonds and functional groups of the graphite, GO, CNTs, FCNTs, GO/FCNTs mixture, and GNFG, FTIR was performed using a FTIR spectrometer (Spectrum Two FTIR spectrometer, PerkinElmer, Waltham, MA, USA) in the wavenumber range of 500 to 4000 cm^{-1}.

A UV–vis absorption spectrophotometer (GENESYS 180, Thermo Fisher Scientific, Waltham, MA, USA) was used to measure the dispersion of the GO, CNTs, FCNTs, GO/FCNTs mixture, and GNFG in aqueous solutions. Before the test, 0.05 g of different nanomaterials were added to 30 mL of distilled water. The concentration of the test solution was the same as that of the solution added to the cement paste, i.e., 1.66 mg/mL, followed by 15 min of sonication with a sonicator (Qsonica, Q700 Sonicator) at an amplitude of 50%. As a rule, in the absorption UV–vis spectrum, the absorption is proportional to the concentration, according to the Beer–Lambert law [44]. However, only the low concentration of the solution obeys the Beer–Lambert Law [45]. Therefore, the solutions after ultrasound treatment were diluted to 0.1 mg/mL [46]. SP was not added to the solutions for this test to exclude the effect of SP on the dispersion of the nanomaterials.

An important aspect of this study is to investigate the role of highly dispersed GNFG in the enhancement of cement properties. Several previous studies have extensively compared the roles of GO, CNTs, FCNTs, and GO/FCNTs mixtures in enhancing the properties of cement [34,47–50]. However, the structure of FCNTs can be damaged by chemical treatment resulting in shorter lengths, smaller diameters, and rough surfaces. Therefore, the mechanical properties of pure CNTs are better than those of FCNTs [51]. Based on the above, in this study, the GO, CNTs, and GO/FCNTs mixture were compared with GNFG.

2.4. Preparation of the GO, CNTs, GO/FCNTs Mixture, and GNFG Cement Pastes

Cement paste containing 0.025 wt.% CNTs/0.025 wt.% GO composites can significantly improve various properties of the cement paste [34]. Therefore, the ratio of nanomaterials selected for this study was also 0.05% of the weight of cement, and the w/c ratio was 0.3. The solution used to prepare specimens was obtained by adding 0.05 g of GO, CNTs, GO/FCNTs mixture, and GNFG to 30 g of distilled water. This solution was treated with ultrasonic waves to uniformly disperse the nanomaterials in water. However, if the energy of ultrasonic waves is too low, the dispersion will be non-uniform. In contrast, if the energy of the ultrasound is too high, the structure of the material will be damaged [52]. Hence, to ensure the same experimental conditions, four different nanomaterials were sonicated (Qsonica, Q700 Sonicator) using the same processing time and amplitude, which were 15 min and 50%, respectively. In addition, the beaker with the solution was placed in an ice-water bath during the sonication process to prevent the solution temperature from increasing during continuous sonication. Five specimens were prepared as shown in Table 5. For each specimen, SP was added at a 0.1 wt.% dosage based on the cement weight.

Table 5. Mix proportions.

Specimen	Cement (g)	Water (g)	SP (g)	GO (g)	CNTs (g)	GO/FCNTs (g)	GNFG (g)
OPC	100	30	0.1	–	–	–	–
0.05 wt.% GO	100	30	0.1	0.05	–	–	–
0.05 wt.% CNTs	100	30	0.1	–	0.05	–	–
0.05 wt.% GO/FCNTs	100	30	0.1	–	–	0.05	–
0.05 wt.% GNFG	100	30	0.1	–	–	–	0.05

Usually, SP is added to cement paste for two reasons. Firstly, SP is added to improve the workability of the cement paste, since the w/c ratio was chosen to be small and the addition of nanomaterials will cause the workability of the cement paste to deteriorate [53,54]. Secondly, SP is used as a surfactant to disperse the nanomaterials in the cement matrix [53]. Although these nanomaterials can be relatively stable dispersed in aqueous solutions after sonication, previous studies have shown that, in the alkaline environment of the cement matrix, the dispersion of CNTs and GO will be reduced, causing agglomeration [55,56]. S. Chuah [55] reported that premixing SP with cement reduces the alkalinity of the cement matrix in the early stages of cement hydration, as well as reducing the effects of alkaline conditions on dispersion. Thus, in this work, cement (100 g) and 0.1 g SP were pre-mixed to obtain a uniform mixture. Then the GO, CNTs, GO/FCNTs mixture, and GNFG solutions were mixed with the cement and SP mixture using a paste mixer (Malcom, SPS–1, Tokyo, Japan) for 10 min, respectively. Subsequently, the fresh cement paste was placed into compressive strength molds ($5 \times 5 \times 10$ mm^3) and splitting tensile strength molds ($\Phi 10 \times 20$ mm^2) and cured for 24 h. After 24 h, the specimens were demolded and stored in a constant temperature-humidity curing cabinet; the curing temperature and humidity were 25 °C and 65%, respectively.

2.5. Compressive and Splitting Tensile Strength Tests

To observe the variation of the mechanical properties of the cement paste with hydration time, the compressive and splitting tensile strength of the specimens were tested after curing 1, 3, 7, and 28 days, respectively [57]. The compressive and splitting tensile strengths of the hydrated cement paste were measured using a micro-compressive machine (Deben, Micro-Compressive Machine, Edmunds, Suffolk, UK) [58]. Figure 2 shows a schematic of the compressive and splitting tensile strength tests [57,59]. The sizes of the samples used for the compressive and splitting tensile strength tests were $5 \times 5 \times 10$ mm^3 and $\Phi 10 \times 20$ mm^2, respectively.

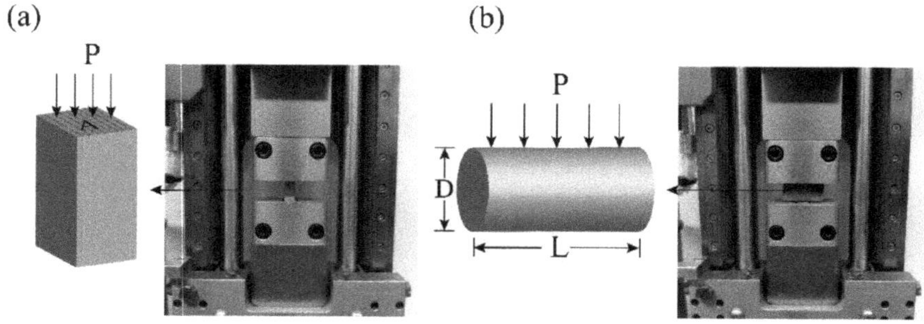

Figure 2. Schematics of strength tests: (**a**) compressive strength and (**b**) splitting tensile strength. Where: (**a**) F = P/A, (**b**) T = 2P/πLD; F: Compressive strength (MPa); T: Splitting tensile strength (MPa); P: Maximum load applied to the specimen (N); A: Cross-sectional area of the specimen (mm^2); L: Length of specimen (mm); D: Diameter of specimen (mm).

2.6. Microstructure and Mineral Analysis

The microstructures of the fracture surfaces of the specimens after curing for 28 days were analyzed by SEM (Thermo Fisher Scientific, Verios G4, Waltham, MA, USA). XRD (Bruker, D2 PHASER, Billerica, MA, USA) and TG analysis (HITACHI, STA7200 Simultaneous Thermogravimetric Analyzer, Tokyo, Japan) were used to study the cement hydration products after curing for 1, 3, 7, and 28 days. The TG experiments were conducted in a temperature range of 20 to 1000 °C with a heating rate of 10 °C/min, under an N_2 atmosphere with a flow rate of 200 mL/min.

2.7. Heat of Hydration

Isothermal conduction calorimetry (TA instrument, TAM-air, New Castle, DE, USA) was conducted to evaluate the effects of GO, CNTs, GO/FCNTs mixture, and GNFG on the hydration process of the cement paste at an early stage (72 h) [60]. According to the w/c ratio, the cement, nanomaterials, SP, and water were mixed to a total mass of 5 g. The aqueous solution containing the nanomaterials was sonicated before mixing with the cement.

3. Results and Discussion

3.1. Characterization of the Graphite, GO, CNTs, FCNTs, GO/FCNTs Mixture, and GNFG

3.1.1. Morphology Investigation

The TEM images of the graphite, GO, CNTs, FCNTs, GO/FCNTs mixture, and synthesized GNFG are shown in Figures 3 and 4. As shown in Figure 3a, CNTs with high aspect ratios are agglomerated due to severe entanglement of CNTs and the absence of hydrophilic functional groups in the structure, which reduced their dispersion in aqueous solutions [61]. Figure 3b shows the FCNTs morphology; the FCNTs are fractured at the ends and do not have closed caps. This is due to the oxidation of the CNTs [62]. Meanwhile, the degree of agglomeration of FCNTs remains relatively high [63]. Figure 3c shows the morphological characteristics of graphite, indicating a multilayer graphite structure with a darker color. Figure 3d shows that the surface morphology of GO is wrinkled, like crumpled paper [64].

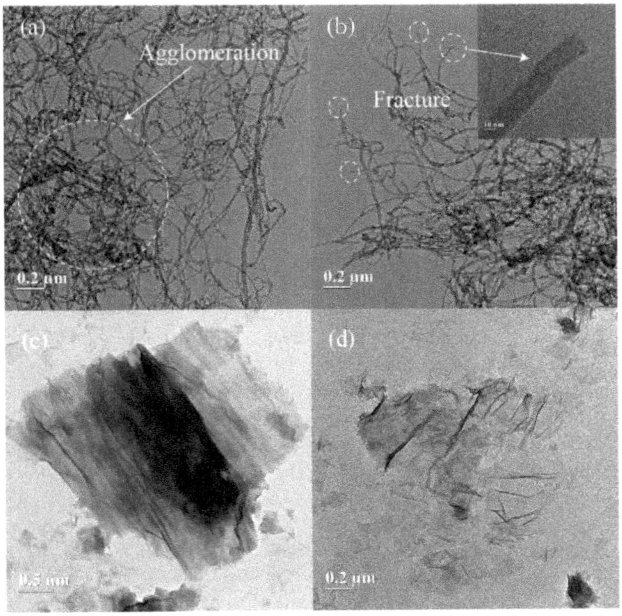

Figure 3. TEM images of the (**a**) CNTs, (**b**) FCNTs, (**c**) graphite, and (**d**) GO.

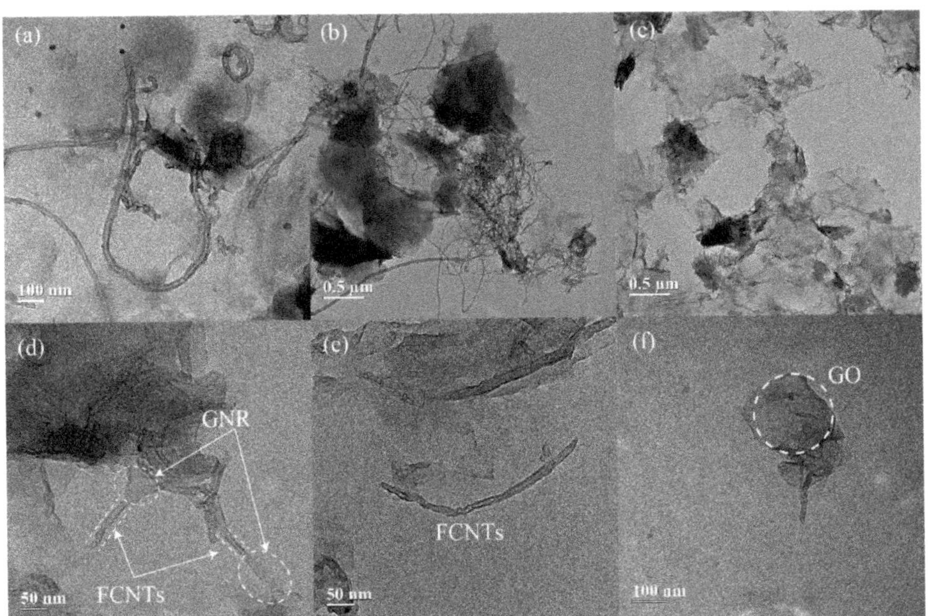

Figure 4. TEM images of the (**a**,**b**) GO/FCNTs mixture and (**c**–**f**) GNFG.

Figure 4a,b display the TEM images of the GO/FCNTs mixture, in which the GO and FCNTs were directly mixed without any chemical treatment. The results show that FCNTs are still being intertwined, with the GO and FCNTs remaining independent. Figure

4c depicts the overall morphological characteristics of GNFG. The FCNTs in the GNFG exhibit a shorter length, and the agglomeration degree is reduced compared with the FCNTs in the GO/FCNTs mixture. Pure CNTs are easily oxidized to FCNTs by strong oxidizing agents, such as sulfuric acid, nitric acid, and potassium permanganate [41,65]. In the synthesis of GNFG, sulfuric acid and potassium permanganate were also used; thus, FCNTs were generated. As the length of the FCNTs in GNFG shortens, the high aspect ratio decreases, reducing the entanglement and agglomeration in water.

Furthermore, as previously reported, when the amount of H_2SO_4 and $KMnO_4$ was sufficiently high, CNTs could unzip and become graphene-oxide nanoribbons (GNR) [66]. However, in the synthesis process of GNFG, the amounts of H_2SO_4 and $KMnO_4$ present are not enough to result in all CNTs unzipping to form GNR. Only part of the CNTs is unzipped into GNR, and the other part is oxidized to FCNTs, resulting in the formation of the unique connected structure of functionalized carbon nanotubes-graphene oxide nanoribbons (FCNTs–GNR), and the novel connected structure presented in Figure 4d. There are significant differences in the structures of GNR and FCNTs. As shown in Figure 4a,b, the walls of the FCNTs can be clearly seen. Unlike the FCNTs, the GNR in FCNTs–GNR does not have multiple layers of walls. In addition to this connected structure of FCNTs–GNR in GNFG, independently existing FCNTs and GO can also be found (Figure 4e,f). In summary, there are three types of components in GNFG: type I is FCNTs–GNR, type II is GO, and type III is FCNTs. To facilitate an interpretation of the images shown in Figure 4d–f, a schematic of the different components of GNFG is shown in Figure 5. The dispersion performance of this novel structure may be improved because the better dispersion of GNR can restrain the strong van der Waals forces of FCNTs connected to it.

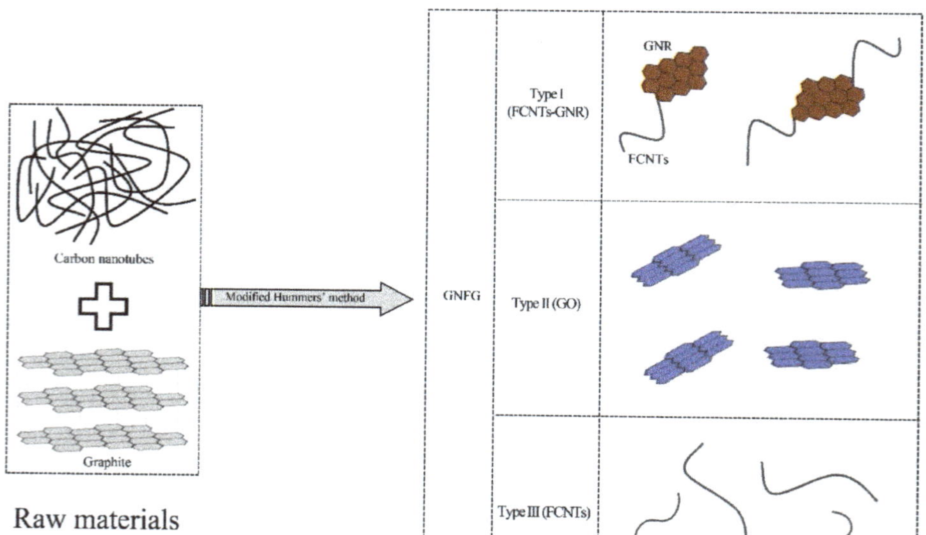

Figure 5. Schematic of the GNFG components.

3.1.2. Determination of Crystal Structure

The XRD patterns of graphite, GO, CNTs, FCNTs, GO/FCNTs mixture, and GNFG are presented in Figure 6. For graphite, a strong diffraction peak is observed at $2\theta = 26.38^\circ$ for the (002) graphitic plane, corresponding to an interlayer spacing of 0.337 nm (calculated using Bragg's equation). However, in the case of GO, which obtained by the chemical treatment of graphite, the XRD peak is observed at a different position. Compared with graphite, the diffraction peak at $2\theta = 26.38°$ disappeared, and a new enlarged peak appeared at approximately $2\theta = 13.1°$ (interlayer spacing 0.675 nm) for the (001) crystal plane. This indicated that the crystal structure changed after the oxidization of graphite, resulting in an increased interlayer spacing. Sharma [47] suggested that an increase in the interlayer spacing was due to the insertion of oxygen-containing functional groups into the carbon atom layer. For the CNTs and FCNTs, a strong diffraction peak at $2\theta = 25.58°$ is observed for the (002) plane with an interlayer spacing of 0.347 nm. The FCNTs, as the oxidation product of the CNTs, did not exhibit an altered structure compared to the CNTs; therefore, there were no significant differences in the XRD patterns. The XRD pattern of GNFG shows two small peaks, a (001) diffraction peak at $2\theta = 10.17°$ with an interlayer spacing of 0.868 nm and a (002) diffraction peak at $2\theta = 22.82°$ corresponding to an interlayer spacing of 0.389 nm. This diffuse peak at $2\theta = 22.82°$ is also similar to the previous study [67]. To highlight the difference in GNFG, the XRD pattern of the GO/FCNTs mixture is also shown. For the GO/FCNTs mixture, unique diffraction peaks were preserved. The (002) diffraction peak of the FCNTs appears at $2\theta = 25.58°$; however, the (001) diffraction peak of GO is shifted to the left and appears at $2\theta = 10.32°$. This indicates that the interlayer spacing of GO in the GO/FCNTs mixture increased. Wang [68] reported that CNTs can be dispersed relatively well between GO sheets through uniform mixing of GO/CNTs, resulting in an increase in the GO layer spacing. Besides the differences in their crystal structures, there are some other differences based on their chemical functional groups. Thus, the functional groups of these materials were analyzed by FTIR and were discussed in detail in the next section.

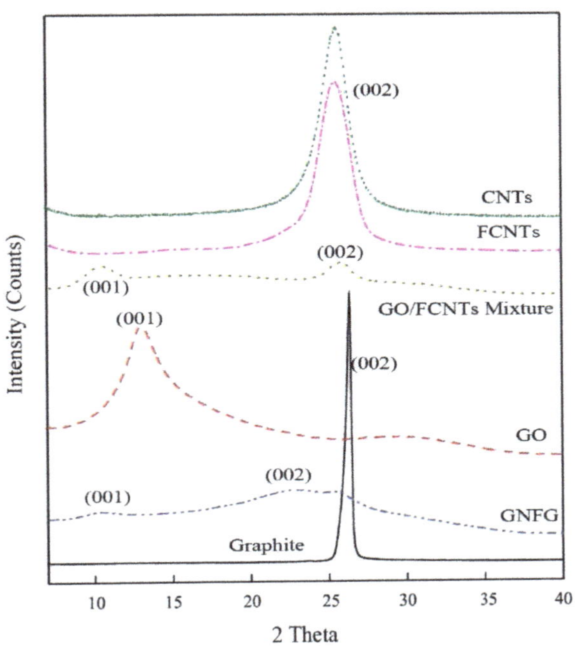

Figure 6. XRD patterns of the graphite, GO, CNTs, FCNTs, GO/FCNTs mixture, and GNFG.

3.1.3. FTIR Spectroscopy

The FTIR spectra of the graphite, GO, CNTs, FCNTs, GO/FCNTs mixture, and GNFG is shown in Figure 7a,b. There is a small peak arising from the C=C stretching vibration of graphite at 1542 cm^{-1} in the graphite spectrum [47,69]. In addition, owing to the purity of the graphite, there is a C=O stretching vibration peak at approximately 1734 cm^{-1} [47,69]. However, there are various functional groups in the GO structure after the graphite-oxidation treatment, including peaks for the –OH group at 3350 cm^{-1} and the C=O at 1702 cm^{-1} [47]. Moreover, an absorption peak for the C=C stretch is observed at approximately 1560 cm^{-1} [47,69]. The peaks at 1153 and 1016 cm^{-1} correspond to the stretching vibrations of C–O–C and C–O, respectively [47,69]. Since the functional groups were inserted into the graphite atomic layer, the layer spacing of GO increased; this was also evidenced by XRD results. Furthermore, a peak is observed in the CNTs spectra at 1517 cm^{-1}, which is related to aromatic C=C bonds [47,69]. After oxidation, although the XRD patterns do not differ significantly, the functional groups of the FCNTs and CNTs are quite distinct. FCNTs have a C=C stretching vibration peak at 1518 cm^{-1}, and –OH and C–O–C peaks at 3339 and 1108 cm^{-1}, respectively. This indicates that after functionalization, oxygen-containing functional groups were inserted into the structure of the CNTs [47,69]. As for the GO/FCNTs mixture, the FTIR spectrum is similar to GO. The similarity in the spectrum is possibly attributed to the presence of GO in the GO/FCNTs mixture. In the case of GNFG, the peak appears at approximately 3353 cm^{-1}, which corresponds to –OH stretching vibrations. Moreover, the peaks observed at 1697, 1548, 1147, 1025, and 848 cm^{-1} are related to the stretching vibrations of C=O, C=C, C–O–C, C–O, and epoxy groups, respectively [47,69]. Due to the graphene oxide nanoribbons in the GNFG complex, the FTIR spectrum of GNFG is also similar to the pure graphene oxide nanoribbons [70]. The properties of the functional groups could also determine the dispersion and stability of the nanomaterials in aqueous solutions, as determined via UV–vis spectroscopy.

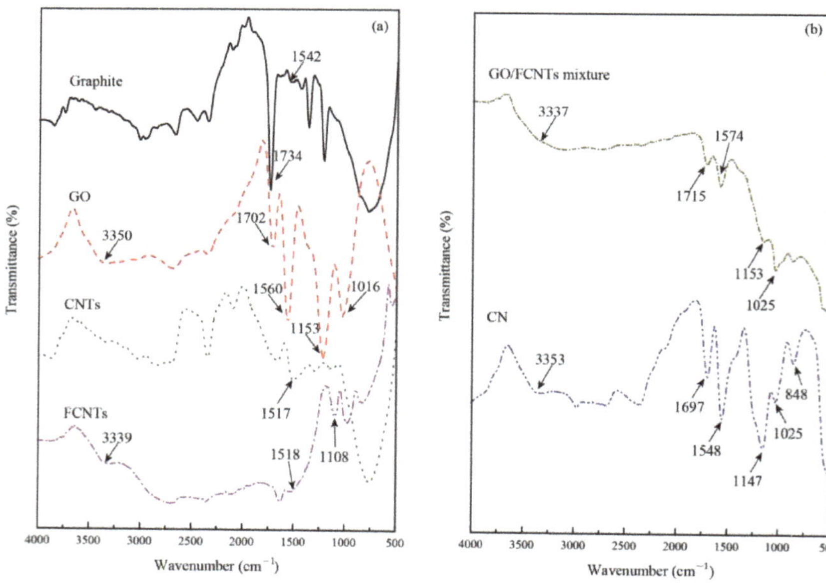

Figure 7. FTIR spectra: (**a**) graphite, GO, CNTs, and FCNTs, (**b**) GO/FCNTs mixture and GNFG.

3.1.4. Dispersion and Stability of Nanomaterials

To investigate the dispersion properties of GNFG in aqueous solutions, UV–vis spectroscopy was conducted on GNFG and the obtained results were compared to those of the CNTs, FCNTs, GO, and GO/FCNTs mixture (Figure 8). The absorbance is proportional to the dispersion of the nanoparticles because only uniformly dispersed nanoparticles can effectively absorb light in the UV–vis region [34]. Thus, the best dispersity in aqueous solutions was observed for GO, followed by GNFG, GO/FCNTs mixture, FCNTs, and CNTs.

Owing to the high aspect ratio and strong van der Waals forces between the molecules, CNTs are likely to agglomerate in aqueous solutions and have poor dispersion [71]. In comparison with CNTs, the insertion of oxygen-containing functional groups results in the surface of the FCNTs being negatively charged due to the ionization, which generates electrostatic repulsion leading to improved dispersion of FCNTs in aqueous solutions [72]. As indicated in the FTIR spectra, GO contains hydrophilic oxygen-containing functional groups (–OH, –COOH), which generate electrostatic repulsion, and there is no strong van der Waals forces between the GO, resulting in the best dispersion [73]. Additionally, the absorbance of the GO/FCNTs mixture is higher than those of the CNTs and FCNTs, also due to the fact that the negatively charged surfaces of the FCNTs and GO generate electrostatic repulsion, increasing the degree of dispersion [34]. However, it should be noted that the absorbance of GNFG as a complex was higher than that of the GO/FCNTs mixture. The excellent dispersion of GNFG is attributed to several characteristics. Firstly, as described in the TEM results, due to the partial unzipping of CNTs to form FCNTs–GNR, the electrostatic repulsion on the GNR surface is able to inhibit the van der Waals forces between FCNTs in FCNTs–GNR. In particular, as the length of independently-existing FCNTs in GNFG shortens and the high aspect ratio decreases, the degree of intertwining is reduced, which results in an improved dispersion of GNFG [74]. Furthermore, the electrostatic repulsion generated by the negatively charged functional groups on the surface of independently existing GO and FCNTs can effectively improve the dispersion [75,76].

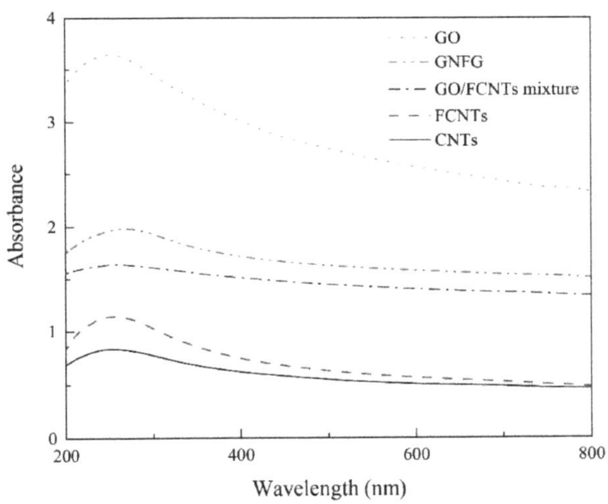

Figure 8. UV–vis spectra of the GO, CNTs, FCNTs, GO/FCNTs mixture, and GNFG.

To further evaluate the dispersion of nanomaterials in aqueous solutions, the stability of dispersion is also investigated. The dispersion stability of the CNTs, FCNTs, GO, GO/FCNTs mixture, and GNFG in aqueous solutions was determined over time after ultrasonic treatment, as shown in Figure 9. Figure 9a shows the results of observations made immediately after the ultrasonication, at which point all the nanomaterials were uniformly dispersed in the solutions and there were no clear agglomerates that precipitated out. After ten minutes (Figure 9b), the CNTs and FCNTs exhibited a different extent of agglomeration and precipitation. Although the absorbance of FCNTs in the UV–vis was higher than that of the CNTs, the stability of FCNTs in an aqueous solution was still unsatisfactory. However, the GO/FCNTs mixture, GO, and GNFG were still uniformly dispersed in the aqueous solution. After 6 h, the FCNTs and CNTs were completely agglomerated and precipitated out of the solution (Figure 9c). Furthermore, the color of the solution containing the GO/FCNTs mixture became lighter, indicating that the mixture of GO/FCNTs almost completely precipitated out of the solution. The GNFG remained partially and steadily dispersed in the solution, while GO remained steadily dispersed. After 24 h, only GO was still partially dispersed in the aqueous solution.

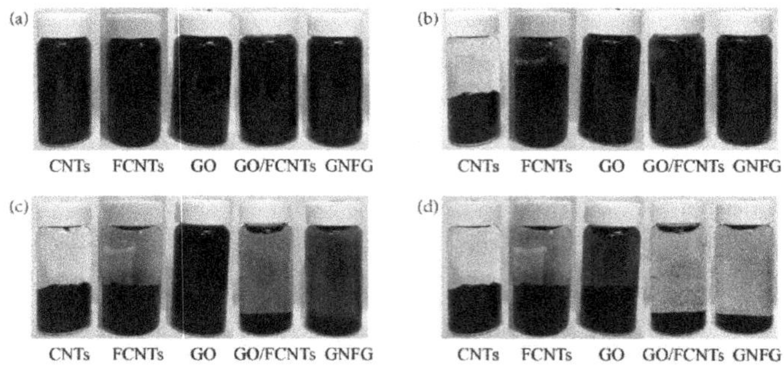

Figure 9. Dispersion of the nanomaterials over time after the ultrasonic treatment: (**a**) 0 min, (**b**) 10 min, (**c**) 6 h, and (**d**) 24 h.

Although the preparation of the GO/FCNTs mixture is simple and the dispersion of the GO/FCNTs mixture is better than that of the CNTs and FCNTs, it is still less stable. However, chemically-treated GNFG not only has better dispersion in aqueous solutions, but also has a better stability. This also provided a guarantee that GNFG would show a reinforcement effect in cement-based materials. The advantages of GNFG as a reinforcing agent for cementitious systems will be further discussed in the following sections.

3.2. *Mechanical Properties*

The effects of different nanomaterials (GO, CNTs, GO/FCNTs mixture, and GNFG) on the mechanical properties of hardened cement paste were assessed after 1, 3, 7, and 28 days. Figure 10a,b represent the compressive and splitting tensile strength of the cement paste with and without GO, CNTs, GO/FCNTs mixture, and GNFG, respectively. In comparison with OPC, after 28 days of curing, it was evidenced that incorporating 0.05 wt.% CNTs led to an increase in both the compressive strength (6.65%) and splitting tensile strength (15.39%) of the cement paste. Moreover, the addition of 0.05 wt.% GO increased the compressive and splitting tensile strengths of the cement paste by 2.98% and 11.76%, respectively, after 28 days of curing. This indicates that CNTs and GO can improve the mechanical properties of cement paste, which agrees with previous reports [77]. Addition of CNTs and GO increase the mechanical strength of the cement paste based on the following effects: (1) the filling effect, in which the CNTs and GO function as fillers in the nanoscale pores of the cement matrix; (2) the bridging effect, in which the nanomaterials connect the microcracks in the cement and inhibit the development of cracks; and (3) the nucleation effect, in which the nanomaterials provide more nucleation sites for hydration of the cement paste [30,78,79]. Further, when 0.05 wt.% of the GO/FCNTs mixture was added to the cement paste, the compressive and splitting tensile strengths were improved by 16.5% and 15.6%, respectively, fully demonstrating the synergistic effect of GO and FCNTs, as previously reported [34].

Notably, after hydration for 28 days, the 0.05 wt.% GNFG specimen showed the highest compressive strength (123.87 MPa) and splitting tensile strength (5.25 MPa), which were improved by 25.39% and 17.31%, respectively, compared to OPC. GNFG appeared to improve the mechanical properties of the cement paste, possibly because of the good dispersion and unique connected structure. Although GO has better dispersion in aqueous solutions than GNFG, the co-effect of GNFG can significantly improve the mechanical properties of the cement paste.

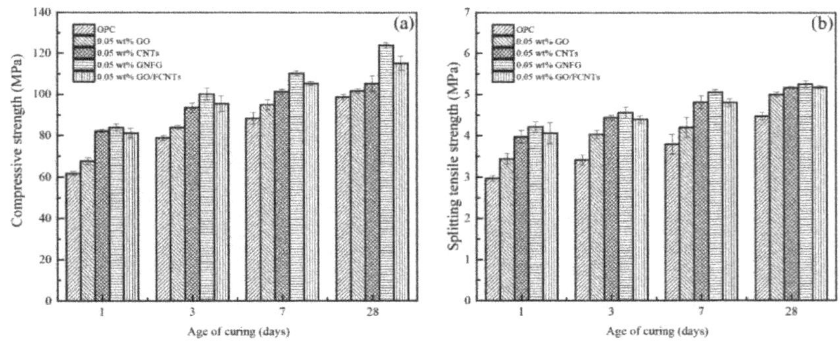

Figure 10. (**a**) Compressive and (**b**) splitting tensile strength of the specimens.

3.3. Microstructure Observations

A previous study [80] showed that the mechanical strength of nanomaterial-incorporated hardened cement paste is related to its microstructure. Therefore, the microstructure of the fractured surface of hardened cement paste with and without nanomaterials, after 28 days of curing, was observed via SEM. Figure 11a,b show the SEM images of OPC as the control sample. For the cement composites without any nanomaterials, the cement matrix structure is not compact and porous, with several microscale pores and cracks that lead to relatively poor mechanical properties for OPC compared to nanomaterial-incorporated cement pastes.

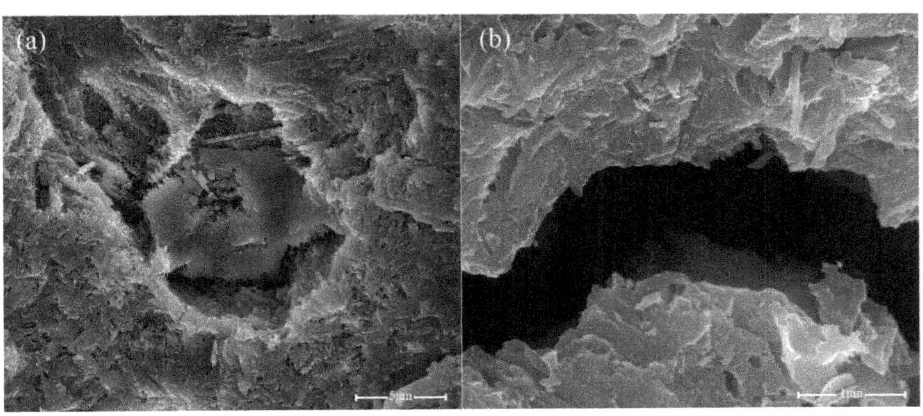

Figure 11. SEM images of OPC (**a**,**b**).

Compared to OPC, the microstructure of the 0.05 wt.% GO specimen (Figure 12a,b) is relatively compact, with reduced microscale cracks. GO can fill the microscale pores and cracks and provide more nucleation sites for cement hydration, thereby improving the microstructure of the cement. The addition of 0.05 wt.% CNTs resulted in a more uniform and compact structure, as the CNTs generated bridges between the cement hydrates, as shown in Figure 13. Previous studies [81–83] have already confirmed that the bridging effect of CNTs improves the load-bearing capacity of the cement matrix. In contrast, CNTs also show outstanding filling effects; they can fill the microscale pores and increase the compactness of the cement matrix. However, although CNTs can improve the microstructure of the cement, CNTs tend to agglomerate in the cement matrix (Figure 13a), which could interfere with its excellent mechanical properties. In addition, Figure 14a,b present the microstructure of cement paste with 0.05 wt.% GO/FCNTs mixture. It can be seen that the cement matrix of 0.05 wt.% GO/FCNTs mixture is compact, and some hydration products are covered on the surface of the cement matrix. Besides, the FCNTs exhibit a lower degree of agglomeration and are filled in the cement matrix.

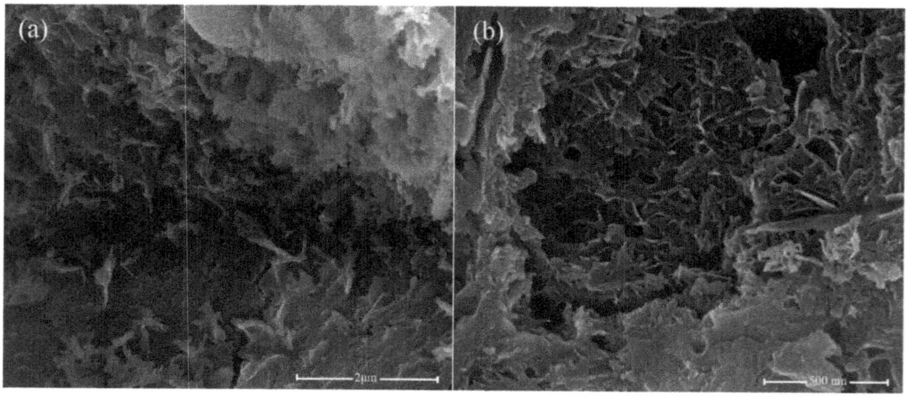

Figure 12. SEM images of 0.05 wt.% GO (**a**,**b**).

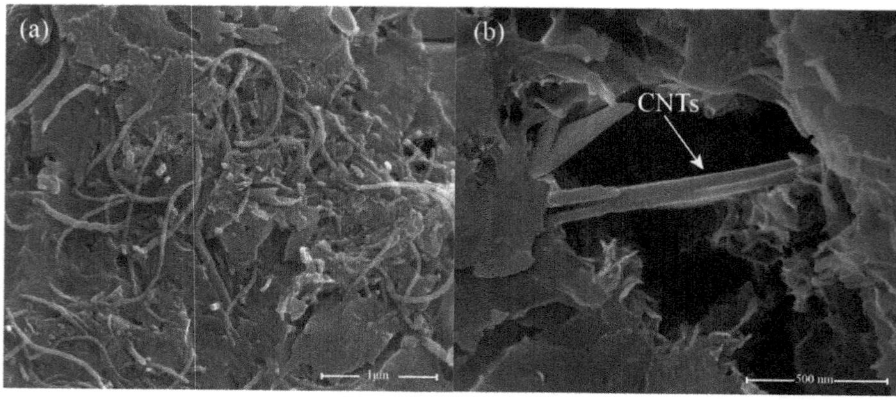

Figure 13. SEM images of 0.05 wt.% CNTs (**a**,**b**).

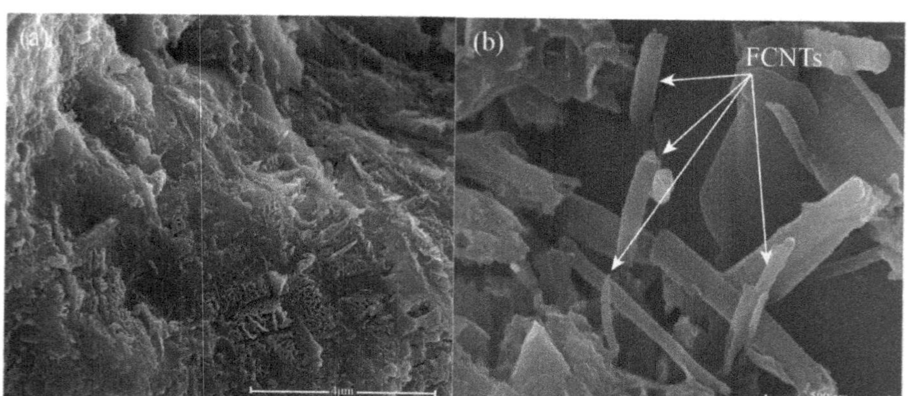

Figure 14. SEM images of 0.05 wt.% GO/FCNTs (**a**,**b**).

The microstructures of the cement hydrates containing 0.05 wt.% GNFG after 28 days of hydration are shown in Figure 15. It is clear that hydration products of calcium–silicate–hydrate (C–S–H) are uniformly distributed in the matrix without any large pores, as shown in Figure 15a. Moreover, in the 0.05 wt.% GNFG specimen, aggregation of FCNTs is not observed in the cement matrix. The GNFG can fill the microscale pores in the cement matrix and play a bridging role (Figure 15b). In addition, GNFG is able to bridge surficial Ca^{2+} of the cement matrix due to its wrinkled surface with oxygen-containing functional groups which enhances the interfacial bonding [84]. This also contributes to the very dense microstructure of cement paste and the 0.05 wt.% GNFG-incorporated specimen exhibiting the highest compressive strength and splitting tensile strength among the five groups.

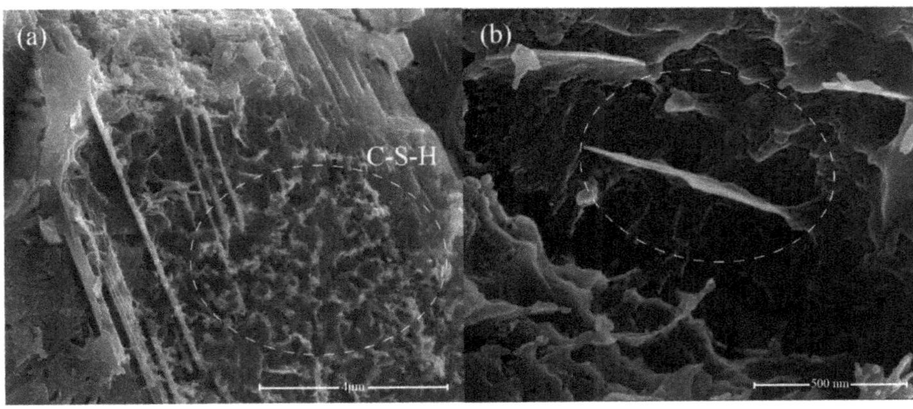

Figure 15. SEM images of 0.05 wt.% GNFG (**a**,**b**).

3.4. *Heat of Hydration*

Nanomaterials have a significant effect on the early hydration process of cementitious systems [85]. To fully observe and compare the effects of the different carbon nanomaterials on the early hydration process (within 72 h) of cement, isothermal calorimetry was used to evaluate the amount of heat released from the OPC and cement pastes containing 0.05 wt.% GO, CNTs, GO/FCNTs mixture, and GNFG (Figure 16a,b). Typically, exothermic reactions with four stages can be observed during the hydration process of the cement: the first stage is the initial period, the second stage is the induction period, the third stage is the acceleration period, and the fourth stage is the deceleration period [86,87]. Previous studies have shown that the SP delays the induction period [88]. The SP, as an anionic surfactant, will be more easily adsorbed on the surface of an oppositely charged cement. Thus, due to adding the SP to the cement paste, the surface of the cement particles was covered, which hindered the exchange of ions in the hydration system and reduced the hydration rate. In addition, the interaction between Ca^{2+} and the SP reduces the concentration of Ca^{2+}, and hinders the nucleation of hydration products [89,90]. However, the five groups of specimens in this study contain the same content of SP, the delayed effect of SP on the early induction period of hydration can be ignored.

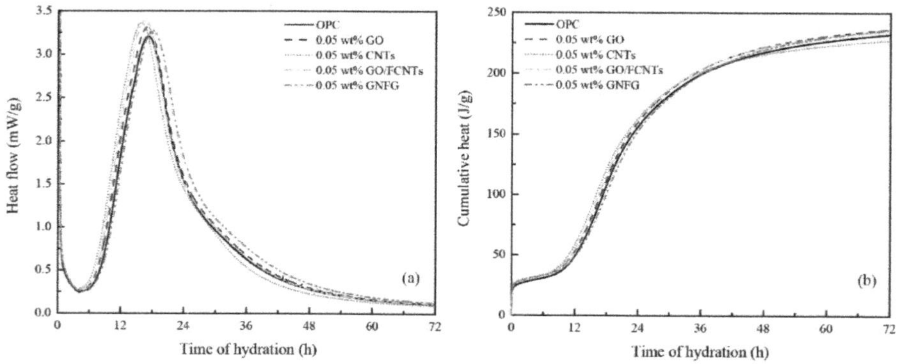

Figure 16. (**a**) Heat flow curves and (**b**) cumulative heat curves of the cement paste specimens.

As shown in Figure 16a, although the addition of different nanomaterials, there was no significant influence on the initial and induction periods of the cement hydration process. However, from about four hours, the hydration reaction enters an acceleration period. It can be clearly observed that the maximum heat flow of the cement paste with nanomaterials is higher than that of OPC. Therefore, these carbon nanomaterials contribute obviously to the hydration of cement, which is also consistent with previous studies [91]. When all samples entered the deceleration period, it was found that the cement paste incorporating GNFG had a higher heat flow during the deceleration period compared to the other samples. Although the maximum heat flow of GNFG is slightly lower than that of specimens with other nanomaterials, the higher heat flow during the deceleration period results in a cumulative heat of 72 h that is not significantly different from that of specimens with other nanomaterials. In addition, it could also be found that the heat flow of the samples containing 0.05 wt.% CNTs was lower than all samples in the deceleration period and resulted in the lowest cumulative heat over 72 h (Figure 16b). This is attributed to the agglomeration of CNTs in the cement matrix as presented in the SEM image (Figure 13a), which has an impact on the deceleration period of the hydration reaction. Overall, owing to the addition of the nanomaterials (besides CNTs), the cumulative heat of the cement hydration increased after 72 h, indicating the enhanced degree of hydration of the cement at early stage. However, although CNTs promote the hydration reaction during the acceleration period, they also have some impact on the hydration reaction during the deceleration period. It is difficult to evaluate the degree of hydration of different samples at various periods by only using calorimetry. Therefore, XRD and TG analysis also were applied to investigate the effects of the CNTs, GO, GO/FCNTs mixture, and GNFG on the hydration reaction of cement pastes at various curing times.

3.5. XRD Analysis of the Hydration Products

XRD was employed to qualitative analyze the hydration products and the consumption rate of the anhydrous phases at different hydration periods [92]. Previous research indicated that owing to the large specific surface area of CNTs and GO, more nucleation sites are present for cement hydration and more hydration products are generated [10,93]. The XRD patterns of OPC and the cement pastes containing 0.05 wt.% GO, 0.05 wt.% CNTs, 0.05 wt.% GO/FCNTs mixture, and 0.05 wt.% GNFG are shown in Figure 17. Usually, as the hydration reaction progresses, the unhydrated calcium silicate content gradually decreases [94]. As reported previously [95–97], the peaks of C_3S and C_2S overlap with those of the other minerals (C–S–H, calcium hydroxide (CH), and calcite). In Figure 17, the peaks of C_3S and C_2S also appear to overlap; thus, it was not easy to determine the changes in reduction.

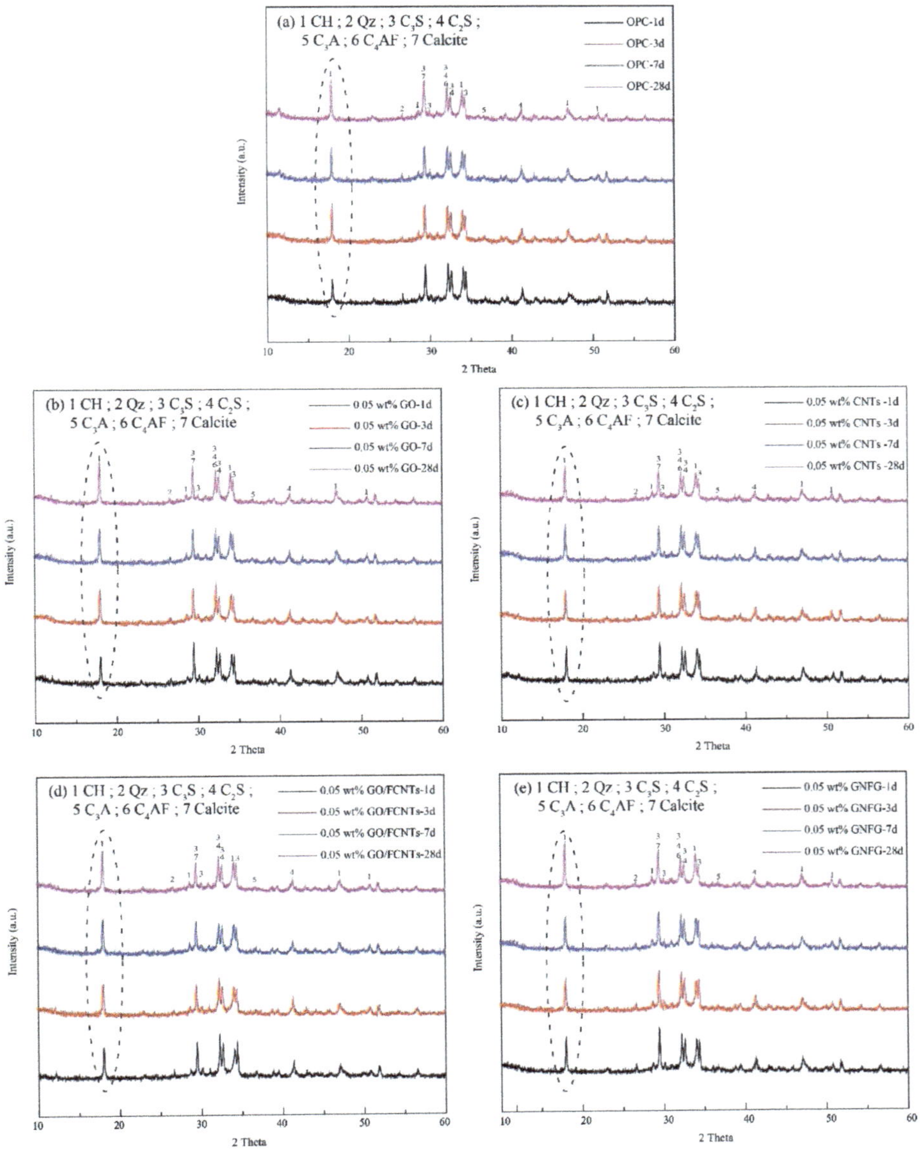

Figure 17. XRD spectra of (**a**) OPC, (**b**) 0.05 wt.% GO, (**c**) 0.05 wt.% CNTs, (**d**) 0.05 wt.% GO/FCNTs. and (**e**) 0.05 wt.% GNFG.

It is well known that C–S–H and CH are produced simultaneously by the hydration of calcium silicate [47,98]. The relative hydration rate can be established by measuring the content of C–S–H or CH. In general, XRD is used to determine the CH content instead of the C–S–H content because CH has a hexagonal crystal structure that can be easily detected by XRD. In contrast, the C–S–H phase forms a semi-crystalline or amorphous phase, which limits the use of XRD as a detection technique [47,99–101]. In the crystallization patterns (Figure 17), the diffraction peaks of CH were observed at 2θ = 17.9°, 28.6°, 34.1°, 47.1°, and 50.1° [47]. For the OPC, 0.05 wt.% GO, 0.05 wt.% CNTs, 0.05 wt.% GO/FCNTs mixture, and 0.05 wt.% GNFG specimens, the CH diffraction peaks show significant differences. The XRD patterns of the five groups show that the 0.05 wt.% GNFG specimen has the highest CH intensity over the different periods, which indicates that GNFG improves the degree of hydration of cement paste and more hydration products are produced compared to OPC. In addition, when 0.05 wt.% CNTs, GO, and GO/FCNTs mixture were added to the cement, the CH content also increased. This indicates that hydration was promoted when these nanomaterials were used. The amounts of CH in each sample at the different periods were quantified by TG analysis.

3.6. Thermogravimetric Analysis

The OPC, 0.05 wt.% GO, 0.05 wt.% CNTs, 0.05 wt.% GO/FCNTs mixture, and 0.05 wt.% GNFG samples were analyzed by TG and DTG at 1, 3, 7, and 28 days, as shown in Figure 18. The mass loss at approximately 100 °C is related to the evaporation of free water. Mass loss in the range of 400–500 °C is due to the decomposition of CH, and between 600 and 700 °C is attributed to calcium carbonate ($CaCO_3$) decomposition [102]. However, as stated in the XRD section, calcium silicates produce both C–S–H and CH when hydrated [47]. Based on the values in Figure 18, the CH content for each sample at the different curing times was calculated according to Equation (1). $WL_{Ca(OH)_2}$ represents the percentage weight loss of CH, and $m_{Ca(OH)_2}$ and m_{H_2O} are the molecular masses of portlandite (74 g/mol) and water (18 g/mol), respectively. The calculation results are listed in Table 6.

$$Ca(OH)_{2,measured} = WL_{Ca(OH)_2} \times m_{Ca(OH)_2} / m_{H_2O}. \tag{1}$$

Table 6. Percentage content (%) of $Ca(OH)_2$ in the hardened cement paste after different curing times.

Specimen	$Ca(OH)_2$ Content (%)			
	1 Day	3 Days	7 Days	28 Days
OPC	5.43%	6.42%	7.23%	7.44%
0.05 wt.% GO	6.17%	6.50%	7.46%	7.62%
0.05 wt.% CNTs	6.29%	6.81%	8.43%	10.03%
0.05 wt.% GO/FCNTs	6.83%	8.54%	9.35%	10.27%
0.05 wt.% GNFG	8.62%	9.49%	9.66%	10.97%

As shown in Figure 18 and Table 6, compared with the samples containing 0.05 wt.% CNTs, GO, GO/FCNTs mixture, and GNFG, OPC showed the lowest relative amount of CH (1 d: 5.43%; 3 d: 6.42%; 7 d: 7.23%; and 28 d: 7.44%) for each period. This indicates that OPC has fewer hydration products and shows a lower degree of hydration, which is also consistent with the XRD results. For the samples with nanomaterials, the hydration of the cement paste was enhanced at each curing time, although the degree of enhancement differed. Using a dosage of 0.05 wt.%, GNFG induces the greatest enhancement in hydration among the four nanomaterials, followed by the GO/FCNTs mixture, CNTs, GO, and OPC. This also proves that GNFG can promote the hydration of cement paste and generate more hydration products, and the promotion effect is better than other nanomaterials.

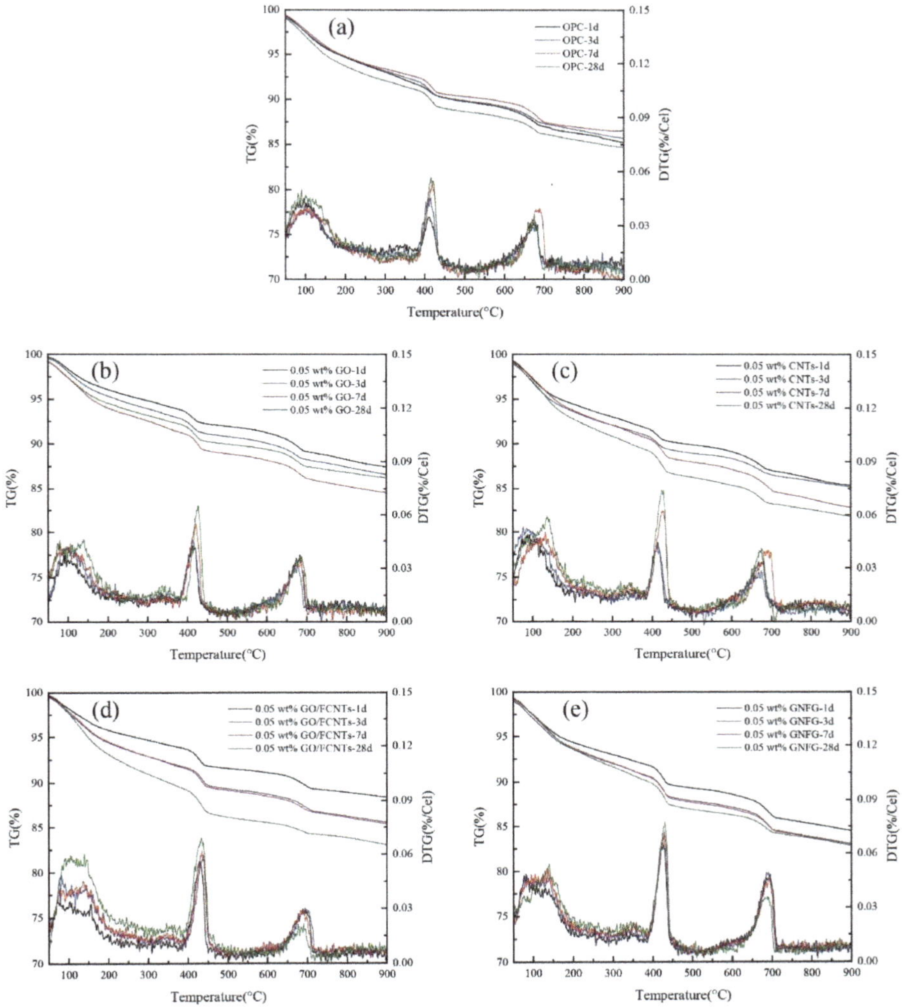

Figure 18. TG/DTG curves of the samples containing (**a**) OPC, (**b**) 0.05 wt.% GO, (**c**) 0.05 wt.% CNTs, (**d**) 0.05 wt.% GO/FCNTs, and (**e**) 0.05 wt.% GNFG.

4. Conclusions

In this study, the highly-dispersed GNFG was hydrothermally synthesized and its effects on the mechanical properties, hydration kinetics, hydration products, and microstructure of cementitious composites were investigated. The results were compared with those using the composites incorporating other typical carbon nanomaterials. Based on the experimental results, the following conclusions can be drawn:

(1) GNFG, as a new carbon nanomaterial, was successfully synthesized via a chemical method, and there are three different components (FCNTs–GNR, FCNTs, and GO) of GNFG due to the treatment conditions and led to better dispersibility.

(2) Although the GO, CNTs, GO/FCNTs mixture, and GNFG improved the compressive and splitting tensile strength of cement paste at the same content (0.05 wt.%), GNFG induced the greatest improvement in the mechanical properties of the cement paste due to the denser microstructure.

(3) GNFG, as a new reinforcing nanomaterial for cementitious systems, similar to other nanomaterials, can fill cracks and play a bridging role in the cement matrix.
(4) GNFG can increase the maximum heat flow and cumulative heat of cement hydration reaction, promote the hydration reaction and generate more hydration products.

These experimental results demonstrate that GNFG exhibits excellent potential as a new nano-reinforcing agent for cementitious materials, especially relating to the improvement in the mechanical properties of cementitious composites.

Author Contributions: Conceptualization, S.L., E.Z.N. and S.B.; formal analysis, P.L. and S.H.; resources, S.B.; writing—original draft preparation, P.L. and J.L.; writing—review and editing, S.L., E.Z.N. and S.B.; supervision, S.L., E.Z.N. and S.B. All authors have read and agreed to the published version of the manuscript.

Funding: This work is supported by the Korea Agency for Infrastructure Technology Advancement(KAIA) grant funded by the Ministry of Land, Infrastructure, and Transport (Grant 21NANO-B156177-02).

Data Availability Statement: Data are contained within the article.

Conflicts of Interest: The authors declare no conflict of interest.

References

1. Gong, K.; Pan, Z.; Korayem, A.H.; Qiu, L.; Li, D.; Collins, F.; Wang, C.M.; Duan, W.H. Reinforcing effects of graphene oxide on portland cement paste. *J. Mater. Civ. Eng.* **2015**, *27*, A4014010. [CrossRef]
2. Li, N.; Jin, Z.; Long, G.; Chen, L.; Fu, Q.; Yu, Y.; Zhang, X.; Xiong, C. Impact resistance of steel fiber-reinforced self-compacting concrete (SCC) at high strain rates. *J. Build. Eng.* **2021**, *38*, 102212. [CrossRef]
3. Singh, N.K.; Rai, B. Assessment of synergetic effect on microscopic and mechanical properties of steel-polypropylene hybrid fiber reinforced concrete. *Struct. Concr.* **2021**, *22*, 516–534. [CrossRef]
4. Małek, M.; Jackowski, M.; Łasica, W.; Kadela, M.; Wachowski, M. Mechanical and material properties of mortar reinforced with glass fiber: An experimental study. *Materials* **2021**, *14*, 698. [CrossRef]
5. Wang, Z.; Ma, G.; Ma, Z.; Zhang, Y. Flexural behavior of carbon fiber-reinforced concrete beams under impact loading. *Cem Concr Compos.* **2021**, *118*, 103910. [CrossRef]
6. Park, T.; Her, S.; Jee, H.; Yoon, S.; Cho, B.; Hwang, S.-H.; Bae, S. Evaluation of orientation and distribution of steel fibers in high-performance concrete column determined via micro-computed tomography. *Constr. Build. Mater.* **2021**, *270*, 121473. [CrossRef]
7. Xue, G.; Yilmaz, E.; Song, W.; Cao, S. Fiber length effect on strength properties of polypropylene fiber reinforced cemented tailings backfill specimens with different sizes. *Constr. Build. Mater.* **2020**, *241*, 118113. [CrossRef]
8. Cao, S.; Zheng, D.; Yilmaz, E.; Yin, Z.; Xue, G.; Yang, F. Strength development and microstructure characteristics of artificial concrete pillar considering fiber type and content effects. *Constr. Build. Mater.* **2020**, *256*, 119408. [CrossRef]
9. Parveen, S.; Rana, S.; Fangueiro, R. A review on nanomaterial dispersion, microstructure, and mechanical properties of carbon nanotube and nanofiber reinforced cementitious composites. *J. Nanomater.* **2013**, *2013*, 710175. [CrossRef]
10. Liu, J.; Fu, J.; Ni, T.; Yang, Y. Fracture toughness improvement of multi-wall carbon nanotubes/graphene sheets reinforced cement paste. *Constr. Build. Mater.* **2019**, *200*, 530–538. [CrossRef]
11. Zhao, L.; Guo, X.; Song, L.; Song, Y.; Dai, G.; Liu, J. An intensive review on the role of graphene oxide in cement-based materials. *Constr. Build. Mater.* **2020**, *241*, 117939. [CrossRef]
12. Liew, K.; Kai, M.; Zhang, L. Carbon nanotube reinforced cementitious composites: An overview. *Compos. Part A Appl. Sci. Manuf.* **2016**, *91*, 301–323. [CrossRef]
13. Iijima, S.; Ichihashi, T. Single-shell carbon nanotubes of 1-nm diameter. *Nature* **1993**, *363*, 603–605. [CrossRef]
14. Siddique, R.; Mehta, A. Effect of carbon nanotubes on properties of cement mortars. *Constr. Build. Mater.* **2014**, *50*, 116–129. [CrossRef]
15. Georgakilas, V.; Perman, J.A.; Tucek, J.; Zboril, R. Broad family of carbon nanoallotropes: Classification, chemistry, and applications of fullerenes, carbon dots, nanotubes, graphene, nanodiamonds, and combined superstructures. *Chem. Rev.* **2015**, *115*, 4744–4822. [CrossRef] [PubMed]
16. Yu, M.-F.; Lourie, O.; Dyer, M.J.; Moloni, K.; Kelly, T.F.; Ruoff, R.S. Strength and breaking mechanism of multiwalled carbon nanotubes under tensile load. *Science* **2000**, *287*, 637–640. [CrossRef]
17. Zhang, Q.; Huang, J.Q.; Qian, W.Z.; Zhang, Y.Y.; Wei, F. The road for nanomaterials industry: A review of carbon nanotube production, post-treatment, and bulk applications for composites and energy storage. *Small* **2013**, *9*, 1237–1265. [CrossRef]
18. Chaipanich, A.; Nochaiya, T.; Wongkeo, W.; Torkittikul, P. Compressive strength and microstructure of carbon nanotubes-fly ash cement composites. *Mater. Sci. Eng. A* **2010**, *527*, 1063–1067. [CrossRef]
19. Assi, L.; Alsalman, A.; Bianco, D.; Ziehl, P.; El-Khatib, J.; Bayat, M.; Hussein, F.H. Multiwall Carbon Nanotubes (MWCNTs) Dispersion & Mechanical Effects in OPC Mortar & Paste: A review. *J. Build. Eng.* **2021**, *43*, 102512.

20. Neto, J.D.S.A.; Santos, T.A.; de Andrade Pinto, S.; Dias, C.M.R.; Ribeiro, D.V. Effect of the combined use of carbon nanotubes (CNT) and metakaolin on the properties of cementitious matrices. *Constr. Build. Mater.* **2021**, *271*, 121903. [CrossRef]
21. Li, Y.; Li, H.; Jin, C. Effect of multi-walled carbon nanotubes on the damping property of cement mortar and mechanism analysis. *Arch. Civ. Mech. Eng.* **2021**, *21*, 1–14. [CrossRef]
22. Xu, S.; Liu, J.; Li, Q. Mechanical properties and microstructure of multi-walled carbon nanotube-reinforced cement paste. *Constr. Build. Mater.* **2015**, *76*, 16–23. [CrossRef]
23. Kharisov, B.I.; Kharissova, O.V.; Méndez, U.O. Methods for dispersion of carbon nanotubes in water and common solvents. *MRS Online Proc. Libr.* **2014**, *1700*, 109–114. [CrossRef]
24. Dyke, C.A.; Tour, J.M. Covalent functionalization of single-walled carbon nanotubes for materials applications. *J. Phys. Chem. A* **2004**, *108*, 11151–11159. [CrossRef]
25. Jeon, I.-Y.; Chang, D.W.; Kumar, N.A.; Baek, J.-B. Functionalization of carbon nanotubes. In *Carbon Nanotubes-Polymer Nanocomposites*; IntechOpen: London, UK, 2011; pp. 91–110.
26. Mousavi, M.A.; Bahari, A. Influence of functionalized MWCNT on microstructure and mechanical properties of cement paste. *Sādhanā* **2019**, *44*, 103. [CrossRef]
27. Chen, J.; Yao, B.; Li, C.; Shi, G. An improved Hummers method for eco-friendly synthesis of graphene oxide. *Carbon* **2013**, *64*, 225–229. [CrossRef]
28. Gong, S.; Ni, H.; Jiang, L.; Cheng, Q. Learning from nature: Constructing high performance graphene-based nanocomposites. *Mater. Today* **2017**, *20*, 210–219. [CrossRef]
29. Zhu, Y.; Murali, S.; Cai, W.; Li, X.; Suk, J.W.; Potts, J.R.; Ruoff, R.S. Graphene and graphene oxide: Synthesis, properties, and applications. *Adv. Mater.* **2010**, *22*, 3906–3924. [CrossRef] [PubMed]
30. Peng, H.; Ge, Y.; Cai, C.; Zhang, Y.; Liu, Z. Mechanical properties and microstructure of graphene oxide cement-based composites. *Constr. Build. Mater.* **2019**, *194*, 102–109. [CrossRef]
31. Pan, Z.; He, L.; Qiu, L.; Korayem, A.H.; Li, G.; Zhu, J.W.; Collins, F.; Li, D.; Duan, W.H.; Wang, M.C. Mechanical properties and microstructure of a graphene oxide–cement composite. *Cem. Concr. Compos.* **2015**, *58*, 140–147. [CrossRef]
32. An, J.; McInnis, M.; Chung, W.; Nam, B.H. Feasibility of using graphene oxide nanoflake (GONF) as additive of cement composite. *Appl. Sci.* **2018**, *8*, 419. [CrossRef]
33. Liu, J.; Suh, H.; Jee, H.; Xu, J.; Nezhad, E.Z.; Choi, C.-S.; Bae, S. Synergistic effect of carbon nanotube/TiO_2 nanotube multi-scale reinforcement on the mechanical properties and hydration process of portland cement paste. *Constr. Build. Mater.* **2021**, *293*, 123447. [CrossRef]
34. Lu, Z.; Hou, D.; Meng, L.; Sun, G.; Lu, C.; Li, Z. Mechanism of cement paste reinforced by graphene oxide/carbon nanotubes composites with enhanced mechanical properties. *RSC Adv.* **2015**, *5*, 100598–100605. [CrossRef]
35. Zhou, C.; Li, F.; Hu, J.; Ren, M.; Wei, J.; Yu, Q. Enhanced mechanical properties of cement paste by hybrid graphene oxide/carbon nanotubes. *Constr. Build. Mater.* **2017**, *134*, 336–345. [CrossRef]
36. Kaur, R.; Kothiyal, N. Positive synergistic effect of superplasticizer stabilized graphene oxide and functionalized carbon nanotubes as a 3–D hybrid reinforcing phase on the mechanical properties and pore structure refinement of cement nanocomposites. *Constr. Build. Mater.* **2019**, *222*, 358–370. [CrossRef]
37. William, S.; Hummers, J.; Offeman, R.E. Preparation of graphitic oxide. *J. Am. Chem. Soc.* **1958**, *80*, 1339.
38. Eigler, S.; Hirsch, A. Chemistry with graphene and graphene oxide-challenges for synthetic chemists. *Angew. Chem. Int. Ed.* **2014**, *53*, 7720–7738. [CrossRef]
39. Yu, H.; Zhang, B.; Bulin, C.; Li, R.; Xing, R. High-efficient synthesis of graphene oxide based on improved hummers method. *Sci. Rep.* **2016**, *6*, 36143. [CrossRef]
40. Marcano, D.C.; Kosynkin, D.V.; Berlin, J.M.; Sinitskii, A.; Sun, Z.; Slesarev, A.; Alemany, L.B.; Lu, W.; Tour, J.M. Improved synthesis of graphene oxide. *ACS Nano* **2010**, *4*, 4806–4814. [CrossRef]
41. Datsyuk, V.; Kalyva, M.; Papagelis, K.; Parthenios, J.; Tasis, D.; Siokou, A.; Kallitsis, I.; Galiotis, C. Chemical oxidation of multiwalled carbon nanotubes. *Carbon* **2008**, *46*, 833–840. [CrossRef]
42. Al Mgheer, T.; Abdulrazzak, F.H. Oxidation of multi-walled carbon nanotubes in acidic and basic Piranha mixture. *Front. Nanosci. Nanotechnol.* **2016**, *2*, 155–158. [CrossRef]
43. Rahman, M.J.; Mieno, T. Water-dispersible multiwalled carbon nanotubes obtained from citric-acid-assisted oxygen plasma functionalization. *J. Nanomater.* **2014**, *2014*, 508192. [CrossRef]
44. Njuguna, J.; Vanli, O.A.; Liang, R. A review of spectral methods for dispersion characterization of carbon nanotubes in aqueous suspensions. *J. Spectrosc.* **2015**, *2015*, 463156. [CrossRef]
45. Rance, G.A.; Marsh, D.H.; Nicholas, R.J.; Khlobystov, A.N. UV-vis absorption spectroscopy of carbon nanotubes: Relationship between the π-electron plasmon and nanotube diameter. *Chem. Phys. Lett.* **2010**, *493*, 19–23. [CrossRef]
46. Elkashef, M.; Abou-Zeid, M.N. Performance of carbon nanotubes in mortar using different surfactants. *Can. J. Civ. Eng.* **2017**, *44*, 619–625. [CrossRef]
47. Sharma, S.; Kothiyal, N.; Chitkara, M. Enhanced mechanical performance of cement nanocomposite reinforced with graphene oxide synthesized from mechanically milled graphite and its comparison with carbon nanotubes reinforced nanocomposite. *RSC Adv.* **2016**, *6*, 103993–104009. [CrossRef]

48. Kaur, R.; Kothiyal, N. Comparative effects of sterically stabilized functionalized carbon nanotubes and graphene oxide as reinforcing agent on physico-mechanical properties and electrical resistivity of cement nanocomposites. *Constr. Build. Mater.* **2019**, *202*, 121–138. [CrossRef]
49. Kaur, R.; Kothiyal, N.; Singh, J. Ultrasonic and superplasticizer assisted dispersion of hybrid carbon nanomaterials (FCNT and GO): Its effect on early stage hydration and physico-mechanical strength of cement mortar. *J. Adhes. Sci. Technol.* **2020**, *34*, 192–218. [CrossRef]
50. Petrunin, S.; Vaganov, V.; Sobolev, K. The effect of functionalized carbon nanotubes on the performance of cement composites. *NANOCON* **2013**, *10*, 16–18.
51. Sahebian, S.; Zebarjad, S.; Vahdati Khaki, J.; Lazzeri, A. A study on the dependence of structure of multi-walled carbon nanotubes on acid treatment. *J. Nanostructure Chem.* **2015**, *5*, 287–293. [CrossRef]
52. Sindu, B.; Sasmal, S. Properties of carbon nanotube reinforced cement composite synthesized using different types of surfactants. *Constr. Build. Mater.* **2017**, *155*, 389–399. [CrossRef]
53. Du, S.; Wu, J.; AlShareedah, O.; Shi, X. Nanotechnology in cement-based materials: A review of durability, modeling, and advanced characterization. *Nanomaterials* **2019**, *9*, 1213. [CrossRef]
54. Zhang, M.-H.; Sisomphon, K.; Ng, T.S.; Sun, D.J. Effect of superplasticizers on workability retention and initial setting time of cement pastes. *Constr. Build. Mater.* **2010**, *24*, 1700–1707. [CrossRef]
55. Chuah, S.; Li, W.; Chen, S.J.; Sanjayan, J.G.; Duan, W.H. Investigation on dispersion of graphene oxide in cement composite using different surfactant treatments. *Constr. Build. Mater.* **2018**, *161*, 519–527. [CrossRef]
56. Chen, S.J.; Qiu, C.Y.; Korayem, A.H.; Barati, M.R.; Duan, W.H. Agglomeration process of surfactant-dispersed carbon nanotubes in unstable dispersion: A two-stage agglomeration model and experimental evidence. *Powder Technol.* **2016**, *301*, 412–420. [CrossRef]
57. ASTM. Standard Test Method for Compressive Strength of Hydraulic Cement Mortars (Using 2–In. or [50 mm] Cube Specimens). Available online: https://compass.astm.org/EDIT/html_annot.cgi?C109+20b (accessed on 21 June 2021).
58. Jee, H.; Park, J.; Zalnezhad, E.; Jeong, K.; Woo, S.M.; Seok, S.; Bae, S. Characterization of Titanium Nanotube Reinforced Cementitious Composites: Mechanical Properties, Microstructure, and Hydration. *Materials* **2019**, *12*, 1617. [CrossRef] [PubMed]
59. ASTM. Standard Test Method for Splitting Tensile Strength of Cylindrical Concrete Specimens. Available online: https://compass.astm.org/EDIT/html_annot.cgi?C496+17 (accessed on 21 June 2021).
60. Sedaghat, A.; Zayed, A.; Sandberg, P. Measurement and prediction of heat of hydration of portland cement using isothermal conduction calorimetry. *J. Test. Eval.* **2013**, *41*, 943–950. [CrossRef]
61. Chung, D.D. *Carbon Composites: Composites with Carbon Fibers, Nanofibers, and Nanotubes*; Butterworth–Heinemann: Oxford, UK, 2016.
62. Avilés, F.; Ponce, A.; Cauich-Rodríguez, J.; Martínez, G. TEM examination of MWCNTs oxidized by mild experimental conditions. *Fuller. Nanotub. Carbon Nanostructures* **2012**, *20*, 49–55. [CrossRef]
63. Musso, S.; Tulliani, J.-M.; Ferro, G.; Tagliaferro, A. Influence of carbon nanotubes structure on the mechanical behavior of cement composites. *Compos. Sci. Technol.* **2009**, *69*, 1985–1990. [CrossRef]
64. Zhao, L.; Guo, X.; Ge, C.; Li, Q.; Guo, L.; Shu, X.; Liu, J. Investigation of the effectiveness of PC@GO on the reinforcement for cement composites. *Constr. Build. Mater.* **2016**, *113*, 470–478. [CrossRef]
65. Zheng, M.; Diner, B.A. Solution redox chemistry of carbon nanotubes. *J. Am. Chem. Soc.* **2004**, *126*, 15490–15494. [CrossRef]
66. Fathi, M.; Saghafi, M.; Mahboubi, F. Graphene oxide nanoribbons and their applications in supercapacitors. *J. Ultrafine Grained Nanostructured Mater.* **2014**, *47*, 71–76.
67. Zhang, Y.; Liu, Y.; Chen, L.; Hu, X.; Zhang, L.; Hu, L.; Chen, Y. One-dimensional graphene nanoribbons hybridized with carbon nanotubes as cathode and anode interfacial layers for high performance solar cells. *RSC Adv.* **2015**, *5*, 49614–49622. [CrossRef]
68. Wang, K.; Pang, J.; Li, L.; Zhou, S.; Li, Y.; Zhang, T. Synthesis of hydrophobic carbon nanotubes/reduced graphene oxide composite films by flash light irradiation. *Front. Chem. Sci. Eng.* **2018**, *12*, 376–382. [CrossRef]
69. Ţucureanu, V.; Matei, A.; Avram, A.M. FTIR spectroscopy for carbon family study. *Crit. Rev. Anal. Chem.* **2016**, *46*, 502–520. [CrossRef]
70. Wong, C.H.A.; Pumera, M. Highly conductive graphene nanoribbons from the reduction of graphene oxide nanoribbons with lithium aluminium hydride. *J. Mater. Chem. C* **2014**, *2*, 856–863. [CrossRef]
71. Cui, H.; Yan, X.; Monasterio, M.; Xing, F. Effects of various surfactants on the dispersion of MWCNTs-OH in aqueous solution. *Nanomaterials* **2017**, *7*, 262. [CrossRef] [PubMed]
72. Osorio, A.; Silveira, I.; Bueno, V.; Bergmann, C. $H_2SO_4/HNO_3/HCl$—Functionalization and its effect on dispersion of carbon nanotubes in aqueous media. *Appl. Surf. Sci.* **2008**, *255*, 2485–2489. [CrossRef]
73. Harun, S.W. Technology and Innovations. In *Handbook of Graphene*; John Wiley & Sons: Hoboken, NJ, USA, 2019; Volume 8.
74. Rubel, R.I.; Ali, M.H.; Jafor, M.A.; Alam, M.M. Carbon nanotubes agglomeration in reinforced composites: A review. *AIMS Mater. Sci.* **2019**, *6*, 756–780. [CrossRef]
75. Li, D.; Müller, M.B.; Gilje, S.; Kaner, R.B.; Wallace, G.G. Processable aqueous dispersions of graphene nanosheets. *Nat. Nanotechnol.* **2008**, *3*, 101–105. [CrossRef] [PubMed]
76. Fan, X.; Peng, W.; Li, Y.; Li, X.; Wang, S.; Zhang, G.; Zhang, F. Deoxygenation of exfoliated graphite oxide under alkaline conditions: A green route to graphene preparation. *Adv. Mater.* **2008**, *20*, 4490–4493. [CrossRef]

77. Korayem, A.; Tourani, N.; Zakertabrizi, M.; Sabziparvar, A.; Duan, W. A review of dispersion of nanoparticles in cementitious matrices: Nanoparticle geometry perspective. *Constr. Build. Mater.* **2017**, *153*, 346–357. [CrossRef]
78. Wang, X.; Dong, S.; Ashour, A.; Zhang, W.; Han, B. Effect and mechanisms of nanomaterials on interface between aggregates and cement mortars. *Constr. Build. Mater.* **2020**, *240*, 117942. [CrossRef]
79. Mahinroosta, M.; Allahverdi, A. A Scoping Review on Integrating Inorganic Nanomaterials into Cement Composites. *Adv. Civ. Eng. Mater.* **2019**, *8*, 526–553. [CrossRef]
80. Lv, S.; Ma, Y.; Qiu, C.; Sun, T.; Liu, J.; Zhou, Q. Effect of graphene oxide nanosheets of microstructure and mechanical properties of cement composites. *Constr. Build. Mater.* **2013**, *49*, 121–127. [CrossRef]
81. Li, G.Y.; Wang, P.M.; Zhao, X. Mechanical behavior and microstructure of cement composites incorporating surface-treated multi-walled carbon nanotubes. *Carbon* **2005**, *43*, 1239–1245. [CrossRef]
82. Han, B.; Yu, X.; Ou, J. Multifunctional and Smart Carbon Nanotube Reinforced Cement-Based Materials. In *Nanotechnology in Civil Infrastructure: A Paradigm Shift*; Gopalakrishnan, K., Birgisson, B., Taylor, P., Attoh-Okine, N.O., Eds.; Springer: Berlin/Heidelberg, Germany, 2011; pp. 1–47.
83. Makar, J.; Margeson, J.; Luh, J. Carbon nanotube/cement composites-early results and potential applications. In Proceedings of the 3rd International Conference on Construction Materials: Performance, Innovations and Structural Implications, Vancouver, BC, USA, 22–24 August 2005; pp. 1–10.
84. Zhao, L.; Hou, D.; Wang, P.; Guo, X.; Zhang, Y.; Liu, J.; Zhang, J. Experimental and molecular dynamics studies on the durability of sustainable cement-based composites: Reinforced by graphene. *Constr. Build. Mater.* **2020**, *257*, 119566. [CrossRef]
85. Makar, J. The effect of SWCNT and other nanomaterials on cement hydration and reinforcement. In *Nanotechnology in Civil Infrastructure*; Springer: Berlin/Heidelberg, Germany, 2011; pp. 103–130.
86. Bullard, J.W.; Jennings, H.M.; Livingston, R.A.; Nonat, A.; Scherer, G.W.; Schweitzer, J.S.; Scrivener, K.L.; Thomas, J.J. Mechanisms of cement hydration. *Cem. Concr. Res.* **2011**, *41*, 1208–1223. [CrossRef]
87. Bae, S.; Kanematsu, M.; Hernández-Cruz, D.; Moon, J.; Kilcoyne, D.; Monteiro, P.J.M. In Situ Soft X-ray Spectromicroscopy of Early Tricalcium Silicate Hydration. *Materials* **2016**, *9*, 976. [CrossRef]
88. Zhao, L.; Guo, X.; Liu, Y.; Ge, C.; Guo, L.; Shu, X.; Liu, J. Synergistic effects of silica nanoparticles/polycarboxylate superplasticizer modified graphene oxide on mechanical behavior and hydration process of cement composites. *RSC Adv.* **2017**, *7*, 16688–16702. [CrossRef]
89. Jansen, D.; Neubauer, J.; Goetz-Neunhoeffer, F.; Haerzschel, R.; Hergeth, W.-D. Change in reaction kinetics of a Portland cement caused by a superplasticizer—Calculation of heat flow curves from XRD data. *Cem. Concr. Res.* **2012**, *42*, 327–332. [CrossRef]
90. Uchikawa, H.; Hanehara, S.; Sawaki, D. The role of steric repulsive force in the dispersion of cement particles in fresh paste prepared with organic admixture. *Cem. Concr. Res.* **1997**, *27*, 37–50. [CrossRef]
91. Cui, H.; Yang, S.; Memon, S.A. Development of carbon nanotube modified cement paste with microencapsulated phase-change material for structural-functional integrated application. *Int. J. Mol. Sci.* **2015**, *16*, 8027–8039. [CrossRef] [PubMed]
92. Snellings, R.; Salze, A.; Scrivener, K. Use of X-ray diffraction to quantify amorphous supplementary cementitious materials in anhydrous and hydrated blended cements. *Cem. Concr. Res.* **2014**, *64*, 89–98. [CrossRef]
93. Chuah, S.; Pan, Z.; Sanjayan, J.G.; Wang, C.M.; Duan, W.H. Nano reinforced cement and concrete composites and new perspective from graphene oxide. *Constr. Build. Mater.* **2014**, *73*, 113–124. [CrossRef]
94. Xie, S.; Cheng, Z.; Wan, L. Hydration And Microstructure Of Astm Type I Cement Paste. *SECM* **2019**, *26*, 215–220. [CrossRef]
95. Mohsen, A.; Aiad, I.; El-Hossiny, F.; Habib, A. Evaluating the Mechanical Properties of Admixed Blended Cement Pastes and Estimating its Kinetics of Hydration by Different Techniques. *Egypt. J. Pet.* **2020**, *29*, 171–186. [CrossRef]
96. Habib, A.; Aiad, I.; El-Hosiny, F.; Abd El-Aziz, A. Development of the fire resistance and mechanical characteristics of silica fume-blended cement pastes using some chemical admixtures. *Constr. Build. Mater.* **2018**, *181*, 163–174. [CrossRef]
97. Tafesse, M.; Kim, H.-K. The role of carbon nanotube on hydration kinetics and shrinkage of cement composite. *Compos. B Eng.* **2019**, *169*, 55–64. [CrossRef]
98. Suh, H.; Jee, H.; Kim, J.; Kitagaki, R.; Ohki, S.; Woo, S.; Jeong, K.; Bae, S. Influences of rehydration conditions on the mechanical and atomic structural recovery characteristics of Portland cement paste exposed to elevated temperatures. *Constr. Build. Mater.* **2020**, *235*, 117453. [CrossRef]
99. Jee, H.; Im, S.; Kanematsu, M.; Suzuki, H.; Morooka, S.; Taku, K.; Machida, A.; Bae, S. Determination of atomistic deformation of tricalcium silicate paste with high-volume fly ash. *J. Am. Ceram. Soc.* **2020**, *103*, 7188–7201. [CrossRef]
100. Bae, S.; Jee, H.; Suh, H.; Kanematsu, M.; Shiro, A.; Machida, A.; Watanuki, T.; Shobu, T.; Morooka, S.; Geng, G. Analysis of atomistic structural deformation characteristics of calcium silicate hydrate in 53-year-old tricalcium silicate paste using atomic pair distribution function. *Constr. Build. Mater.* **2020**, *237*, 117714. [CrossRef]
101. Bae, S.; Jee, H.; Kanematsu, M.; Shiro, A.; Machida, A.; Watanuki, T.; Shobu, T.; Suzuki, H. Pair distribution function analysis of nanostructural deformation of calcium silicate hydrate under compressive stress. *J. Am. Ceram. Soc.* **2018**, *101*, 408–418. [CrossRef]
102. Scrivener, K.; Snellings, R.; Lothenbach, B. *A Practical Guide to Microstructural Analysis of Cementitious Materials*; CRC Press: Boca Raton, FL, USA, 2018.

nanomaterials

MDPI

Article

pH-Sensitive Chitosan Nanoparticles for Salivary Protein Delivery

Yi Zhu [1], Lina M. Marin [2], Yizhi Xiao [3], Elizabeth R. Gillies [1,4] and Walter L. Siqueira [2,*]

[1] School of Biomedical Engineering, The University of Western Ontario, 1151 Richmond Street, London, ON N6A 3K7, Canada; yzhu336@uwo.ca (Y.Z.); egillie@uwo.ca (E.R.G.)

[2] College of Dentistry, University of Saskatchewan, 105 Wiggins Rd, Saskatoon, SK S7N 5E4, Canada; lina.marin@usask.ca

[3] Schulich Medicine and Dentistry, The University of Western Ontario, 1151 Richmond Street, London, ON N6A 5C1, Canada; yxiao32@uwo.ca

[4] Department of Chemistry, Department of Chemical and Biochemical Engineering, The University of Western Ontario, 1151 Richmond Street, London, ON N6A 5B7, Canada

* Correspondence: walter.siqueira@usask.ca

Abstract: Salivary proteins such as histatins (HTNs) have demonstrated critical biological functions directly related to tooth homeostasis and prevention of dental caries. However, HTNs are susceptible to the high proteolytic activities in the oral environment. Therefore, pH-sensitive chitosan nanoparticles (CNs) have been proposed as potential carriers to protect proteins from enzymatic degradation at physiological salivary pH. Four different types of chitosan polymers were investigated and the optimal formulation had good batch to batch reproducibility, with an average hydrodynamic diameter of 144 ± 6 nm, a polydispersity index of 0.15 ± 0.04, and a zeta potential of 18 ± 4 mV at a final pH of 6.3. HTN3 encapsulation and release profiles were characterized by cationic polyacrylamide gel electrophoresis. The CNs successfully encapsulated HTN3 and selectively swelled at acidic pH to facilitate HTN3 release. Protection of HTN3 against enzymatic degradation was investigated in diluted whole saliva. HTN3 encapsulated in the CNs had a prolonged survival time compared to the free HTN3. CNs with and without HTN3 also successfully reduced biofilm weight and bacterial viability. The results of this study have demonstrated the suitability of CNs as potential protein carriers for oral applications, especially for complications occurring at acidic conditions.

Keywords: protein carrier; drug delivery; polymeric nanoparticles; biopolymers

1. Introduction

Saliva is a complex fluid composed of proteins, enzymes, and a variety of electrolytes. Many physiological functions, such as chewing, digestion, wetting, and lubrication are regulated by saliva [1]. The concentration of proteins present in the saliva may be used to monitor oral health, as the severity and occurrence of oral diseases have been associated with the composition and quantitative changes in salivary proteins [2]. Many of the salivary proteins are active in the regulation of tooth homeostasis, either by directly exerting antimicrobial effects or interfering with microbial colonization [3]. Examples include statherin, histatins (HTNs), defensins, lactoferrin, and mucin [4]. In particular, the HTNs are of great interest because these proteins are multifunctional. They have demonstrated biological functions including the inhibition of calcium and phosphate precipitation on enamel and antimicrobial activities, which are directly related to the regulation of oral homeostasis [5–7]. The HTNs mainly consist of HTN1, HTN3, and HTN5, and each supplies about 20–30% of the total HTN pool [8]. HTN1 is the only phosphorylated HTN able to reduce bacterial colonization on tooth surfaces [9]. HTN5 has the most potent antifungal activity against the pathogenic yeast *Candida albicans* [10]. HTN3 was demonstrated to be the most effective in killing against *Streptococcus mutans* (*S. mutans*) [11], which is a significant contributor towards dental caries.

These proteins are often present in low concentrations inside the oral cavity due to the high proteolytic activity of saliva [12,13]. Therefore, we proposed a delivery system to encapsulate and protect these proteins for their use as inhibitors of tooth decay. The use of stimulus-responsive polymeric nanoparticles synthesized from natural polymers has recently gained substantial attention, as such smart delivery systems have the potential to improve the overall colloidal stability of the encapsulated molecules and modulate pharmacokinetics, often resulting in reduced toxicity and enhanced efficacy [14–17]. The encapsulated cargo can be released in a stimulus-responsive manner. Examples of stimuli include changes in pH, ionic strength, temperature, UV light, magnetic field, or the presence of specific biological molecules [18–21]. In particular, for delivering drugs to the oral cavity, it would be ideal for the delivery system to remain stable at physiological salivary pH and selectively release the encapsulated cargo under acidic conditions. The pH-responsive property is crucial because the oral environment acidifies following carbohydrate intake as a result of the consumption of food and beverages. These conditions promote the development of oral diseases such as dental caries and dental erosion [22,23].

Many polymer-based materials have been studied for applications in controlled drug delivery, including naturally occurring biopolymers. For instance, chitosan is a copolymer composed of *N*-acetyl-D-glucosamine and β(1–4) linked D-glucosamine [24]. It is mainly obtained through the deacetylation of chitin under alkaline chemicals [25]. Chitin is a biopolymer found in the exoskeletons of crustaceans, insects, and some fungi [26]. As the only known naturally occurring polycationic polysaccharide and with its ability to interact with anionic molecules, chitosan and its derivatives have been studied extensively for applications in the agricultural, medicinal, and pharmaceutical industries [27–29].

Previous studies have utilized chitosan in toothpaste as an antimicrobial agent due to its broad antimicrobial spectrum, covering both Gram-negative and Gram-positive bacteria and fungi [30,31]. Aside from its antimicrobial effect, chitosan also offers other advantages including biodegradability [32] and biocompatibility [33]. It also exhibits pH-dependent solubility due to the presence of amino groups on the polymer chains [34]. The functional amino groups can serve as platforms for interactions with other anionic molecules, where the choice of anionic molecules depends on the intended applications.

Chitosan nanoparticles (CNs) have been commonly used to deliver therapeutic agents, such as insulin [35], tretinoin [36], and doxorubicin [37]. In the current work, to the best of our knowledge, for the first time we propose pH-sensitive chitosan nanoparticles for salivary protein delivery. We hypothesize that salivary protein-loaded CNs can selectively release the encapsulated proteins under low pH environments, offer protection against proteolysis at the physiological salivary pH, and reduce *S. mutans* biofilm formation on hydroxyapatite. To test these hypotheses, HTN3 was selected as the target protein, and our objectives were to: (1) optimize CN formulation via ionic gelation with four different types of chitosan polymers; (2) characterize both blank and HTN3-loaded CNs; (3) quantify encapsulation and release profiles of HTN3; (4) assess the protection offered by encapsulation against enzymatic degradation in human saliva; and (5) evaluate the effectiveness of HTN3-loaded CNs in reducing *S. mutans* biofilm formation.

2. Materials and Methods

2.1. Materials

Ultra-low molecular weight (MW) chitosan (20 kg/mol, 91% deacetylated) and low MW chitosan (1) (250 kg/mol, 91% deacetylated) were purchased from Glentham Life Sciences (Corsham, UK). The other low MW chitosan (2) (50–190 kg/mol, 75–85% deacetylated), medium MW chitosan (190–310 kg/mol, 75–85% deacetylated), and ZipTip C18 pipette tips were acquired from Millipore Sigma (Oakville, ON, Canada). All filters, including 0.45 and 0.22 µm syringe filters and 10k Nanosep filters were purchased from Pall Corporation (Mississauga, ON, Canada). HTN3 was acquired from Synpeptide Co., Ltd. (Shanghai, China). *S. mutans* UA159 was kindly donated by Dr. Yoav Finer from the University of Toronto (Toronto, ON, Canada). Ceramic hydroxyapatite discs with 5 mm diameter and 2 mm thickness were obtained from Clarkson Chromatography Products Inc. (South Williamsport, PA, USA). Bicinchoninic acid assay kit was acquired from Thermo Fisher Scientific (Mississauga, ON, Canada). All other chemicals, including sodium tripolyphosphate (TPP), were purchased from Millipore Sigma (Oakville, ON, Canada).

2.2. Optimized CNs Preparation

The ionic gelation procedure described previously was modified to optimize CNs synthesis [38]. Briefly, preliminary experiments were done to assess the effect of chitosan molar mass and degree of deacetylation, pH of chitosan/TPP solutions, and chitosan to TPP mass ratios on CN characteristics. These particle preparations were performed in triplicate. After determining the optimal conditions, unloaded CNs were synthesized by dissolving 100 mg of ultra-low MW chitosan in 100 mL of 0.4% *v/v* concentrated HCl. The pH of the chitosan solution was adjusted to 5.9 and the solution was filtered through a 0.45 µm syringe filter. Then, 100 mg of TPP was dissolved in 100 mL of Milli-Q water, its pH was adjusted to 6.0 and solution was filtered through a 0.22 µm syringe filter. Finally, 0.36 mL of TPP solution was added dropwise into 2.5 mL of chitosan solution under constant stirring at 700 rpm, such that the mass ratio of chitosan to TPP was 6.94:1. HTN3-loaded CNs were synthesized by mixing HTN3 with the chitosan solution before ionic gelation with TPP. Briefly, 0.29 mL of the 0.1% *w/v* TPP solution was introduced dropwise into 2 mL of 0.1% *w/v* chitosan and 0.0025% *w/v* HTN3 under constant stirring at 700 rpm. The loading ratio of chitosan to HTN3 by mass was 40:1, and 0.45 mL of the suspension contained 10 µg of HTN3. The final pH of the nanoparticle solution was 6.3.

2.3. Characterizations of CNs

Unloaded CNs and HTN3-loaded CNs were initially characterized by measuring the Z-average particle diameter, polydispersity index (PDI), and zeta potential by dynamic light scattering (DLS, Malvern Zetasizer Nano ZS instrument, Malvern Instruments Ltd, Malvern, UK) at a wavelength of 630 nm and a constant temperature of 25 °C. To determine the morphology of CNs, an aliquot from each CNs suspension was firstly desalted by dialyzing it against 1 L of deionized water for 3 h under constant stirring, using a 50 kg/mol MW cut-off regenerated cellulose dialysis tubing (Spectrum Labs, New Brunswick, NJ, USA). Then, samples were placed onto Formvar®-coated copper grid, dried overnight under an air atmosphere, and the morphology of the CNs was visualized by transmission electron microscopy (TEM, Philips CM10, Philips, Amsterdam, The Netherlands) at 80 kV. To investigate the pH-dependence of CN size, initial (time = 0 min) particle size measurements were taken at a starting pH of 6.3. The pH of the suspension was then adjusted to 3.0, 4.0, or 5.0. After the pH adjustments, particle size, PDI, and zeta potential were assessed by DLS at 10, 20, 30, 60, 120, and 240 min. To evaluate colloidal stability, CNs were stored at 4 °C for 61 days. Particle size and zeta potential were measured by DLS on days 1, 10, 15, 22, 30, 35, 45, 52, and 61.

2.4. Encapsulation and Release of HTN3

To assess the pH-dependent release of HTN3 from HTN3-loaded CNs, 0.05 mL of buffer (25 mM phosphate buffer/500 mM NaCl solution, pH 6.8), was added to 0.45 mL of HTN3-loaded CNs suspension. The 0.5 mL of suspension was then filtered with a 10 kDa cut-off Nanosep filter by centrifuging the suspension at 14,000× *g* for 10 min to separate free HTN3 (4063 Dalton) from the encapsulated CNs. The filtrate was collected to determine encapsulation efficiency by cationic polyacrylamide gel electrophoresis (cationic-PAGE) [39]. The pH of the retentate on top of the filter was adjusted to pH 3.0, 4.0, or 5.0 and incubated at 37 °C for 30 min. The suspension was centrifuged again, allowing the passage of released HTN3 through the filter. Subsequently, the filtrate was collected to determine the extent of protein release by cationic-PAGE, using pure HTN3 as a reference control. HTN3 encapsulation efficiency was tested for loading ratios of 2%, 5%, and 10% *w/w* (HTN3 to chitosan). Encapsulation samples were prepared as previously described with modifications to the mass ratio of HTN3 to chitosan. The encapsulation efficiency was determined by Image Lab (BioRad) via the relative quantity tool, corresponding to the ratio of the band volume and intensity divided by the reference (HTN3 standard) band volume and intensity. To evaluate the cumulative release of HTN3, HTN3-loaded CNs were prepared as previously described. Then, the retentate on top of the filter was incubated for 30, 60, 90, 120, 150, 180, and 210 min at 37 °C in buffer solutions with different pH values (6.8, 3.0, 4.0, or 5.0). Then, samples were centrifuged at 14,000× *g* for 10 min to separate the released protein from the CNs. Subsequently, the filtrate was collected. This procedure was repeated for each time point and the filtrate and a HTN3 standard were analyzed by cationic-PAGE as previously described. The ability of the HTN3-loaded CNs to release HTN3 under pH-cycling conditions, reflecting the pH changes that occur in the oral cavity, was determined by incubating the CNs in buffer with pH of 6.8 and then in buffer with pH 4.0. This cycle was repeated for a total of four times, and the amount of HTN3 released at each pH in function of time was determined by cationic-PAGE, following the protocol previously described.

2.5. Protein Degradation Study

Considering that proteins are susceptible to proteolytic degradation inside the oral cavity [40], the protection offered by encapsulation was evaluated in diluted whole saliva. Stimulated whole saliva was collected by chewing a 5 cm × 5 cm parafilm piece from three healthy individuals, between 9:00 am to 11:00 am, to minimize the effect of circadian rhythm [13], and at least two hours after breakfast. Whole saliva was kept on ice during the collection and then was immediately centrifuged for 10 min 14,000× *g* at 4 °C to separate the bacteria, cells, and other debris from the supernatant containing salivary proteins. The concentration of salivary proteins in the supernatant, referred as whole saliva supernatant (WSS), was quantified using a bicinchoninic acid assay kit. The degradation of free HTN3 was assessed in 10-fold diluted WSS, following the protocol previously reported [6]. Briefly, 50 µg of free HTN3 was added to diluted WSS (100 µg salivary protein/mL) in a final volume of 1 mL. The pH was adjusted to 6.8 using 1 M HCl or 1 M NaOH, depending on the pH of the WSS. Then, an aliquot of 100 µL was incubated for 0, 0.5, 1, 2, 3, or 6 h at 37 °C. At each time point, samples were boiled for 5 min to terminate proteolytic activity, dried, and desalted. The extent of degradation was quantified after cationic-PAGE, using Image Lab relative quantity tool. Subsequently, the degradation of HTN3-loaded CNs in WSS was assessed by mixing 0.5 mL of the HTN3-loaded CNs (containing 50 µg of HTN3) with diluted WSS (100 µg salivary protein/mL) to reach a final volume of 1 mL. The final pH was standardized to 6.8 with 1 M HCl or 1 M NaOH, depending on the initial pH of saliva. An aliquot of 100 µL was collected after incubation at 37 °C at the following time points: 0, 0.5, 1, 2, 3, and 6 h. The pH was adjusted to 3 to release all of the HTN3 within the delivery system. After centrifugation with a 10 kDa Nanosep filter, the filtrate was retrieved and boiled for 5 min to terminate proteolytic activity. Samples were dried and desalted, followed by cationic-PAGE. The extent of degradation was quantified by Image Lab through the relative quantity tool.

2.6. *Streptococcus mutans* Killing Assay and Biofilm Formation

S. mutans (UA 159) colonies were spiked from blood agar and grown for 14 h in 10 mL of tryptone yeast extract broth (TYEB) supplemented with 1% glucose at 37 °C and 10% CO_2. After incubation, the bacterial suspension was washed twice with 0.9% NaCl and the pellet was resuspended in 1.2 mL of phosphate-buffered saline (PBS). The optical density (OD) at 600 nm was adjusted to 1.5, corresponding to a bacterial concentration of 1×10^9 CFU/mL, based on a growth curve previously done. An aliquot of 0.1 mL of the bacterial suspension with 1.5 OD reading was added to 9.9 mL of PBS to obtain a bacterial concentration of 10^7 CFU/mL. This suspension was then added to an equal volume of a serial dilution series of HTN3 from 0 to 200 μM in a 96-well polypropylene microtiter plate. The samples were then incubated at 37 °C for 1.5 h. After incubation, each sample was diluted 10^3 and 10^4-fold in PBS, and 20 μL of each dilution was plated onto Todd Hewitt broth (THB) agar plates. Bacterial viability was assessed by colony counting using comparisons against control samples incubated without HTN3. The IC50 value was calculated based on the dose–response curve and used in the biofilm study described below. To assess the capability of HTN3-loaded CNs to reduce *S. mutans* biofilm formation on hydroxyapatite (HA) discs, the discs were fixated on the interior of a 24-well polypropylene microtiter plate lid, where the discs were carefully positioned so that the plate lid fits both 24 and 96-well plates. The plate lid was placed to immerse all discs in a 96-well plate filled with 200 μL of 5 different treatment solutions (n = 6/group): control (PBS), 7.4 μM HTN3, 0.1% *w/v* unloaded CNs, 0.1% *w/v* CNs containing 7.4 μM of encapsulated HTN3, and 12,300 ppm fluoride solution. Treatments were done under constant stirring for 2 h at 37 °C to allow the formation of a single-component pellicle onto HA, simulating the formation of the acquired enamel pellicle in the oral cavity. Subsequently, the plate cover was placed on a 24-well plate with 2 mL of 0.9% *w/v* NaCl to wash the discs. Afterwards the discs were immersed into 2 mL of TYEB supplemented with 1% *w/v* glucose and 10^7 CFU/mL of bacteria. The discs were then incubated at 37 °C in 10% CO_2 for 8 h. After incubation, the discs were washed again with NaCl, followed by further incubation for 16 h immersed in 2 mL of TYEB supplemented with 0.1 mM glucose. Over the next 4 days, the discs were incubated in a repeating cycle between solution 1 and 2 where solution 1 was TYEB supplemented with 1% *w/v* sucrose for 8 h, and solution 2 was TYEB supplemented with 0.1 mM glucose for 16 h. On the 6th day, all discs were transferred to separate tubes filled with 1 mL of PBS, followed by sonication to remove the biofilms off the discs. An aliquot of 0.5 mL of the suspension was transferred to preweighted tubes and centrifuged at 14,000× *g* for 5 min. The supernatant was removed, and the biofilm wet weight was measured. The same suspension was diluted 10^6 and 10^7 times in PBS, and 20 μL of each dilution was plated onto THB agar plates, incubated for 48 h at 37 °C and 10% CO_2, and then bacterial viability was assessed by colony counting.

2.7. Statistical Analyses

Statistical analyses were performed using software Prism 8.0 GraphPad. Biofilm mass and bacterial viability were analyzed by ordinary one-way ANOVA followed by Tukey's multiple comparisons test between each treatment group. The level of significance (α) was set at 0.05 (95% confidence interval).

3. Results

3.1. Results

3.1.1. Optimization of CNs Formulation

Four different types of chitosan were investigated to select the best formulation. As shown in Figure 1a, in general the particles were unstable and aggregated at low chitosan to TPP mass ratios. At higher ratios, stable particles were formed and the diameter decreased. Upon further increasing the chitosan to TPP mass ratio, the particle diameter steadily increased for most systems. PDI followed a similar trend as demonstrated in Figure 1b. Zeta potential increased steadily with a corresponding increase in chitosan to TPP ratio. Based on these results, CNs prepared from ultra low MW chitosan were selected as the optimal formulation, because they had the smallest particle diameter (Figure S1), lowest PDI, with a fairly high zeta potential. The particle diameter, PDI, and zeta potential were stable at 4 °C for at least 60 days (Figure S2).

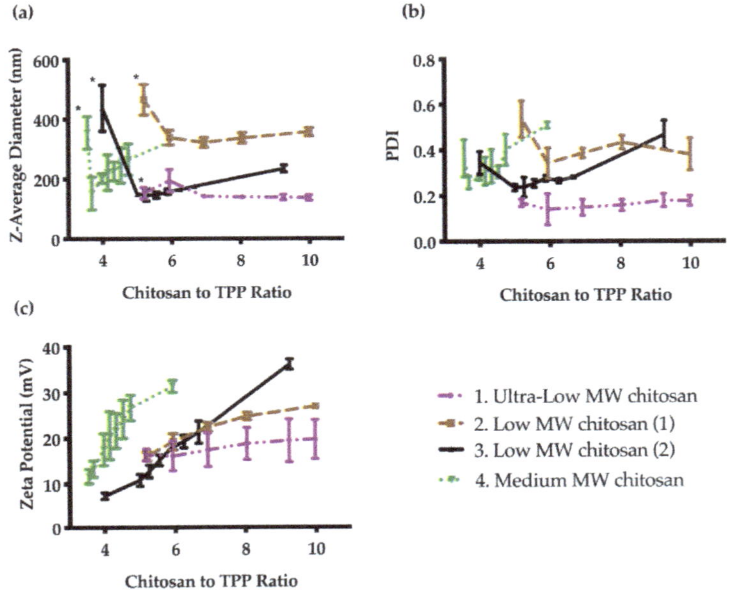

Figure 1. Effect of chitosan to TPP mass ratio on particle properties as measured by DLS: (**a**) Z-average diameter; (**b**) PDI; and (**c**) Zeta potential. Error bars correspond to the standard deviations on triplicate particle preparations. The particles were unstable and sedimented below a certain ratio (indicated by *), then the diameter generally decreased as the ratio was increased, and finally increased as the ratio was further increased. PDI followed a similar trend, while the zeta potential continually increased due to the increasing cationic charge with increasing chitosan to TPP ratio.

3.1.2. TEM Images of Unloaded and HTN3-Loaded CNs

The selected ultra-low CNs were then visualized with TEM. As shown in Figure 2, the CNs had a spherical morphology, and the observed diameters were in good agreement with the data obtained from DLS. There were no obvious differences in size or dispersity among unloaded CNs and those with different loading ratios (Figure S3).

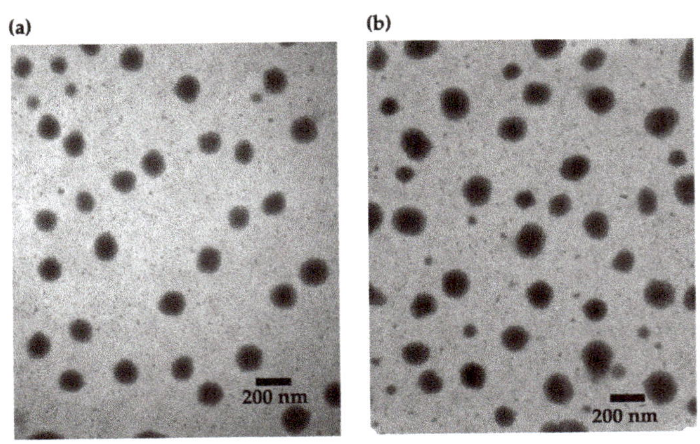

Figure 2. TEM images of CNs: (**a**) unloaded and (**b**) HTN3-loaded at a 2% *w/w* loading ratio. There are no obvious differences between the samples.

3.1.3. pH-Dependent Swelling CNs

The pH-responsive properties of the unloaded CNs were examined first by adjusting the pH from the initial pH of 6.3 for the formulation to pH 3, 4, or 5. DLS measurements of the Z-average diameters at each pH indicated greater swelling at lower pH values (Figure 3a). The CNs swelled from 146 nm at pH 6.3 to 260 nm at pH 3. The swelling was rapid at each pH, reaching an equilibrium diameter at 10 min. We also subjected the CNs suspension to pH-cycling between 6.3 and 4. Measurements of the Z-average diameter suggested that the CNs were able to selectively swell under acidic conditions and then to contract when the pH was increased upon removal of the stimulus (Figure 3b). However, there was a general trend towards larger diameters with repeated swelling cycles.

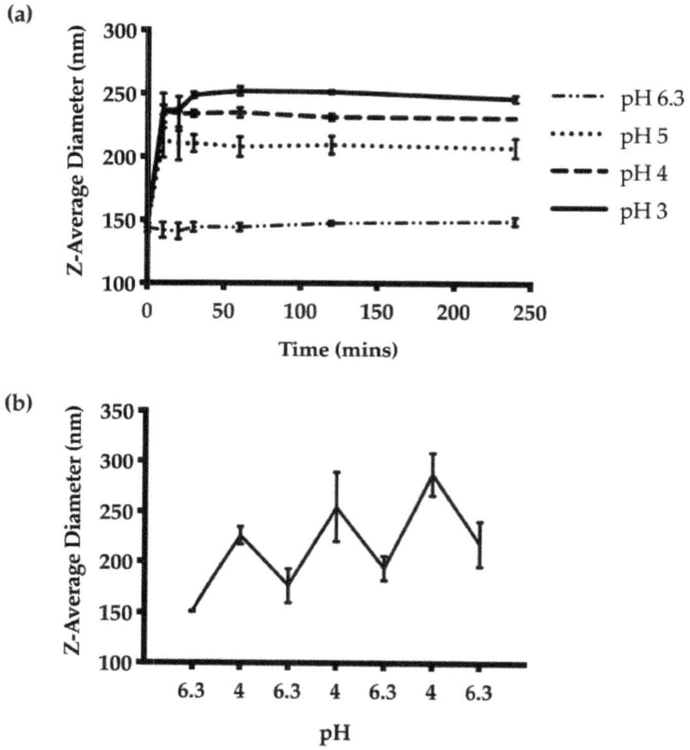

Figure 3. pH-responsive behaviour of the CNs: (**a**) Z-average diameters over time at different pH values from 3 to 6.3. The results suggest the degree of swelling is proportional to the acidity of the suspension. The swelling behaviour was also rapid, and reached equilibrium 10 min after pH adjustment. (**b**) The pH-responsive behaviour was further examined through pH cycling between pH 6.3 and 4. The nanoparticles were able to swell and contract accordingly based on the pH.

3.1.4. pH-Dependent Release of HTN3 from CNs

The release kinetics of HTN3 encapsulated CNs were investigated at different pH values. The highest cumulative amount of HTN3 release was observed at pH 3, and there was minimal release at pH 6.8. On average, at pH 3, the CNs were able to release 84% ± 7% of the encapsulated HTN3, 58% ± 9% at pH 4, 36% ± 3% at pH 5, and 2% ± 2% at pH 6.8 over seven separate releases (Figure 4a). Thus, the CNs were able to respond to the environmental pH changes and release the protein selectively at acidic pH. pH-cycling was also performed to reflect the pH changes that happen many times in the oral cavity throughout the day. HTN3 loaded CNs were subjected to pH treatments in the following sequence: 6.8, 4, 6.8, 4, 6.8, 4, 6.8, and 4. After isolation of released protein and cationic-PAGE at each pH change, protein bands of HTN3 were only observed at pH 4 and were absent at pH 6.8 (Figure 4b). This result shows the pH selective release for the CNs.

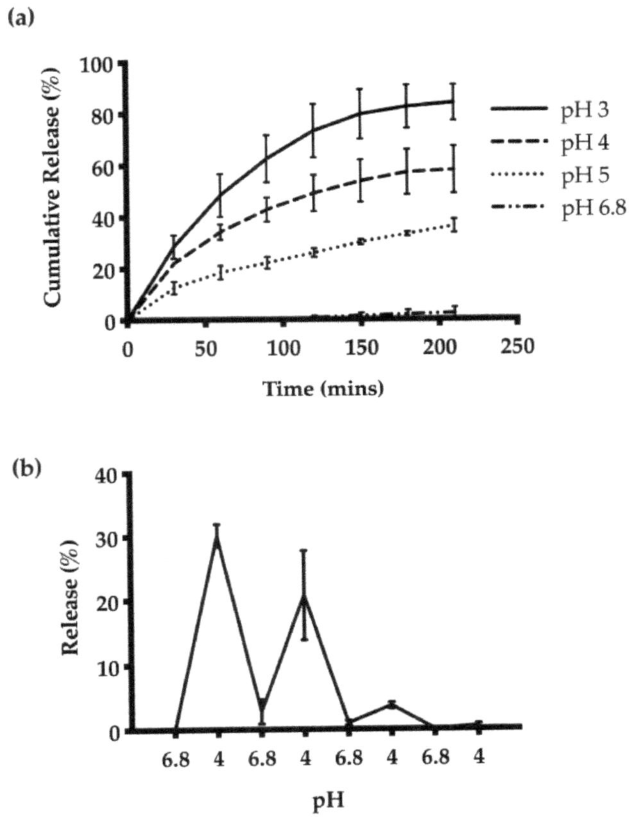

Figure 4. (**a**) Release of HTN3 from CNs at different pH values. The highest extent of release and most rapid release was observed at pH 3, and there was minimal release at pH 6.8. On average, at pH 3, the CNs were able to release 84% ± 7% of the encapsulated protein, 58% ± 9% at pH 4, 36% ± 3% at pH 5, and 2% ± 2% at pH 6.8. (**b**) pH cycling release results show that the CNs can adapt to environmental pH changes and can release protein selectively at pH 4 over 3 pH cycles.

3.1.5. HTN3 Degradation Study in WSS

To assess the protection offered through encapsulation against enzymatic degradation in the oral cavity, a protein degradation study was conducted in diluted human saliva to compare the degradation kinetics of free HTN3 and HTN3 loaded CNs. The degradation over time was quantified for both free HTN3 and HTN3 loaded CNs (Figure 5 and Table S1). For free HTN3, only 6% ± 5% of the free HTN3 remained after 2 h, whereas 47% ± 8% of HTN3 was intact when it was encapsulated in the CNs delivery system.

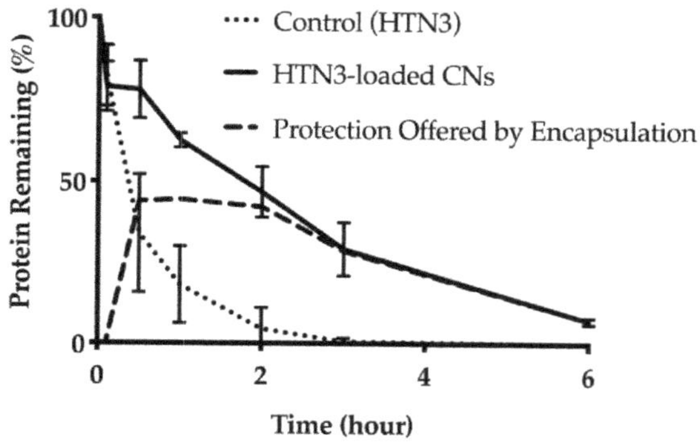

Figure 5. Protein degradation over time for free HTN3 and HTN3-loaded into CNs. For free HTN3, most protein was degraded in 2 h. At the same time point, about half of HTN3 was still intact in the CNs delivery system. The dashed line represents the degree of protection offered through CNs encapsulation.

3.1.6. Biofilm Formation

Wet biofilm mass was measured to test the effectiveness of four different treatment groups including 7.4 µM HTN3, 0.1% *w/v* unloaded CNs, 7.4 µM HTN3 in CNs, and the gold standard 12,300 ppm fluoride solution. These groups were compared against PBS as a control group. The control group had an average wet biofilm mass of 15 ± 2 mg, while the mass was 12 ± 1 mg for HTN3, 8 ± 2 mg for fluoride, 7 ± 1 mg for unloaded CNs, and 6 ± 1 for HTN3-loaded CNs (Figure 6).

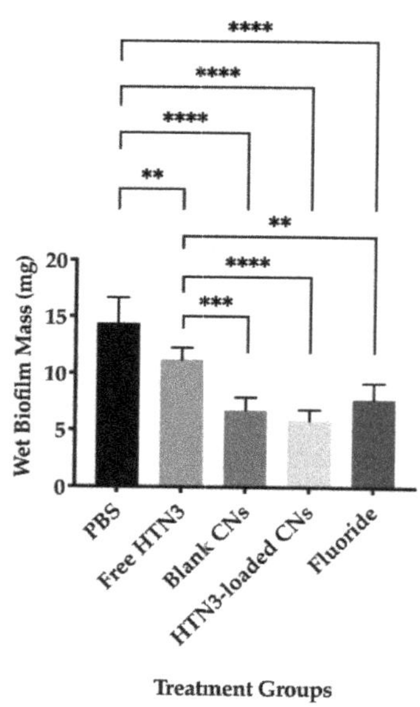

Figure 6. Wet biofilm masses for four treatment groups were compared against the control group PBS. All treatment groups led to significantly lower biofilm mass than PBS. Fluoride, blank, and histatin encapsulated CNs performed significantly better than free HTN3 in minimizing biofilm formation, but no significant difference in biofilm mass was observed between these treatment groups (** $p \leq 0.01$, *** $p \leq 0.001$, **** $p \leq 0.0001$).

Bacterial cell viability was also evaluated and presented in Figure 7. The PBS control led to $5.0 \times 10^9 \pm 0.4 \times 10^8$ CFU/mL of viable bacteria, while free HTN3 had $4.6 \times 10^9 \pm 0.3 \times 10^9$ CFU/mL, fluoride had $3.8 \times 10^9 \pm 0.5 \times 10^9$ CFU/mL, unloaded CNs had $2.7 \times 10^9 \pm 0.4 \times 10^9$ CFU/mL, and HTN3-loaded CNs had $2.3 \times 10^9 \pm 0.3 \times 10^9$ CFU/mL.

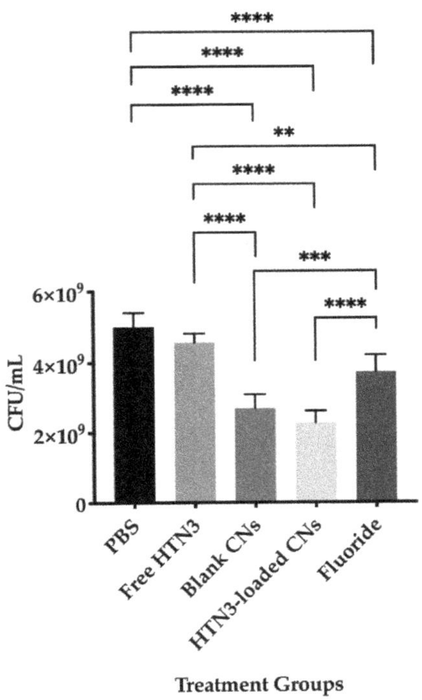

Figure 7. Bacterial viability for the four treatment groups were compared against the control group treated with PBS only. While free HTN3 did not control bacterial growth, fluoride treatment significantly decreased bacterial growth. Blank and HTN3-loaded CNs were the most effective at decreasing bacterial population but are insignificantly different from each other (** $p \leq 0.01$, *** $p \leq 0.001$, **** $p \leq 0.0001$).

4. Discussion

Salivary proteins such as statherin and HTN have demonstrated several functions that are directly related to the inhibition of dental caries. They inhibit enamel demineralization, promote enamel remineralization, and modulate growth of *S. mutans* [12,41]. However, given the oral cavity can be hostile to proteins due to the high proteolytic activity, these proteins cannot be readily applied as therapeutics for protein-mediated homeostasis. In this study, we produced pH-sensitive CNs, which are able to encapsulate our protein of interest, HTN3, and release it selectively under acidic conditions. The observed pH-sensitivity is important because major oral complications, such as dental caries and dental erosion, occur under acidified conditions. The encapsulation also prolonged the lifetime of HTN3 inside 10-fold diluted human saliva.

The chitosan to TPP mass ratio was first studied systematically to optimize the formulation. At low ratios of chitosan to TPP, the particles were unstable, due to charge neutralization, resulting in visible sedimentation. They became stable as the mass ratio exceeded values ranging from 2:1 to 5:1, depending on the form of chitosan. In general, the smallest diameters and lowest PDI values were observed as a cation:anion molar ratio of about 2:1. As the ratio was further increased, the Z-average diameter increased, which can perhaps be attributed to more chitosan per particle or aggregation of the particles. For the same reasons, the PDI also followed a similar trend. For the zeta potential, a steady increase was observed with increasing chitosan to TPP ratios, since chitosan is positively charged. The findings are in agreement with many studies in the literature [42,43].

The optimal formulation was selected based on particle diameter, PDI, and zeta potential. Ideally, it is best to have a formulation with a sub-200 nm diameter nanoparticles, with a relatively low PDI and a high zeta potential. Therefore, the ultra-low MW chitosan formulation at a ratio of 6.94:1 (chitosan to TPP) was chosen as it has the smallest particle diameter of 144 ± 6 nm, with the lowest PDI of 0.15 ± 0.04 and a zeta potential of 18 ± 4 mV. Through a colloidal stability study by time-course of particle size and PDI measurements by DLS, we found that these nanoparticles maintain their average diameter and PDI when the zeta potential is above 15 mV. These optimized CNs were visualized by TEM and the average particle diameter estimated from the TEM images was in good agreement with the average particle diameter obtained from DLS. Since HTN3 is a low MW salivary protein, its encapsulation in the CNs did not significantly affect the particle diameter. As shown by the TEM images, there were no significant differences in size or dispersity between the unloaded CNs and those with different HTN3 loading ratios.

The pH-responsive properties of the optimized CNs were also investigated. The extent of swelling under different pH conditions was studied. The highest degree of swelling was observed at pH 3 and the CNs remained unchanged at pH 6.3. The results also demonstrated that the degree of swelling depended on the acidity of the suspension. The swelling response elicited by the pH stimulus was rapid, suggesting that CNs were able to quickly respond to pH changes. The versatility of this pH-responsive property was further examined through pH-cycling between 6.3 and 4, and the results suggested the CNs delivery system was capable of swelling selectively at acidic pH and reversed the swelling process upon removal of the pH stimulus.

The encapsulation loading efficiency of HTN3 was studied at 2%, 5%, and 10% *w/w* loading ratios of HTN3 to chitosan. The absence of unloaded protein suggested that HTN3 was quantitatively encapsulated in the delivery system at all ratios, which is beneficial, as the loading can therefore be tuned. Cumulative release studies at different pH values including 3, 4, 5, and 6.8 were also conducted to assess the extent of release of encapsulated HTN3 from the delivery system. These values were chosen because they reflect the cariogenic conditions promoting different oral diseases. For instance, dental caries initiate at pH 5 and dental erosion occurs at pH 3. In accordance with the demonstration of the pH-responsive property shown previously, the extent of HTN3 release was proportionate to the acidity of the environment. A maximum cumulative release of 84% ± 7% was achieved at pH 3 over 210 min, while a minimum release of 2% ± 2% was observed at pH 6.8 over the same time period. To better mimic the pH changes in the oral cavity throughout the day, pH-cycling release was performed between pH 6.8 and 4. HTN3-loaded CNs were able to selectively release encapsulated HTN3 at pH 4, and retain the HTN3 at the salivary pH of 6.8. The ability to selectively release HTN3 under acidic conditions is crucial because throughout the day the salivary pH fluctuates many times from the consumption of food or beverages [23,44]. As saliva becomes acidic following carbohydrate intake, the drop in pH could trigger the release of HTN3, which promotes oral homeostasis by inhibiting the demineralization process. Salivary pH would gradually recover to its physiological value due to the buffer capacity of saliva, which would halt HTN3 release from the CNs. Further release of HTN3 would require another sugar challenge, which would drop salivary pH.

Salivary proteins are susceptible to the high proteolytic activity in the oral cavity, which prevent them from being used as potential therapeutics on their own. To evaluate the protection offered through CN encapsulation against enzymatic degradation, a protein degradation study was performed in 10-fold diluted human saliva. Saliva was diluted to better capture the degradation kinetics of proteins, since at the original concentration the degradation would have happened too rapidly to allow easy measurement of the process. Only 6% ± 5% of the free HTN3 remained at the 2-h mark. At the same time point, 47% ± 8% of HTN3 was still present in the CN delivery system. At the 6-h mark, only 7% ± 1% of HTN3 remained in the delivery system, which could be due to the potential breakdown of the CNs by other salivary proteins. Nevertheless, the delivery system increased the protein survival time significantly. The increase in survival time is important because it allows less frequent administration of the formulation.

Ultimately, a biofilm model was applied to evaluate the therapeutic effect of HTN3-loaded CNs against *S. mutans* biofilm formation on hydroxyapatite discs. *S. mutans* was chosen because it is a major contributor responsible for the initiation and development of tooth decay [23,45]. It metabolizes sucrose to produce sticky polysaccharides that allow the bacteria to aggregate and adhere to the tooth enamel, forming a biofilm. The biofilm, together with frequent sugar intake promotes fermentation of dietary sugar into acidic products. Persistence of the resulting acidic conditions favours the proliferation of acidogenic bacteria. The low pH environment in the biofilm initiates the dental caries process [23,46]. Hydroxyapatite is a mineral composed of calcium phosphate, which greatly resembles human hard tissues including bone and tooth enamel in composition and morphology [47]. It is also the most stable calcium phosphate mineral under physiological conditions [48]. Therefore it has been extensively used in oral applications, such as pellicle formation and biofilm formation [49,50].

Two key factors used to evaluate biofilm growth are biofilm mass and bacterial cell viability. Effective treatment should result in low biofilm mass and reduced bacterial cell viability. For biofilm mass measurements, based on the ANOVA analysis and Tukey's multiple comparisons test, all four treatment groups resulted in significantly reduced biofilm mass compared to the control group. Fluoride together with unloaded and HTN3-loaded CNs were significantly more effective than free HTN3. However, no significant difference was observed between the HTN3-loaded CNs and unloaded CNs. Both of these treatments successfully lowered the biofilm mass by at least half. Bacterial cell viability was also determined for all treatment groups. Free HTN3 was not effective at reducing bacterial cell viability and did not perform better than the control group. Fluoride, unloaded, and HTN3-loaded CNs significantly reduced cell viability when compared against the control group. Fluoride was significantly more effective when compared against HTN3. Unloaded and HTN3-loaded CNs were the most competent treatment groups. They reduced cell viability in half, but their performance in reducing bacterial population were not significantly different from one another. While these results suggest that CNs alone can potentially control bacterial population growth, further studies that encapsulate greater HTN3 concentrations within CNs, or reduce the amount of CNs while maintaining HTN3 concentrations are required. In addition, CNs are known to exhibit antimicrobial effects against *S. mutans* [51]. Another possible explanation would be CNs alone already prevented the initiation of demineralization, therefore there was no steep drop in pH to trigger HTN3 release from the CNs.

5. Conclusions

In conclusion, the results of this study demonstrated the pH-responsive properties of the CNs. The CNs were also able to offer protection against enzymatic degradation in comparison to free HTN3. While HTN3-loaded CNs had successfully reduced biofilm growth as reflected by reduced biofilm mass and bacterial cell viability, these particles did not significantly outperform unloaded CNs. Future studies should include use of the biofilm model to characterize the contribution of HTN3 within CNs and CNs alone in controlling bacterial population. Nonetheless, this work has shown that CNs can be used as a protein carrier for oral applications, especially for complications involving acidic environments. This delivery system can also be applied to encapsulate other salivary proteins for oral delivery. The ultimate goal of studying protein encapsulation within CNs is to integrate multifunctional salivary proteins into daily dental hygiene products like toothpaste and mouthwash to provide a preventative approach to address dental caries. Altogether, our results show that a pH-sensitive delivery system that can release salivary proteins under oral conditions can be achieved by utilizing a biodegradable, biocompatible, and naturally derived polymer like chitosan.

Supplementary Materials: The following are available online at https://www.mdpi.com/2079-4991/11/4/1028/s1, Figure S1: Volume distribution of the ultra-low MW CNs at chitosan to TPP mass ratio of 6.94. There is a small batch to batch variability; Figure S2: CNs were stored at 4 °C, and three parameters including (a) Z-average diameter, (b) PDI, and (c) Zeta potential were measured by DLS routinely over the course of 60 days. The consistency of obtained data suggest the nanoparticles are stable; Figure S3: TEM images of (a) blank chitosan nanoparticles, (b) HTN3 loaded chitosan nanoparticles with a 2% *w/w* loading ratio, (c) HTN3 loaded chitosan nanoparticles with a 5% *w/w* loading ratio, and (d) HTN3 loaded chitosan nanoparticles with a 10% *w/w* loading ratio. No significant variance in size or dispersity was observed among blank chitosan nanoparticles and those with different loading ratios; Table S1: Protein degradation over time for free HTN3 and HTN3-loaded into CNs.

Author Contributions: Conceptualization, E.R.G. and W.L.S.; methodology, Y.Z., L.M.M., Y.X., E.R.G. and W.L.S.; validation, Y.Z.; formal analysis, Y.Z.; investigation, Y.Z.; resources, L.M.M. and Y.X.; data curation, Y.Z.; writing—original draft preparation, Y.Z.; writing—review and editing, Y.Z., L.M.M., E.R.G. and W.L.S.; visualization, Y.Z.; supervision, E.R.G. and W.L.S.; project administration, Y.Z., L.M.M. and Y.X.; funding acquisition, E.R.G. and W.L.S. All authors have read and agreed to the published version of the manuscript.

Funding: This research was funded by Canadian Institutes of Health Research—CIHR, grant numbers 106657 and 400347; Canada Foundation for Innovation—Leaders Opportunity Fund, grant number 25116.

Institutional Review Board Statement: The study was conducted according to the guidelines of the Declaration of Helsinki, and approved by the Human Research Ethics Committee at The University of Western Ontario (protocol number 16181E, 6 June 2016).

Informed Consent Statement: Informed consent was obtained from all subjects involved in the study.

Data Availability Statement: The data presented in this study are available on request from the corresponding author.

Acknowledgments: The authors acknowledge the staffs from Biotron's Imaging and Microscopy Facility at The University of Western Ontario for technical assistance with TEM imaging. This paper is based on a thesis submitted by the first author to the School of Biomedical Engineering, The University of Western Ontario, London ON, Canada, in partial fulfillment of the requirements for the Master's program in biomedical engineering [52].

Conflicts of Interest: The authors declare no conflict of interest.

References

1. Siqueira, W.L.; Dawes, C. The salivary proteome: Challenges and perspectives. *Proteom. Clin. Appl.* **2011**, *5*, 575–579. [CrossRef]
2. Scarano, E.; Fiorita, A.; Picciotti, P.M.; Passali, G.C.; Calò, L.; Cabras, T.; Inzitari, R.; Fanali, C.; Messana, I.; Castagnola, M.; et al. Proteomics of saliva: Personal experience. *Acta Otorhinolaryngol.* **2010**, *30*, 125–130.
3. Mandel, I.D. The Functions of Saliva. *J. Dent. Res.* **1987**, *66*, 623–627. [CrossRef] [PubMed]
4. Siqueira, W.L.; Zhang, W.; Helmerhorst, E.J.; Gygi, S.P.; Oppenheim, F.G. Identification of Protein Components ininvivoHuman Acquired Enamel Pellicle Using LC−ESI−MS/MS. *J. Proteome Res.* **2007**, *6*, 2152–2160. [CrossRef] [PubMed]
5. Oppenheim, F.G.; Salih, E.; Siqueira, W.L.; Zhang, W.; Helmerhorst, E.J. Salivary Proteome and Its Genetic Polymorphisms. *Ann. N. Y. Acad. Sci.* **2007**, *1098*, 22–50. [CrossRef] [PubMed]
6. Xiuli, S.; Erdjan, S.; Oppenheim, F.G.; Helmerhorst, E.J. Kinetics of histatin proteolysis in whole saliva and the effect on bioactive domains with metal-binding, antifungal, and wound-healing properties. *FASEB J.* **2009**, *23*, 2691–2701. [CrossRef] [PubMed]
7. Siqueira, W.L.; Lee, Y.H.; Xiao, Y.; Held, K.; Wong, W. Identification and characterization of histatin 1 salivary complexes by using mass spectrometry. *Proteomics* **2012**, *12*, 3426–3435. [CrossRef] [PubMed]
8. Gusman, H.; Leone, C.; Helmerhorst, E.J.; Nunn, M.; Flora, B.; Troxler, R.F.; Oppenheim, F.G. Human salivary gland-specific daily variations in histatin concentrations determined by a novel quantitation technique. *Arch. Oral Biol.* **2004**, *49*, 11–22. [CrossRef]
9. Shimotoyodome, A.; Kobayashi, H.; Tokimitsu, I.; Matsukubo, T.; Takaesu, Y. Statherin and Histatin 1 Reduce Parotid Sali-va-Promoted Streptococcus Mutans Strain MT8148 Adhesion to Hydroxyapatite Surfaces. *Caries Res.* **2006**, *40*, 403–411. [CrossRef]

10. Puri, S.; Edgerton, M. How Does It Kill? Understanding the Candidacidal Mechanism of Salivary Histatin *Eukaryot. Cell* **2014**, *13*, 958–964. [CrossRef]
11. Basiri, T.; Johnson, N.; Moffa, E.; Mulyar, Y.; Nunes, P.S.; Machado, M.; Siqueira, W. Duplicated or Hybridized Peptide Functional Domains Promote Oral Homeostasis. *J. Dent. Res.* **2017**, *96*, 1162–1167. [CrossRef] [PubMed]
12. Helmerhorst, E.; Alagl, A.; Siqueira, W.; Oppenheim, F. Oral fluid proteolytic effects on histatin 5 structure and function. *Arch. Oral Biol.* **2006**, *51*, 1061–1070. [CrossRef]
13. McDonald, E.E.; Goldberg, H.A.; Tabbara, N.; Mendes, F.M.; Siqueira, W.L. Histatin 1 Resists Proteolytic Degradation when Adsorbed to Hydroxyapatite. *J. Dent. Res.* **2010**, *90*, 268–272. [CrossRef]
14. Mihu, M.R.; Sandkovsky, U.; Han, G.; Friedman, J.M.; Nosanchuk, J.D.; Martinez, L.R. The use of nitric oxide releasing nanoparticles as a treatment againstAcinetobacter baumanniiin wound infections. *Virulence* **2010**, *1*, 62–67. [CrossRef]
15. Yang, X.; Grailer, J.J.; Pilla, S.; Steeber, U.A.; Gong, S. Tumor-Targeting, pH-Responsive, and Stable Unimolecular Micelles as Drug Nanocarriers for Targeted Cancer Therapy. *Bioconj. Chem.* **2010**, *21*, 496–504. [CrossRef]
16. Shen, M.; Huang, Y.; Han, L.; Qin, J.; Fang, X.; Wang, J.; Yang, V.C. Multifunctional drug delivery system for targeting tumor and its acidic microenvironment. *J. Control. Release* **2012**, *161*, 884–892. [CrossRef] [PubMed]
17. Xiong, M.-H.; Li, Y.-J.; Bao, Y.; Yang, X.-Z.; Hu, B.; Wang, J. Bacteria-Responsive Multifunctional Nanogel for Targeted Antibiotic Delivery. *Adv. Mater.* **2012**, *24*, 6175–6180. [CrossRef]
18. Lee, E.S.; Shin, H.J.; Na, K.; Bae, Y.H. Poly(l-histidine)–PEG block copolymer micelles and pH-induced destabilization. *J. Control. Release* **2003**, *90*, 363–374. [CrossRef]
19. Basel, M.T.; Shrestha, T.B.; Troyer, D.L.; Bossmann, S.H. Protease-Sensitive, Polymer-Caged Liposomes: A Method for Making Highly Targeted Liposomes Using Triggered Release. *ACS Nano* **2011**, *5*, 2162–2175. [CrossRef]
20. Chen, K.-J.; Chaung, E.-Y.; Wey, S.-P.; Lin, K.-J.; Cheng, F.; Lin, C.-C.; Liu, H.-L.; Tseng, H.-W.; Liu, C.-P.; Wei, M.-C.; et al. Hyperthermia-Mediated Local Drug Delivery by a Bubble-Generating Liposomal System for Tumor-Specific Chemotherapy. *ACS Nano* **2014**, *8*, 5105–5115. [CrossRef] [PubMed]
21. Fan, B.; Gillies, E.R. Poly(ethyl glyoxylate)-Poly(ethylene oxide) Nanoparticles: Stimuli-Responsive Drug Release via End-to-End Polyglyoxylate Depolymerization. *Mol. Pharm.* **2017**, *14*, 2548–2559. [CrossRef]
22. Nikaido, T.; Moriya, K.; Hiraishi, N.; Ikeda, M.; Kitasako, Y.; Foxton, R.M.; Tagami, J. Surface Analysis of Dentinal Caries in Primary Teeth Using a PH-Imaging Microscope. *Dent. Mater. J.* **2004**, *23*, 628–632. [CrossRef] [PubMed]
23. Bowen, W.H.; Tenuta, L.M.A.; Koo, H.; Cury, J.A. Dental Caries: Etiology and Pathogenesis. In *Oral Microbiology and Immunology*, 3rd ed.; Lamont, R.J., Hajishengalis, G.M., Koo, H., Jenkinson, H.F., Eds.; ASM Press: Washington, DC, USA, 2019; pp. 251–265.
24. Dash, M.; Chiellini, F.; Ottenbrite, R. Chitosan—A versatile semi-synthetic polymer in biomedical applications. *Prog. Polym. Sci.* **2011**, *36*, 981–1014. [CrossRef]
25. Raafat, D.; Sahl, H.-G. Chitosan and its antimicrobial potential—a critical literature survey. *Microb. Biotechnol.* **2009**, *2*, 186–201. [CrossRef] [PubMed]
26. Tang, W.J.; Fernandez, J.G.; Sohn, J.J.; Amemiya, C.T. Chitin Is Endogenously Produced in Vertebrates. *Curr. Biol.* **2015**, *25*, 897–900. [CrossRef]
27. Souza, M.P.; Vaz, A.F.; Correia, M.T.S.; Cerqueira, M.A.; Vicente, A.A.; Carneiro-Da-Cunha, M.G. Quercetin-Loaded Lecithin/Chitosan Nanoparticles for Functional Food Applications. *Food Bioprocess Technol.* **2014**, *7*, 1149–1159. [CrossRef]
28. Bugnicourt, L.; Ladavière, C. Interests of chitosan nanoparticles ionically cross-linked with tripolyphosphate for biomedical applications. *Prog. Polym. Sci.* **2016**, *60*, 1–17. [CrossRef]
29. Olivera, S.; Muralidhara, H.B.; Venkatesh, K.; Guna, V.K.; Gopalakrishna, K. Potential applications of cellulose and chitosan nanoparticles/composites in wastewater treatment: A review. *Carbohydr. Polym.* **2016**, *153*, 600–618. [CrossRef] [PubMed]
30. Carvalho, T.S.; Lussi, A. Combined effect of a fluoride-, stannous-and chitosan-containing toothpaste and stannous-containing rinse on the prevention of initial enamel erosion–abrasion. *J. Dent.* **2014**, *42*, 450–459. [CrossRef]
31. Costa, E.; Silva, S.; Madureira, A.; Cardelle-Cobas, A.; Tavaria, F.; Pintado, M. A comprehensive study into the impact of a chitosan mouthwash upon oral microorganism's biofilm formation in vitro. *Carbohydr. Polym.* **2014**, *101*, 1081–1086. [CrossRef]
32. Kean, T.; Thanou, M. Biodegradation, biodistribution and toxicity of chitosan. *Adv. Drug Deliv. Rev.* **2010**, *62*, 3–11. [CrossRef]
33. Thandapani, G.; Prasad, S.; Sudha, P.N.; Sukumaran, A. Size optimization and in vitro biocompatibility studies of chitosan nanoparticles. *Int. J. Biol. Macromol.* **2017**, *104*, 1794–1806. [CrossRef]
34. Cha, J.; Lee, W.B.; Park, C.R.; Cho, Y.W.; Ahn, C.-H.; Kwon, I.C. Preparation and characterization of cisplatin-incorporated chitosan hydrogels, microparticles, and nanoparticles. *Macromol. Res.* **2006**, *14*, 573–578. [CrossRef]
35. Wang, X.; Zheng, C.; Wu, Z.; Teng, D.; Zhang, X.; Wang, Z.; Li, C. Chitosan-NAC nanoparticles as a vehicle for nasal absorption enhancement of insulin. *J. Biomed. Mater. Res. Part B Appl. Biomater.* **2009**, *88*, 150–161. [CrossRef]
36. Ridolfi, D.M.; Marcato, P.D.; Justo, G.Z.; Cordi, L.; Machado, D.; Durán, N. Chitosan-solid lipid nanoparticles as carriers for topical delivery of tretinoin. *Colloids Surf. B Biointerfaces* **2012**, *93*, 36–40. [CrossRef]
37. Janes, K.A.; Fresneau, M.P.; Marazuela, A.; Fabra, A.; Alonso, M.J. Chitosan nanoparticles as delivery systems for doxorubicin. *J. Control. Release* **2001**, *73*, 255–267. [CrossRef]
38. Calvo, P.; Vila-Jato, J.L.; Alonso, M.J. Novel hydrophilic chitosan-polyethylene oxide nanoparticles as protein carriers. *J. Appl. Polym. Sci.* **1997**, *63*, 125–132. [CrossRef]

39. Siqueira, W.L.; Margolis, H.C.; Helmerhorst, E.J.; Mendes, F.M.; Oppenheim, F.G. Evidence of Intact Histatins in thein vivoAcquired Enamel Pellicle. *J. Dent. Res.* **2010**, *89*, 626–630. [CrossRef]
40. Nakamura, M.; Slots, J. Salivary enzymes. *J. Periodontal Res.* **1983**, *18*, 559–569. [CrossRef] [PubMed]
41. Siqueira, W.; Custodio, W.; McDonald, E. New Insights into the Composition and Functions of the Acquired Enamel Pellicle. *J. Dent. Res.* **2012**, *91*, 1110–1118. [CrossRef]
42. Fan, W.; Yan, W.; Xu, Z.; Ni, H. Formation mechanism of monodisperse, low molecular weight chitosan nanoparticles by ionic gelation technique. *Colloids Surf. B Biointerfaces* **2012**, *90*, 21–27. [CrossRef]
43. Jonassen, H.; Kjøniksen, A.-L.; Hiorth, M. Stability of Chitosan Nanoparticles Cross-Linked with Tripolyphosphate. *Biomacromolecules* **2012**, *13*, 3747–3756. [CrossRef]
44. Bowen, W.H. The Stephan Curve revisited. *Odontology* **2012**, *101*, 2–8. [CrossRef]
45. Garcia, S.; Blackledge, M.; Michalek, S.; Su, L.; Ptacek, T.; Eipers, P.; Morrow, C.; Lefkowitz, E.; Melander, C.; Wu, H. Targeting of Streptococcus mutans Biofilms by a Novel Small Molecule Prevents Dental Caries and Preserves the Oral Microbiome. *J. Dent. Res.* **2017**, *96*, 807–814. [CrossRef] [PubMed]
46. Argimón, S.; Caufield, P.W. Distribution of Putative Virulence Genes in Streptococcus mutans Strains Does Not Correlate with Caries Experience. *J. Clin. Microbiol.* **2011**, *49*, 984–992. [CrossRef]
47. Wei, G.; Ma, P.X. Structure and properties of nano-hydroxyapatite/polymer composite scaffolds for bone tissue engineering. *Biomaterials* **2004**, *25*, 4749–4757. [CrossRef] [PubMed]
48. Kalita, S.J.; Bhardwaj, A.; Bhatt, H.A. Nanocrystalline calcium phosphate ceramics in biomedical engineering. *Mater. Sci. Eng. C* **2007**, *27*, 441–449. [CrossRef]
49. Smith, A.M.V.; Bowen, W.H. In situ studies of pellicle formation on hydroxyapatite discs. *Arch. Oral Biol.* **2000**, *45*, 277–291. [CrossRef]
50. Takeshita, T.; Yasui, M.; Shibata, Y.; Furuta, M.; Saeki, Y.; Eshima, N.; Yamashita, Y. Dental plaque development on a hydroxyapatite disk in young adults observed by using a barcoded pyrosequencing approach. *Sci. Rep.* **2015**, *5*, 8136. [CrossRef] [PubMed]
51. De Paz, L.E.C.; Resin, A.; Howard, K.A.; Sutherland, D.S.; Wejse, P.L. Antimicrobial Effect of Chitosan Nanoparticles on Streptococcus mutans Biofilms. *Appl. Environ. Microbiol.* **2011**, *77*, 3892–3895. [CrossRef] [PubMed]
52. Zhu, Y. A pH-Sensitive Delivery System for the Prevention of Dental Caries Using Salivary Proteins. Master's thesis, The University of Western Ontario, London, ON, Canada, 2019. Available online: https://ir.lib.uwo.ca/etd/6227 (accessed on 15 April 2021).

nanomaterials

Article

Designing pH-Dependent Systems Based on Nanoscale Calcium Carbonate for the Delivery of an Antitumor Drug

Victoriya Popova , Yuliya Poletaeva, Inna Pyshnaya , Dmitrii Pyshnyi and Elena Dmitrienko *

Institute of Chemical Biology and Fundamental Medicine (ICBFM), Siberian Branch of Russian Academy of Sciences (SB RAS), 8 Lavrentiev Avenue, 630090 Novosibirsk, Russia; fom.nin198@mail.ru (V.P.); fabaceae@yandex.ru (Y.P.); pyshnaya@niboch.nsc.ru (I.P.); pyshnyi@niboch.nsc.ru (D.P.)
* Correspondence: elenad@niboch.nsc.ru; Tel.: +7-(383)-363-5102

Abstract: Materials based on calcium carbonate ($CaCO_3$) are widely used in biomedical research (e.g., as carriers of bioactive substances). The biocompatibility of $CaCO_3$ and dependence of its stability on pH make these materials promising transporters of therapeutic agents to sites with low pH such as a tumor tissue. In this work, we developed an approach to the preparation of nanoscale particles based on $CaCO_3$ (CaNPs) up to 200 nm in size by coprecipitation and analyzed the interaction of the nanoparticles with an anticancer drug: *DOX*orubicin (*DOX*). We also showed a prolonged pH-dependent release of *DOX* from a CaNP nanocarrier and effective inhibition of cancer cell growth by a $CaCO_3$-and-*DOX*–based composite ($CaNP_7$-*DOX*) in in vitro models.

Keywords: calcium carbonate nanoparticle; *DOX*orubicin; prolonged release

1. Introduction

Calcium carbonate ($CaCO_3$) is one of the most common inorganic materials with a wide range of applications in biomedical fields, for example, as the basis for delivery systems of biologically active substances or for biosensor construction [1]. Studies on the use of $CaCO_3$ particles in biomedicine date back to the 1990s. The advantages of $CaCO_3$-based materials include good availability of reagents, the absence of toxicity, and the gradual biodegradation of $CaCO_3$ nanoparticles (CaNPs) [2]. The biocompatibility of CaNPs toward cells has also been reported, as has the suitability of such materials as safe drug carriers [3].

A large number of studies have been published aimed at the design of Ca^{2+}-based delivery systems for anticancer drugs (including *DOX*orubicin; *DOX*) [4,5]. Most of the studies on the preparation and characterization of nanomaterials based on $CaCO_3$ involve porous particles in the micrometer size; the advantages of such materials include the large surface area (due to the porosity of such structures) available for interaction with the drug; however, the large particle size imposes significant limitations in the use of such materials in vivo [6,7]. The literature indicates that for the development of cancer drug delivery vehicles, promising strategies are those that combine both an active targeting modality (for example, transferrin) and a passive one (i.e., a targeting system based on the enhanced permeability and retention effect in deformed cancerous tissues). In the case of passive targeting, nanomaterials with sizes <200 nm are optimal; such nanoparticles have a better ability to penetrate into (and accumulate at) the tumor site due to defects in endothelial cells [8].

There are significantly fewer scientific publications about the preparation of CaNP materials with sizes less than 200 nm than publications about their micrometer scale analogs because of difficulties (i) with the development of the methods for constructing such materials and (ii) with their subsequent stabilization. Zhao Y et al. presented a technique for the fabrication of nanoscale monodisperse CaNPs, but the material was not stable in an aqueous solution, therefore, it was functionalized with a silicate shell for further applications. This modification significantly lowered the efficiency of drug encapsulation and reduced the ability of *DOX* to desorb from the CaNP-based particles, namely, those authors failed to achieve a release of more than 25% of the loaded drug at a physiological temperature [9]. Hamidu A. et al. conducted a lot of work on the preparation and characterization of CaNPs loaded with *DOX*, but were unable to prevent aggregation of the nanomaterials into macro objects [10]. Wenliang Fu and colleagues [11] worked with similar nanomaterials. The poor monodispersity significantly limits prospective applicability of CaNPs to biomedicine due to a substantial increase in the size of the particles because they tend to adhere to each other.

Despite the current absence of FDA-approved $CaCO_3$-based delivery vehicles (carriers) of cancer drugs for clinical use, this material undoubtedly holds promise. Scientists continue to develop protocols of CaNP construction for subsequent biomedical applications including in vivo experiments [12,13].

Wenliang F. et al. demonstrated the potential utility of CaNPs (a size distribution of 20–60 nm; prepared from seashells) as a drug delivery system for the treatment of osteosarcoma in an orthotopic rat model of this cancer. After measuring body weight and analyzing serum biochemical parameters and histopathological data, the authors showed that *DOX* administration caused its accumulation (and the manifestation of toxic side effects) in major organs such as the heart, liver, and kidneys. In contrast, during $CaCO_3$-*DOX* therapy, the manifestation of the adverse effects was weaker and the survival rate of the rats was higher compared to the rat group receiving free *DOX*. These researchers argued that this benefit was due to increased passive targeting to tumor tissues because of the enhanced permeability and retention effect (EPR) and pH sensitivity of $CaCO_3$, namely, that this sensitivity reduced systemic toxicity and unintended exposure of normal tissues to *DOX* while increasing the chemotherapeutic action on solid tumors [14]. Maleki Dizaj S. and colleagues investigated the efficacy of *DOX*-loaded pH-dependent CaNPs (<600 nm) in dogs with tumors including osteosarcoma in a single-center open-label clinical trial; they registered osteoid matrix production, death of osteogenic cells, and inhibition of their proliferation [4]. Ghaji M. S. et al. studied CaNPs (20–50 nm) as a carrier of cytarabine acting against human leukemic HL-60 cells and as an antitumor therapy in SCID mice with leukemia; half-maximal inhibitory concentrations (IC_{50}) of cytarabine and CaNP-cytarabine were 5 and 2.5 μg/mL, respectively, after 72 h. CaNP-cytarabine was more effective at inducing apoptosis than free cytarabine [15]. Nonetheless, due to a high polydispersity index (PDI), heterogeneity and aggregation of CaNPs, protocols for the synthesis of such nanomaterials require modification. Nanoparticles can be entrapped in some organs and tissues, this problem is relevant for all nanoparticles. However, the advantages of nanoparticles based on calcium salts are their non-toxicity and biodegradability, which are confirmed by studies in the literature [16]. Moreover, for calcium phosphate nanoparticles of a submicron size, the distribution over organs and the subsequent excretion including through the decomposition of nanoparticles has been shown [17]. Thus, we expect that the biodegradability of nanoparticles based on calcium carbonate will avoid the negative effects of the accumulation of nanoparticles in organs.

Natural properties of $CaCO_3$-based materials such as biocompatibility, biodegradability, and pH sensitivity make $CaCO_3$ a promising candidate carrier for the delivery of various biologically active substances, especially anticancer drugs [18,19]. In the present work, we focused on the materials science part of this topic to develop a fundamental approach to the fabrication of nanomaterials with the special characteristics that will increase the efficacy of $CaCO_3$ applications in biomedicine.

$CaCO_3$ particles that meet several criteria—monodispersity, stability under physiological conditions, partial or complete degradation at low pH levels, and size <200 nm—are attractive carriers for pH-dependent drug delivery [20]. In the literature, there are several basic approaches to the preparation of $CaCO_3$ particles. Depending on the synthesis technique, it is possible to obtain particles of various shapes and sizes. Options for constructing CaNPs are subdivided into three main categories: aeration or carbon dioxide barbotage, coprecipitation of appropriate salts, and microemulsion approaches [21]. The easiest option to implement is the coprecipitation method because it does not require special equipment and reagents. Despite the apparent simplicity of the existing techniques and numerous studies aimed at obtaining CaNPs [22–24], there is no simple and reproducible way to synthesize monodisperse $CaCO_3$ particles of the nanometer size (up to 200 nm) that preserves their stability in solution for prospective in vivo applications. Thus, the main aim of this study was to find a way to prepare a suspension of nanoscale CaNPs including complete screening of various synthesis conditions and assessing the effects of reaction mixture composition, solvents, pH, and various additives on the morphological properties of the formed particles, with subsequent testing of the interaction of the particles with a drug (*DOX*). Additionally, we aimed to study the efficiency of conjugation of CaNPs with the drug, the kinetic release profile of *DOX* from the matrix (CaNPs) depending on pH, and the effectiveness of the inhibition of cancer cell proliferation by CaNP carrying *DOX* in vitro.

In this work, we developed and described a simple approach to the fabrication of nanocomposites based on $CaCO_3$ and researched the physical and chemical properties of these nanoparticles. We noted the absence of toxicity and proved the effectiveness of CaNPs as a drug carrier.

2. Materials and Methods

2.1. Materials

Sodium hydrogen carbonate, calcium chloride, magnesium chloride, and Tween 20 were purchased from Sigma-Aldrich, Co (St. Louis, MO, USA) whereas *DOX* from Teva Pharmaceutical Industries Ltd. (Petach Tikva, Israel). 3-(4,5-Dimethyl-2-thiazolyl)-2,5-diphenyl-2H-tetrazolium bromide (MTT) was acquired from Panreac Química (Barcelona, Spain), and polyethylene glycol (PEG) 1000 Da, 2000 Da, and 6000 Da from Carl Roth (Karlsruhe, Germany), FBS (fetal bovine serum), DMEM (Dulbecco's modified Eagle medium), and antimycotic antibiotic solution from GIBCO, Life Technologies (Carlsbad, CA, USA).

2.2. Synthesis of CaNPs by Coprecipitation

This procedure was performed by mixing two salts on an ultrasonic bath or dispersant: 100 µL of a $CaCl_2$ solution (0.007–0.100 M) was added dropwise to 1 mL of a $NaHCO_3$ (0.1 M) aqueous solution in the absence and presence of additives such as PEG (MW = 1000, 2000, or 6000) at 0.1 mg/mL, detergents (Tween 20, Triton X-100, sodium dodecyl sulfate, or cetyltrimethylammonium bromide) at 0.1 vol.%, DMEM at 0.2–10.0 vol.%, and/or $MgCl_2$ (0.005–0.010 M) [25]. The additives were introduced both individually and as a mixture. The reaction of CaNP formation in a mixture of Tween 20 and PEG-2000 was performed similarly to the reaction in an aqueous solvent replaced by isopropanol [1].

2.3. Characterization of CaNPs

This was carried out by dynamic light scattering (DLS) methods on a Malvern Zetasizer nano particle characterization system in water at room temperature. Suspensions of CaNPs were analyzed by transmission electron microscopy (TEM). For this purpose, a drop of a sample was allowed to adsorb for 1 min on a copper grid covered with formvar film; the excess liquid was then removed, and the grids were examined under a Jem1400 microscope (Jeol, Tokyo, Japan). Images were captured by a side-mounted Veleta digital camera (EM SIS, Muenster, Germany).

2.4. Reversible Binding of CaNPs to DOX

Conjugation was performed with stirring (700 rpm, 25 °C, 12 h) in 10 mM borate buffer (pH 8.5) containing 0.1–1.0 mg/mL CaNPs and 50–500 µg/mL *DOX*. Next, the CaNPs were washed with 10 mM borate buffer pH 8.5 (3 times × 1 mL), and the supernatant was separated by centrifugation (10 min, 13,400 rpm, miniSpin from Eppendorf).

The residual concentration of *DOX* in the supernatant was determined spectrophotometrically in the wavelength range of 400–600 nm at room temperature.

The amount of the drug bound to the nanoparticles was determined as capacity (E) of CaNPs for the drug and calculated according to the formula:

$$E = \frac{DOX_0 - DOX}{N}$$

where E is particle capacity (μg *DOX*/mg CaNPs); DOX_0 represents the initial amount of *DOX* (μg); *DOX* is the amount of *DOX* in the supernatant (μg); and *N* denotes the amount of CaNPs (mg).

The release of *DOX* was investigated at room temperature (unless stated otherwise; in some cases, the temperature was varied from 15 to 45 °C) in 1 mL of 100 mM acetate buffer (pH of 3.0 to 7.0) containing CaNP-*DOX* (132 μg of *DOX* with 0.2 mg of $CaNP_7$) with constant stirring (750 rpm). The amount of *DOX* released into the solution was determined by means of optical density and/or fluorescence intensity of the solution.

2.5. The Cytotoxicity of CaNPs, CaNP-DOX, and DOX

These properties were evaluated on A549 and HEK293 cells in a standard MTT assay. For this purpose, cells at (2.0 ± 0.5) × 10^3 per well were seeded in a 96-well plate containing the culture medium and incubated for 24 h at 37.0 ± 1.0 °C in an atmosphere containing 5.0% ± 0.5% of CO_2. HEK293 and/or A549 cells were then incubated with CaNPs (0.2–22.5 mg/mL), CaNPs carrying *DOX* (0.1–10.0 μM), or free *DOX* (0.1–10.0 μM) in the culture medium for 48 h at 37.0 ± 1.0 °C in an atmosphere containing 5.0% ± 0.5% of CO_2. After that, the medium was removed from all wells, and 200 μL of an MTT solution (0.25 mg/mL in the culture medium containing 1% of an antimycotic antibiotic solution) was added and incubated for 4 h under the same conditions. Next, the medium and MTT were removed from the wells, and 100 μL of dimethyl sulfoxide (DMSO) was added, and optical density was measured on a multichannel plate reader at wavelengths of 570 and 450 nm. The percentage of surviving cells was calculated from the obtained optical density values for each concentration of a tested agent.

2.6. Statistical Evaluation of Experimental Error

This was performed on at least three parallel biological samples. Data obtained in each experimental series are presented as mean ± standard deviation calculated by Excel software.

3. Results and Discussion

3.1. Synthesis and Assessment of $CaCO_3$ Properties

The synthesis of CaNPs by coprecipitation is a simple, inexpensive, and efficient method for producing nanomaterials for biomedical applications. On the other hand, micrometer scale particles (DLS data: 2950 ± 400 nm hydrodynamic diameter [d], PDI = 0.1 ± 0.05) are obtained by equimolar mixing of $CaCl_2$ and $NaHCO_3$ in the absence of additional reagents and do not meet our criteria. To obtain stable monodisperse particles up to 200 nm, we investigated the influence of the reaction mixture composition on the morphological characteristics of CaNPs. The workflow of CaNP fabrication is shown in Figure 1.

An ultrasonic bath (for small-scale synthesis up to 1.5 mL) and a dispersant (up to 50 mL) were employed in the synthesis setup for all CaNPs.

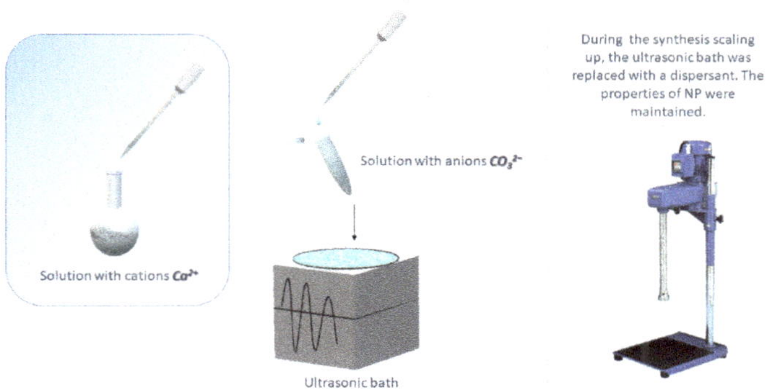

Figure 1. An outline of the fabrication of CaNPs.

The effect of stoichiometric ratios of the reagents on the size of the resulting particles was evaluated first. By varying the calcium chloride concentration (0.007–0.100 M) at a constant concentration of the carbonate anion and vice versa, we found that the size of formed particles at all tested concentrations was greater than 1 micron. According to DLS data, the use of a 10-fold excess of CO_3^{2-} over Ca^{2+} allows particles of the smallest size in the examined range to be obtained (d = 1470 ± 180 nm, PDI = 0.635 ± 0.002). Therefore, the optimal concentrations (0.1 M $NaHCO_3$ and 0.01 M $CaCl_2$) were chosen for further experiments. The addition of surfactants and high-molecular-weight compounds is widely utilized for preparing nanoparticles (up to 200 nm) [26]. The effects of adding the following surfactants to the reaction mixture were studied here: sodium dodecyl sulfate (anionic detergent), cetyltrimethylammonium bromide (cationic detergent), Tween 20 and Triton X-100 (nonionic detergents), and biocompatible high-molecular-weight compounds (PEG-1000, -2000, and -6000) [26]. According to the DLS data, only the addition of detergent Tween 20 led to a significant reduction in particle size (d = 450 ± 30 nm, PDI = 0.11 ± 0.03). Supplementation with PEG, irrespective of molecular weight (MW), yielded particles larger than 700 nm, which were still smaller than the particle sizes obtained in the absence of additives. Accordingly, the effects of the combined addition of high-molecular-weight PEG (1000, 2000, or 6000) and detergent Tween 20 were investigated next. The combined addition of these compounds reduced the size of CaNPs in comparison with the nanomaterials obtained above by separate supplementation with either compound alone. The smallest hydrodynamic size (d = 340.2 ± 0.3 nm, PDI = 0.177 ± 0.003) of CaNPs at this stage of the work was achieved upon the combined addition of PEG-2000 and Tween 20 to the reaction mixture. To determine the influence of codoping with PEG-2000 and Tween 20 on the CaNP formation process, both additives and their combination were examined by TEM. Figure 2 presents the fine structure of the initial solutions of Tween 20 (Figure 2A) and PEG-2000 (Figure 2B) as well as their combination (Figure 2C).

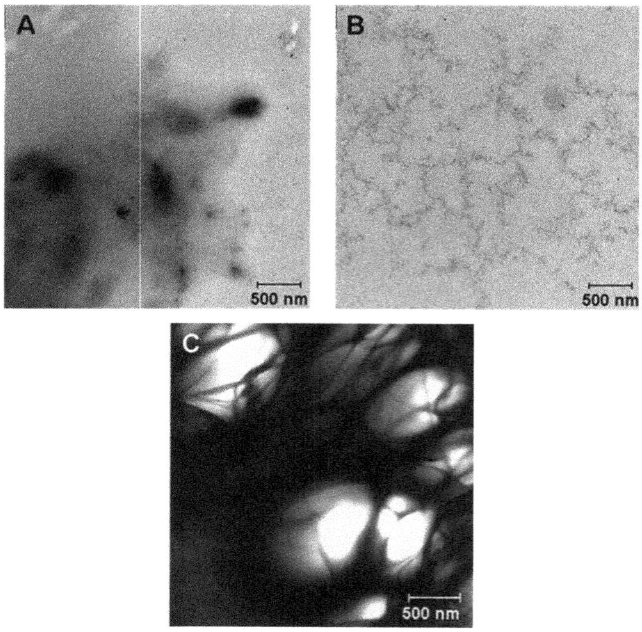

Figure 2. TEM images of Tween 20 (**A**), PEG-2000 (**B**), and the mixture of Tween 20 and PEG-2000 (**C**).

According to the figure, each individual additive has an ultrastructure that is different from that seen when they are mixed. From the TEM data (Figure 2), it was concluded that PEG-2000 and Tween 20 together formed a polymeric structure that acts as a matrix limiting particle growth during nucleation.

The CaNPs obtained in the presence of PEG-2000 and Tween 20 (called $CaNP_1$) were also characterized by TEM (Figure 3).

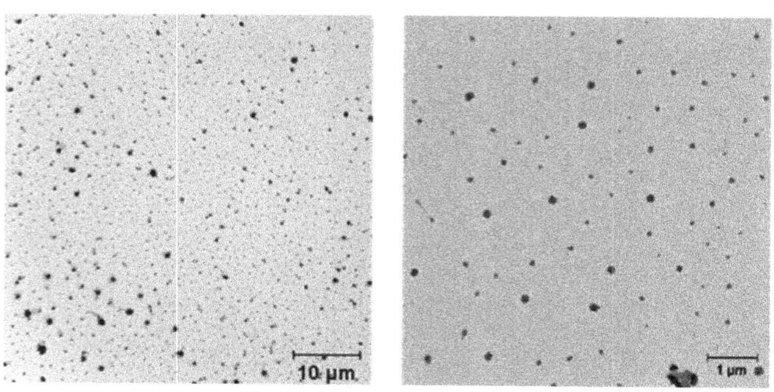

Figure 3. TEM images of the suspensions of the nanomaterial called $CaNP_1$, which was produced in the presence of Tween 20 and PEG-2000.

An analysis of the TEM micrographs (Figure 3) revealed that the obtained CaNPs were heterogeneous in size and shape, and polydispersity was considerable: particle size ranged from 135 to 700 nm.

To increase the stability and monodispersity of the CaNPs and to reduce their size, we then screened other conditions for CaNP fabrication. It is known that the addition of a nutrient medium, DMEM, consisting mainly of amino acids, inorganic salts, and vitamins, and Mg^{2+} cations, to the reaction mixture results in the formation of small CaNPs and in their stabilization [25]. The reason is presumably competing processes and the slowing of crystal formation. Accordingly, we investigated the effects of adding DMEM, $MgCl_2$, or their combination—to the reaction mixture consisting of a salt solution with a 10:1 ratio of CO_3^{2-} to Ca^{2+} and containing PEG-2000 and Tween 20—on the physicochemical properties of the CaNPs formed. Aside from the additives, the influence of the solvent on the characteristics of the obtained CaNPs [1] was also evaluated (Table 1).

Table 1. The influence of changing a single parameter in the $CaNP_1$ fabrication workflow.

# (Figure 4 Panel)	Sample	Special Conditions	D, nm	PDI
1 (A)	$CaNP_1$	-	339 ± 4	0.20 ± 0.01
2 (B)	$CaNP_2$	Solvent: isopropanol	2637 ± 125	0.33 ± 0.05
3 (C)	$CaNP_3$	$MgCl_2$, 0.01 M	278 ± 4	0.15 ± 0.01
4 (D)	$CaNP_4$	DMEM, 2 w.%	333 ± 2	0.26 ± 0.01
5 (E)	$CaNP_5$	$MgCl_2$, 0.01 M; DMEM, 2 w.%	326 ± 6	0.26 ± 0.01
6 (F)	$CaNP_6$	$MgCl_2$, 0.05 M; DMEM, 2 w.%	332 ± 1	0.19 ± 0.01
7 (G)	$CaNP_7$	$MgCl_2$, 0.01 M; DMEM, 10 w.%	249 ± 1	0.10 ± 0.01
8 (H)	$CaNP_8$	$MgCl_2$, 0.05 M; DMEM, 10 w.%	276 ± 4	0.10 ± 0.01

All of the obtained samples were stable in solution (hereafter stability is defined as the preservation of the physical and chemical properties of CaNPs) and were analyzed by TEM to examine the fine structure of the materials (Figure 4).

It was obvious that the composition of the reaction mixture during the CaNP synthesis significantly affected the morphological properties of the nanoparticles: size and ultrastructure (Table 1 and Figure 4). As above-mentioned, the presence of PEG-2000 and Tween 20 ($CaNP_1$, Figure 4A) yielded reproducible fabrication of CaNPs with a wide size distribution beyond the range optimal for biomedical applications, and this phenomenon is related to the tendency of CaNPs to aggregate. Replacing the aqueous solvent with an organic one caused an increase in crystal growth efficiency, which significantly enlarged the particles ($CaNP_2$, Figure 4B). The samples obtained in the presence of Mg^{2+} ($CaNP_3$, Figure 4C) were visibly different from those above, according to the DLS and TEM data. This is because the DLS method cannot determine the size of individual particles; instead, it only quantifies the clusters arising via aggregation of the material, as demonstrated in the TEM images. Because it was not possible to overcome the problem of particle aggregation by either chemical or physical methods, these CaNPs were not used in further work, despite their small size. The addition of DMEM in the absence of magnesium chloride did not cause cardinal differences from the sample obtained in the presence of Tween 20 and PEG (Figure 4A,D). In the comparison of panels E–H (Figure 4), the combined supplementation with 10% DMEM and 0.01 M $MgCl_2$ ($CaNP_7$, Figure 4G) was found to be optimal. Stable monodisperse spherical 249.0 ± 0.8 nm CaNPs were obtained, which were closest to the requirements of the task in question.

Storage and scalability tests of the nanoscale CaNPs fabricated by the newly developed procedure were performed on $CaNP_7$. The stability of these CaNPs was investigated by remeasuring the hydrodynamic dimensions of the CaNPs three months after the synthesis. The physical characteristics of the samples did not change much. The scalability of the proposed approach was assessed by increasing the reaction mixture volume by 30-fold, which made it necessary to replace the ultrasonic bath with a dispersant. According to the

DLS (d = 200 ± 20 nm, PDI = 0.10 ± 0.04) and TEM data (Figure 5), the larger synthesis volume did not increase the hydrodynamic size of the CaNPs.

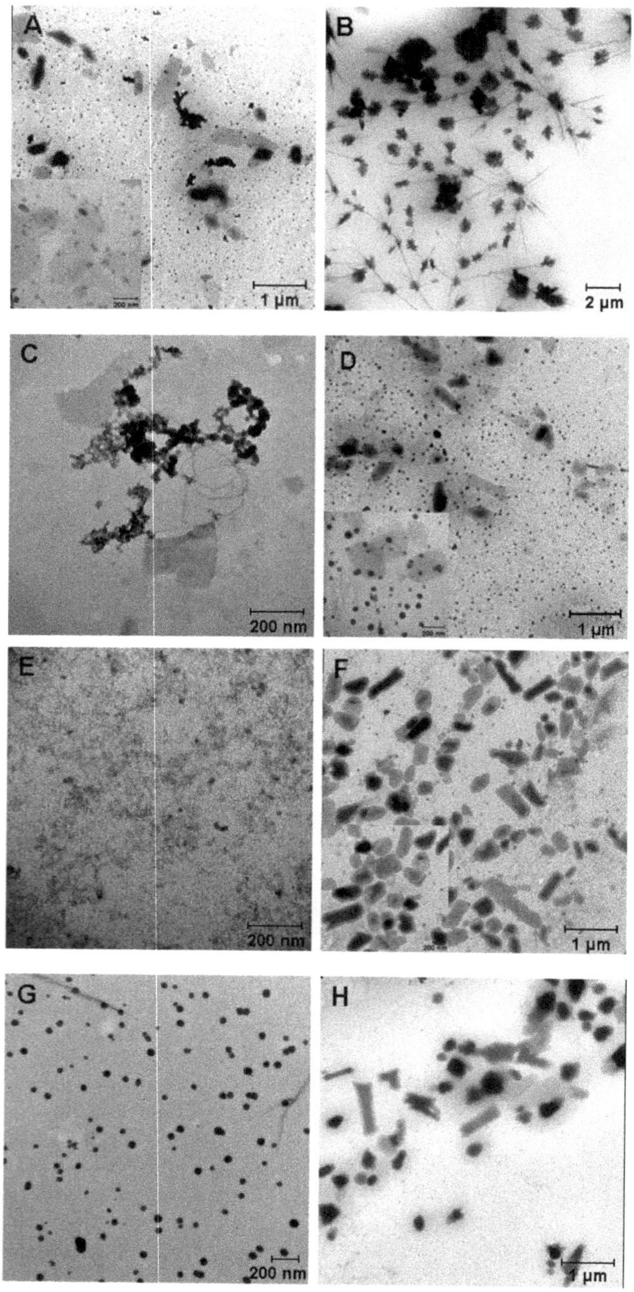

Figure 4. TEM micrographs of suspensions of $CaNP_1$ (**A**), $CaNP_2$ (**B**), $CaNP_3$ (**C**), $CaNP_4$ (**D**), $CaNP_5$ (**E**), $CaNP_6$ (**F**), $CaNP_7$ (**G**), and $CaNP_8$ (**H**).

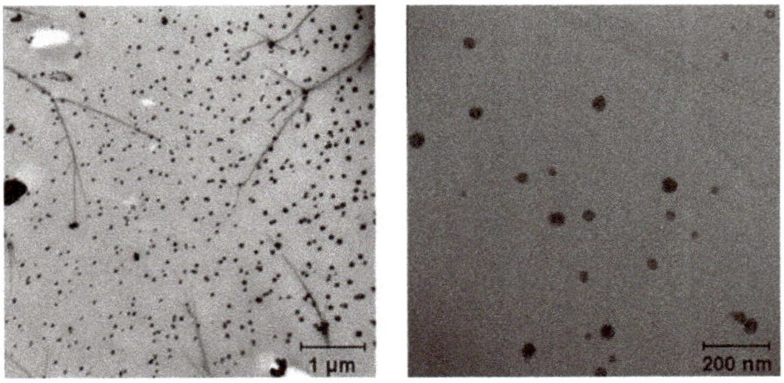

Figure 5. TEM images of suspensions of a $CaNP_7$ preparation obtained in 30× reaction volume.

This nanomaterial was also assayed for stability. During storage for three months, the hydrodynamic size did not change significantly (Figure 6).

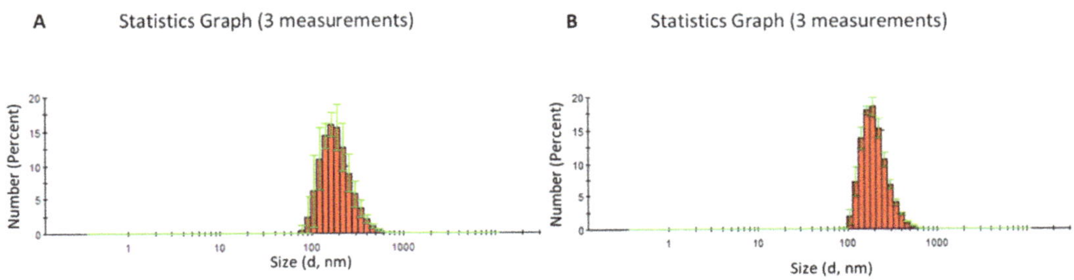

Figure 6. DLS size distribution of the $CaNP_7$ nanoparticles. The particle size was determined by the DLS method to be 204 ± 8 nm, with a polydispersity index (PDI) of 0.14 ± 0.02 (**A**). DLS size distribution of $CaNP_7$ nanoparticles after 90 days of storage. The particle size was determined by the DLS method to be 207 ± 4 nm, with a PDI of 0.11 ± 0.01 (**B**).

As discussed elsewhere, obtaining stable CaNPs in an aqueous solution without additional modifications of the procedure is not an easy task [9], and an aqueous solution is a prerequisite for subsequent successful application of the nanomaterials. The stability of our CaNPs in aqueous solutions, the scalability of our synthesis methodology, and good availability of the reagents will allow for the convenient use of these nanomaterials in the chemical and pharmaceutical industries: the synthesized nanomaterial called $CaNP_7$ can be stored while retaining its original properties for >3 months.

To evaluate the characteristics and stability of $CaNP_7$ in the bloodstream, we conducted a model experiment on the storage of the nanoparticles in a 50% serum solution (Figure 7).

The increase in the hydrodynamic particle size, in comparison with the initial one (t = 0 and t = 5 min, Figure 7), can be primarily explained by the change in the solution in which the DLS analysis was performed; this alteration can affect the obtained hydrodynamic radius. Because the particles between the time points of 5 min and nine days retained their size within the margin of error, it can be said that their stability in serum is sufficient for prospective biomedical applications [27].

Accordingly, the impact of solution composition (addition of a surfactant, DMEM, $MgCl_2$, and a change of the solvent) on the characteristics of the obtained materials was investigated experimentally based on the literature data. As a result, several simple and accessible approaches to obtaining monodisperse CaNPs of different shapes (spherical and rod-shaped) and sizes (from 40 to 200 nm) were developed here. Then, a methodology yielding CaNPs suitable for further biomedical research ($CaNP_7$) was selected. The obtained nanomaterial was found to have the parameters necessary for further investigation of the potential usefulness of CaNPs as a carrier of model antitumor agents.

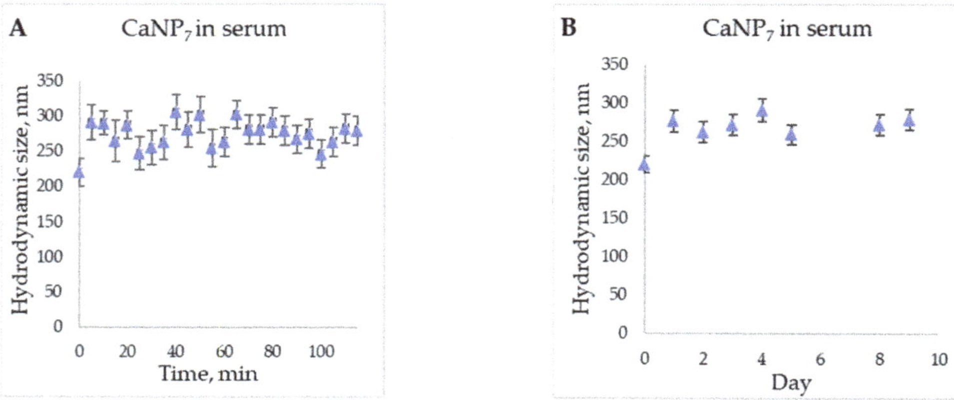

Figure 7. Stability of $CaNP_7$ in 50% serum (FBS) according to DLS data during the first hours (**A**); within 10 days (**B**).

3.2. Interaction of DOX with $CaNP_7$

There are three techniques for conjugating a biologically active compound with a carrier. Water-soluble therapeutic agents can be coprecipitated during nanomaterial synthesis [28]. Another approach is based on the impregnation of the prepared particles with a solution of the drug with constant stirring or shaking [29]. In this case, the binding of the obtained nanoparticles to the drug can be achieved either via adsorption or encapsulation. The third option for the conjugation of the nanomaterial with the drug is based on evaporation of a solvent containing the drug under reduced pressure. The latter two methods are suitable for our purpose because they can load drugs that are poorly soluble in water, aside from other advantages. In our study, we chose the method of postsynthetic impregnation of CaNPs with the drug.

It is likely that the efficiency of adsorption of macromolecules inside $CaCO_3$ particles or onto their surface is determined by electrostatic interactions. In heterogeneous systems (solid–liquid), the relevant factors are steric effects, molecular weight of the drug being immobilized, and its affinity for the CaNP surface. Via adjustment of pH in the reaction mixture, it is possible to change electrostatic interactions and, as a consequence, to manage the adsorption–desorption of the drug by controlling the number of its released/loaded molecules [30].

For example, drug-loaded CaNPs are applicable to the controlled release and delivery of various biomolecules.

DOX is an antibiotic that is also a common broad-spectrum antitumor agent used in chemotherapy. *DOX* was chosen here as the model therapeutic agent for the evaluation of CaNPs as drug carriers. The disadvantages of this drug are several serious adverse effects, ineffective penetration through the cell membrane, rapid excretion from the body, and poor water solubility due to hydrophobicity [31]. The development of a selective prolonged delivery system for a given drug is a highly relevant task. Maximal absorption of *DOX* in the visible spectral region (480 nm) enables quantitative analysis of the effectiveness of *DOX* conjugation with CaNPs by spectrophotometry without the introduction of additional labels.

The efficiency of $CaNP_7$ conjugation with *DOX* was evaluated by means of the drug capacity index, which was calculated as the amount of *DOX* (in μg) bound to 1 mg of $CaNP_7$. The amount of the drug bound to the nanoparticles was calculated as the difference between added *DOX* and *DOX* remaining in solution (supernatant) after incubation with $CaNP_7$. The $CaNP_7$ capacity index was found to be 659 ± 5 μg/mg.

On the basis of the literature data, we believe that the nanocomposites obtained in this work have good prospects for practical applications because they are more than threefold superior to most analogs in terms of *DOX* encapsulation efficiency [7,32].

One of the advantages of materials based on $CaCO_3$ is pH-dependent stability of the particles: with a decrease in pH, the rate of hydrolysis of the nanocarrier matrix increases, thereby facilitating the release of the encapsulated drug. In the next step, we investigated this phenomenon.

The drug release efficiency of the $CaNP_7$–*DOX* nanocomposite as a function of medium acidity was assessed in 100 mM acetate buffer at pH from 3.0 to 7.0. To study the release of *DOX* from the $CaNP_7$–*DOX* complex, the amounts of the components ($CaNP_7$ and *DOX*) were identical among the assays with different pH levels: 132 μg of *DOX* per 0.2 mg of $CaNP_7$.

$CaCO_3$ is considered unstable under acidic conditions; a similar property was expected in CaNPs. Figure 8 shows a pH-dependent release of the drug: as pH of the medium diminished, the proportion of released *DOX* increased, partly due to complete or partial degradation of the nanocarrier matrix. At pH 3.0, complete liberation of loaded *DOX* was achieved, whereas at pH close to physiological, the degree of *DOX* liberation was the lowest: at pH 7.0, the drug liberation rate was 25% after 180 min. Therefore, the pH-dependent release of *DOX* from its complex with $CaNP_7$ was demonstrated successfully.

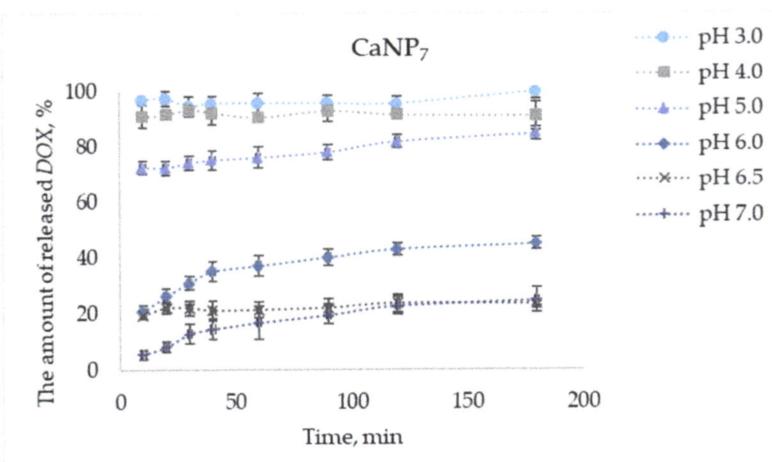

Figure 8. The proportion of *DOX* released from $CaNP_7$–*DOX* with time at pH 3.0–7.0.

High efficiency of a drug release at endosomal pH has been reported elsewhere [33]. In the present study, at physiological pH values, $CaNP_7$ retained more than 75% of the loaded drug within the nanocomposite, and this property may ensure high selectivity of drug distribution in the context of a site-activated therapy.

It is worth mentioning that the nanomaterials developed by us here had a high release efficiency of >80% when pH was lowered. This effect can help to implement passive targeting of the drug to tumor tissues, which are known to have more acidic pH than that of healthy tissues [34]. We expect that in future applications, the sensitivity to the weakly acidic pH of the tumor microenvironment will ensure effective highly specific delivery and a sustained release of anticancer drugs in in vivo experiments.

Next, at pH values close to physiological (pH 6.5), where only low drug release efficiency was seen (Figure 8), we examined the impact of temperature (15, 25, 37, and 45 °C) on the release efficiency of *DOX* from its complex with $CaNP_7$ (Figure 9).

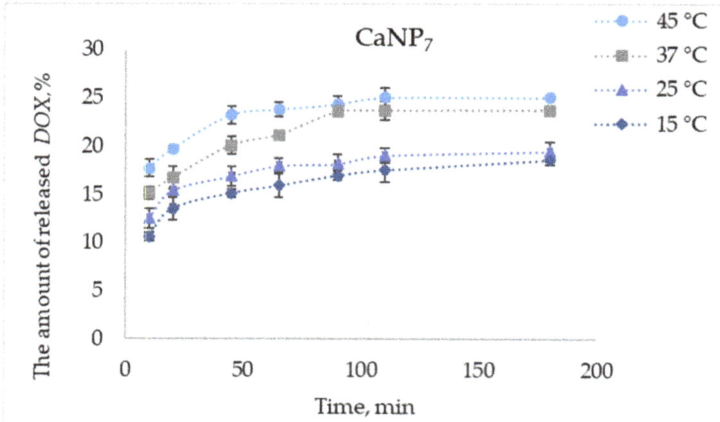

Figure 9. The proportion of released *DOX* (from $CaNP_7$-*DOX*) as a function of time at pH 6.5 and different temperatures.

The efficiency of *DOX* liberation increased with the increasing temperature: the difference between 45 and 15 °C was ~7% at each data point in the experiment (Figure 9).

Thus, it was successfully demonstrated that the obtained $CaCO_3$-based materials can form the basis of drug delivery systems. $CaNP_7$ effectively binds to the model therapeutic agent. During our investigation into the kinetic profiles of the *DOX* release from $CaNP_7$-*DOX*, it was found that CaNPs are pH-sensitive: the drug liberation increases with decreasing pH. This is a promising property for drug delivery to low-acidity sites including malignantly transformed tissues. When the temperature was raised from 15 to 45 °C, the drug release from the tested CaNPs increased. These results suggest that $CaNP_7$ can be further investigated as a nanocarrier for drug delivery. On the other hand, the main prerequisite for the use of CaNPs in vivo is the absence of toxicity.

3.3. Cytotoxicity Assays of $CaNP_7$ and Its Composites with DOX

The cytotoxicity of the obtained CaNPs and nanocomposites before and after complexation with *DOX* ($CaNP_7$-*DOX*) was assessed in the standard MTT assay, which estimates the percentage of surviving cells after exposure of the cells to the agent being tested. To evaluate the effectiveness of cell growth inhibition by the $CaNP_7$–*DOX* nanocomposite, a comparison was made with *DOX* alone and CaNPs alone in an assay involving a lung carcinoma cell line (A549). The absence of toxicity of CaNPs was successfully proven in an assay involving a human embryonic kidney 293 cell line (HEK293) and A549: at CaNP concentrations up to 22.5 μg/mL, the viability of the treated cells did not diminish below 98%. The literature data confirm the safety of $CaCO_3$ micro- and nanoparticles according to MTT assays [6,35]. Nonetheless, we needed to confirm this for the nanomaterial prepared by our method: $CaNP_7$.

Figure 10 indicates the effective inhibition of the growth of cancer cells (A549) by the $CaCO_3$-based nanocomposite containing an antitumor agent (*DOX*) compared to the drug alone.

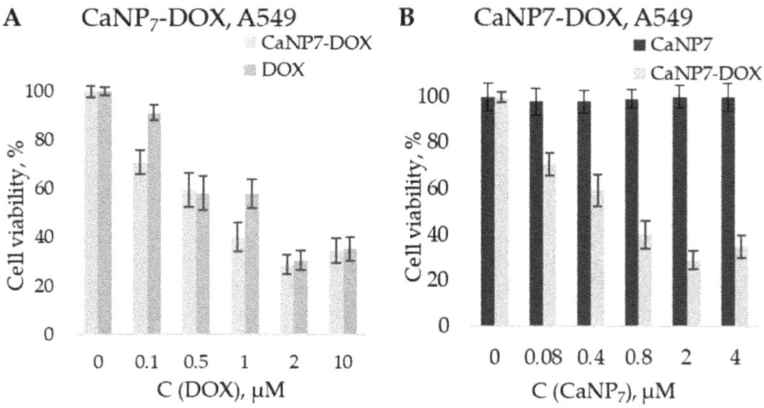

Figure 10. Cytotoxic activity of $CaNP_7$, $CaNP_7$–*DOX*, *DOX* on A549 cells. Cells were incubated with equimolar amounts of *DOX*, either soluble or loaded on nanoparticles (**A**) as well as with equivalent amounts of nanoparticles (**B**).

The $CaNP_7$–*DOX* composite proved to be comparable in efficacy to free *DOX*. This was confirmed by determining IC_{50}: the concentration of the inhibitory agent (either $CaNP_7$-*DOX* or *DOX*) required to reduce the cell proliferation rate by 50% (Table 2).

Table 2. IC_{50} values.

Sample	IC_{50}, μM
$CaNP_7$–*DOX*	0.97 ± 0.04
DOX	2.41 ± 0.02

The non-monotonicity of the MTT test data may be explained by the following: we expect that at a low concentration of *DOX* and $CaNP_7$–*DOX*, the efficiency of penetration of both the pure drug and the composition of nanoparticles affects cell survival. *DOX* in nanocomposites ($CaNP_7$–*DOX*) most probably has better penetration and so is more effective at a concentration of 0.1 μM. With an increase in concentration, it has less effect, because nanostructures release *DOX* gradually and 0.5 μM is the maximum concentration of *DOX* in solution in the case of 100% release, which does not occur immediately, when free *DOX* is immediately added to this solution in full concentration. With a further increase in the concentration, *DOX* tends to be sorbed onto the surface (plate side) because of its hydrophobic properties [36–38]. In this case, *DOX* orubicin in the composition of nanocomposites is more effective. With a further increase in concentration, all these effects have less effect, since the concentration of *DOX*orubicin is high.

This experiment was primarily aimed at proving that the therapeutic properties of *DOX* are preserved when it is encapsulated in $CaNP_7$. These CaNPs can serve as a delivery vehicle for a prolonged pH-dependent release of *DOX*, as depicted in the model experiments in Figure 7. Unfortunately, the MTT assay does not permit testing a pH-dependent release of *DOX* on the same cell type. Nonetheless, there is evidence in the literature that pH-sensitive materials manifesting high efficiency in vitro can be even more effective in vivo [4,12–15].

$CaNP_7$–*DOX* turned out to be more cytotoxic than the free drug. Accordingly, we advanced a hypothesis that the greater reduction in the percentage of surviving cells is due to the gradual release of *DOX* from the $CaNP_7$–*DOX* composite. In contrast, free *DOX* was added to the cells at a single time point at a final concentration and probably precipitated due to its hydrophobicity.

Thus, the $CaNP_7$ nanocarrier proposed in this work was found to be nontoxic in vitro. The efficiency of inhibiting cancer cell growth by the $CaNP_7$–*DOX* composite was not inferior to that of the free therapeutic agent. Nevertheless, due to the pH-dependent stability of the $CaNP_7$ carrier, according to the kinetic profiles of the *DOX* release from $CaNP_7$–*DOX* (Figure 8), a therapeutic efficacy is expected to be higher in vivo due to the selective dosed accumulation of the drug at a tumor site.

Our findings warrant further research into the application of $CaCO_3$-based nanostructures in vivo.

4. Conclusions

Methods were developed for the preparation of monodisperse inorganic nanoparticles based on a $CaCO_3$ (e.g., $CaNP_7$) stable in suspension having sizes of 200 ± 20 nm. The synthesized nanoparticles were characterized by TEM and DLS. Due to their monodispersity and small size, the obtained materials were optimally suited for biomedical applications including intravenous drug delivery. Efficient conjugation of the obtained nanoparticles with an anticancer drug (*DOX*) was demonstrated, as was a pH-dependent profile of a *DOX* release from $CaNP_7$–*DOX*. It is expected that the high efficiency of *DOX* encapsulation and the pH-dependent profile of its release will make it possible to reduce the amount of the administered drug in future in vivo experiments while increasing the effectiveness of the therapy; this is because the CaNPs provides a more efficient release of the drug from the nanoparticles in a site with lowered pH corresponding to a tumor macroenvironment. In the concentration range of 0.5–50.0 μg/mL, no toxicity of the CaNPs to the tested cell lines (HEK293, A549) was detectable. The properties of CaNPs should help to prevent unwanted accumulation of these nanomaterials in major organs such as the liver, heart, and kidneys, due to CaNP biodegradation. The effectiveness of cell growth inhibition by $CaNP_7$–*DOX* turned out to be comparable to that of free *DOX* in the *DOX* concentration range of 0.1–10.0 μM. By means of the model drug *DOX* in the MTT assay, we proved that when our methodology is employed for constructing a pH-dependent transporter with a prolonged release ability, therapeutic efficacy of the drug is preserved. Our results indicate that the newly developed CaNPs hold great promise for further in vivo experiments.

Author Contributions: Methodology, investigation, formal analysis, visualization, writing—original draft preparation, V.P.; investigation, formal analysis, visualization (TEM), Y.P.; data validation, project administration, funding acquisition, I.P.; conceptualization, methodology, data validation, supervision, writing—review and editing, D.P.; data validation, project administration, conceptualization, methodology, writing—review and editing, E.D. All authors have read and agreed to the published version of the manuscript.

Funding: This study was funded by the Russian Science Foundation (grant no. 21-64-00017); synthesis and characterization of CaNPs was funded by the Ministry of Science and Higher Education of the Russian Federation (state registration no. 121031300042-1).

Institutional Review Board Statement: Not applicable.

Informed Consent Statement: Not applicable.

Data Availability Statement: Not applicable.

Acknowledgments: The cell lines were kindly provided by Nushtaeva A and Koval O from the Laboratory of Biotechnology of ICBFM SB RAS. The TEM was performed in the microscopy research group of the ICBFM SB RAS under the supervision of Ryabchikova E.

Conflicts of Interest: The authors declare no conflict of interest.

References

1. Shan, D.; Zhu, M.; Xue, H.; Cosnier, S. Development of amperometric biosensor for glucose based on a novel attractive enzyme immobilization matrix: Calcium carbonate nanoparticles. *Biosens. Bioelectron.* **2007**, *22*, 1612–1617. [CrossRef]

2. Bala, H.; Zhang, Y.; Ynag, H.; Wang, C.; Li, M.; Lv, X.; Wang, Z. Preparation and characteristics of calcium carbonate/silica nanoparticles with core-shell structure. *Colloids Surf. A Physicochem. Eng. Asp.* **2007**, *294*, 8–13. [CrossRef]
3. Zhang, Y.; Ma, P.; Wang, Y.; Du, J.; Zhou, Q.; Zhu, Z.; Yang, X.; Yuan, J. Biocompatibility of Porous Spherical Calcium Carbonate Microparticles on Hela Cells. *World J. Nano Sci. Eng.* **2012**, *2*, 25–31. [CrossRef]
4. Maleki Dizaj, S.; Barzegar-Jalali, M.; Zarrintan, M.H.; Adibkia, K.; Lotfipour, F. Calcium carbonate nanoparticles as cancer drug delivery system. *Expert Opin. Drug Deliv.* **2015**, *12*, 1649–1660. [CrossRef] [PubMed]
5. Iafisco, M.; Delgado-Lopez, J.M.; Varoni, E.M.; Tampieri, A.; Rimondini, L.; Gomez-Morales, J.; Prat, M. Cell surface receptor targeted biomimetic apatite nanocrystals for cancer therapy. *Small* **2013**, *9*, 3834–3844. [CrossRef] [PubMed]
6. Guo, Y.; Li, H.; Shi, W.; Zhang, J.; Feng, J.; Yang, X.; Wang, K.; Zhang, H.; Yang, L. Targeted delivery and pH-responsive release of doxorubicin to cancer cells using calcium car-bonate/hyaluronate/glutamate mesoporous hollow spheres. *J. Colloid Interface Sci.* **2017**, *502*, 59–66. [CrossRef]
7. Li, L.; Yang, Y.; Lv, Y.; Yin, P.; Lei, T. Porous calcite $CaCO_3$ microspheres: Preparation, characterization and release behavior as doxoru-bicin carrier. *Colloids Surf. B Biointerfaces* **2020**, *186*, 110720. [CrossRef] [PubMed]
8. Dai, Y.; Xu, C.; Sun, X.; Chen, X. Nanoparticle design strategies for enhanced anticancer therapy by exploiting the tumour mi-croenvironment. *Chem. Soc. Rev.* **2017**, *46*, 3830–3852. [CrossRef]
9. Zhao, Y.; Luo, Z.; Li, M.; Qu, Q.; Ma, X.; Yu, S.-H.; Zhao, Y. A Preloaded Amorphous Calcium Carbonate/Doxorubicin@Silica Nanoreactor for pH-Responsive Delivery of an Anticancer Drug. *Angew. Chem. Int. Ed.* **2015**, *54*, 919–922. [CrossRef]
10. Hamidu, A.; Mokrish, A.; Mansor, R.; Razak, I.S.A.; Danmaigoro, A.; Jaji, A.Z.; Bakar, Z.A. Modified methods of nanoparticles synthesis in pH-sensitive nano-carriers produc-tion for doxorubicin delivery on MCF-7 breast cancer cell line. *Int. J. Nanomed.* **2019**, *14*, 3615.
11. Fu, W.; Mohd Noor, M.H.; Yusof, L.M.P.; Ibrahim, T.A.T.; Keong, Y.S.; Jaji, A.Z.; Zakaria, M.Z.A.B. In vitro evaluation of a novel pH sensitive drug delivery system based cockle shell-derived aragonite nanoparticles against osteosarcoma. *J. Exp. Nanosci.* **2017**, *12*, 166–187. [CrossRef]
12. Ibiyeye, K.M.; Idris, S.B.; Zuki, A.B.Z. Cockle shell-derived aragonite calcium carbonate nanoparticle for targeting cancer and breast cancer stem cells. *Cancer Nanotechnol.* **2020**, *11*, 1–15. [CrossRef]
13. Danmaigoro, A.; Selvarajah, G.T.; Mohd Noor, M.H.; Mahmud, R.; Abu Bakar, M.Z. Toxicity and safety evaluation of doxorubicin-loaded cockle-shell-derived calcium carbonate nanoparticle in dogs. *Adv. Pharmacol. Sci.* **2018**, *2018*, 4848602.
14. Wenliang, F.; Rameli, M.A.B.P.; Ibrahim, T.A.T.; Noor, M.H.M.; Yusof, L.M.; Zakaria, M.Z.A.B. In vivo evaluation of anticancer efficacy of drug loaded cockle shell-derived aragonite nanoparticles. *J. Biomed. Mater. Res. Part B Appl. Biomater.* **2019**, *107*, 1898–1907. [CrossRef] [PubMed]
15. Ghaji, M.S.; Zakaria, Z.A.B.; Shameha, A.R.; Noor, M.H.M.; Hazilawati, H. Novel synthesis of nanoparticles from cockle shells via mechanical method for cytarabine drug release. *J. Comput. Theor. Nanosci.* **2018**, *15*, 1128–1136. [CrossRef]
16. Som, A.; Raliya, R.; Tian, L.; Akers, W.; Ippolito, J.E.; Singamaneni, S.; Biswas, P.; Achilefu, S. Monodispersed calcium carbonate nanoparticles modulate local pH and inhibit tumor growth in vivo. *Nanoscale* **2016**, *8*, 12639–12647. [CrossRef]
17. Kollenda, S.A.; Klose, J.; Knuschke, T.; Sokolova, V.; Schmitz, J.; Staniszewska, M.; Costa, P.F.; Herrmann, K.; Westendorf, A.M.; Fendler, W.P.; et al. In vivo biodistribution of calcium phosphate nanoparticles after intravascular, in-tramuscular, intratumoral, and soft tissue admin-istration in mice investigated by small animal PET/CT. *Acta Biomater.* **2020**, *109*, 244–253. [CrossRef]
18. Ma, X.; Zhang, X.; Yang, L.; Wang, G.; Jiang, K.; Wu, G.; Cui, W.; Wei, Z. Tunable construction of multi-shelled hollow carbonate nanospheres and their potential ap-plications. *Nanoscale* **2016**, *8*, 8687–8695. [CrossRef] [PubMed]
19. Guo, Y.; Zhang, J.; Jiang, L.; Shi, X.; Yang, L.; Fang, Q.; Fang, H.; Wang, K.; Jiang, K. Facile one-pot preparation of calcite mesoporous carrier for sustained and targeted drug re-lease for cancer cells. *Chem. Commun.* **2012**, *48*, 10636–10638. [CrossRef] [PubMed]
20. Peer, D.; Karp, J.M.; Hong, S.; Farokhzad, O.C.; Margalit, R.; Langer, R. Nanocarriers as an emerging platform for cancer therapy. *Nat. Nanotechnol.* **2007**, *2*, 751–760. [CrossRef] [PubMed]
21. Boyjoo, Y.; Pareek, V.K.; Liu, J. Synthesis of micro and nano-sized calcium carbonate particles and their applications. *J. Mater. Chem. A* **2014**, *2*, 14270–14288. [CrossRef]
22. Krajewska, B. Urease-aided calcium carbonate mineralization for engineering applications: A review. *J. Adv. Res.* **2018**, *13*, 59–67. [CrossRef]
23. Avaro, J.T.; Ruiz-Agudo, C.; Landwehr, E.; Hauser, K.; Gebauer, D. Impurity-free amorphous calcium carbonate, a preferential material for pharmaceutical and medical applications. *Eur. J. Mineral.* **2019**, *31*, 231–236. [CrossRef]
24. Xu, H.; Peng, X.; Bai, S.; Ta, K.; Yang, S.; Liu, S.; Bin Jang, H.; Guo, Z. Precipitation of calcium carbonate mineral induced by viral lysis of cyanobacteria: Evidence from laboratory experiments. *Biogeosciences* **2019**, *16*, 949–960. [CrossRef]
25. Martel, J.; Young, J.D.E. Purported nanobacteria in human blood as calcium carbonate nanoparticles. *Proc. Natl. Acad. Sci. USA* **2008**, *105*, 5549–5554. [CrossRef] [PubMed]
26. Khouzani, M.F.; Chevrier, D.M.; Güttlein, P.; Hauser, K.; Zhang, P.; Hedin, N.; Gebauer, D. Disordered amorphous calcium carbonate from direct precipitation. *CrystEngComm* **2015**, *17*, 4842–4849. [CrossRef]
27. Zelepukin, I.V.; Yaremenko, A.V.; Yuryev, M.V.; Mirkasymov, A.B.; Sokolov, I.L.; Deyev, S.M.; Nikitin, P.I.; Nikitin, M.P. Fast processes of nanoparticle blood clearance: Comprehensive study. *J. Control. Release* **2020**, *326*, 181–191. [CrossRef] [PubMed]

28. Zyuzin, M.V.; Díez, P.; Goldsmith, M.; Carregal-Romero, S. Comprehensive and systematic analysis of the immunocompatibility of polyelectrolyte capsules. *Bioconjug. Chem.* **2017**, *28*, 556–564. [CrossRef]
29. Qiu, N.; Yin, H.; Ji, B.; Klauke, N.; Glidle, A.; Zhang, Y.; Song, H.; Cai, L.; Ma, L.; Wang, G.; et al. Calcium carbonate microspheres as carriers for the anticancer drug camptothecin. *Mater. Sci. Eng. C* **2012**, *32*, 2634–2640. [CrossRef]
30. Tang, H.; Guo, J.; Sun, Y.; Chang, B.; Ren, Q.; Yang, W. Facile synthesis of pH sensitive polymer-coated mesoporous silica nanoparticles and their application in drug delivery. *Int. J. Pharm.* **2011**, *421*, 388–396. [CrossRef] [PubMed]
31. Matyszewska, D. Drug delivery systems in the transport of doxorubicin. *Surf. Innov.* **2014**, *2*, 201–210. [CrossRef]
32. Wu, J.L.; Wang, C.Q.; Zhuo, R.X.; Cheng, S.X. Multi-drug delivery system based on alginate/calcium carbonate hybrid nanoparticles for combination chemotherapy. *Colloids Surf. B* **2014**, *123*, 498–505. [CrossRef]
33. Sushko, A.D.; Yaminskii, I.V.; Gavryushina, E.S.; Drygin, Y.F. Release of ribonucleic acid from human common cold HRV 2 virus in acidic medium. *Colloid J.* **2010**, *72*, 559–564. [CrossRef]
34. Swietach, P.; Vaughan-Jones, R.D.; Harris, A.L.; Hulikova, A. The chemistry, physiology and pathology of pH in cancer. *Philos. Trans. R. Soc. B* **2014**, *369*, 20130099. [CrossRef] [PubMed]
35. Idris, S.B.; Arifah, A.K.; Jesse, F.F.; Ramanoon, S.Z.; Basit, M.A.; Zakaria, M.Z. In Vitro Cytotoxicity Evaluation of Oxytetracycline Loaded Cockle Shell Derived Calcium Carbonate Aragonite Nanoparticles. *Nanosci. Nanotechnol.-Asia* **2021**, *11*, 237–242. [CrossRef]
36. Yamada, Y. Dimerization of Doxorubicin Causes Its Precipitation. *ACS Omega* **2020**, *5*, 33235–33241. [CrossRef] [PubMed]
37. Thedrattanawong, C.; Manaspon, C.; Nasongkla, N. Controlling the burst release of doxorubicin from polymeric depots via ad-justing hydrophobic/hydrophilic properties. *J. Drug Deliv. Sci. Technol.* **2018**, *46*, 446–451. [CrossRef]
38. King, H.D.; Dubowchik, G.M.; Mastalerz, H.; Willner, D.; Hofstead, S.J.; Firestone, R.A.; Lasch, S.J.; Trail, P.A. Monoclonal antibody conjugates of doxorubicin prepared with branched pep-tide linkers: Inhibition of aggregation by methoxytriethyleneglycol chains. *J. Med. Chem.* **2002**, *45*, 4336–4343. [CrossRef] [PubMed]

nanomaterials

MDPI

Article

Novel Experimental In-Office Bleaching Gels Containing Co-Doped Titanium Dioxide Nanoparticles

Matheus Kury [1], Rochelle D. Hiers [2], Yan D. Zhao [3], Mayara Z. D. Picolo [1], Jessica Hsieh [2], Sharukh S. Khajotia [2], Fernando L. Esteban Florez [2,*] and Vanessa Cavalli [1,*]

1 Division of Operative Dentistry, Department of Restorative Dentistry, Piracicaba School of Dentistry, University of Campinas, Piracicaba 13414-903, SP, Brazil
2 Division of Dental Biomaterials, Department of Restorative Sciences, College of Dentistry, University of Oklahoma Health Sciences Center, Oklahoma City, OK 73117, USA
3 Department of Biostatistics and Epidemiology, Hudson College of Public Health, University of Oklahoma Health Sciences Center, Oklahoma City, OK 73104, USA
* Correspondence: fernando-esteban-florez@ouhsc.edu (F.L.E.F.); cavalli@unicamp.br (V.C.)

Abstract: The present study reports on the development and testing of novel bleaching agents containing co-doped metaloxide nanoparticles (NP; 0%, 5%, 10% *v/w*) and hydrogen peroxide (HP, 0%, 6%, 15%, and 35%). Bovine blocks (n = 200, A = 36 mm^2) were obtained and randomly distributed into experimental groups (n = 10/group). NPs were incorporated into gels before bleaching (3 sessions, 7 days apart, 30 min/session, irradiated with violet light-LT). Color changes (ΔE_{00}, ΔWI_D), mineral content (CO_3^{2-}, PO_4^{3-}), and topography were assessed (spectrophotometer, ATR-FTIR, and AFM) before and after bleaching procedures (14 days). Metabolic status and three-dimensional components of non-disrupted *Streptococcus mutans* biofilms were investigated using a multimode reader and confocal microscopy. The results indicate that ΔE_{00} and ΔWI_D significantly increased with NPs' concentrations and LT. The enamel's mineral ratio was adversely impacted by HP, but alterations were less pronounced when using NP-containing gels. The enamel's topography was not damaged by the bleaching protocols tested. The bioluminescence results show that bleaching protocols do not render latent antibacterial properties to enamel, and the confocal microscopy results demonstrate that the 3-dimensional distribution of the components was affected by the protocols. The proposed nanotechnology improved the bleaching efficacy of experimental materials independent of hydrogen peroxide or irradiation and did not adversely impact the enamel's surface properties or its chemical content.

Keywords: tooth bleaching; hydrogen peroxide; light irradiation; nanoparticles

1. Introduction

In-office power bleaching (IPB) is considered an ultraconservative and minimally invasive treatment capable of resolving dental discolorations (low to moderate) in as short as one clinical session [1]. The IPB treatment typically involves three clinical sessions (45 min each; 7 days apart) using hydrogen peroxide-containing bleaching gels (HP, 35% to 45%), in combination with visible light irradiation or not [2], to promote the attainment of immediate esthetic outcomes. IPB's underlying mechanism of action revolves around the generation of reactive oxygen species (ROS). Upon generation, these short-lived and highly reactive free radicals must be efficiently transported from the gel to the dentin-enamel junction (DEJ). Once at the DEJ, free radicals will then break conjugated double bonds present in large organic molecules (chromophores) through a non-specific oxidative process [3].

Even though several reports have demonstrated the bleaching efficacy of IPB [4–6], other studies have indicated that the utilization of these highly caustic bleaching agents may result in the occurrence of adverse effects (short- and long-term), including irreversible changes in enamel topography [7] and chemical make-up [8–10], decreased surface microhardness [11], increased surface roughness [12], diminished bond strength [13], and reduced fracture resistance [14]. From the clinical standpoint, the most prevalent adverse effect reported by patients and clinicians is mild to severe dentin hypersensitivity (DH) [4,15,16]. According to previous studies, there is a strong and positive correlation between dentin hypersensitivity, HP concentration, and pulpal cytotoxicity [17,18], where the higher the HP concentration, the stronger the dentin hypersensitivity [16], and the effects are more durable.

In this critical context, several research groups have tried to overcome the limitations described by adding calcium or fluorine ions in the formulation of highly concentrated bleaching gels. Even though the reported results have demonstrated that adverse effects such as decreased enamel microhardness and rougher surfaces were less pronounced with the utilization of calcium- or fluorine-containing gels [8], subsequent studies have shown that the promising results initially reported were limited to the outermost layers of enamel and did not prevent the loss of minerals at subsurface levels, thereby restricting the therapeutic effect of the novel formulations proposed [9]. Follow-up studies investigated the efficacy of experimental protocols modulated by low-concentrated bleaching gels (6–15%) and near-UVA wavelengths (405 $\pm$ 15 nm) as an alternative approach to reducing the incidence of dentin hypersensitivity while trying to achieve desirable whitening outcomes [19–21]. Even though the utilization of low-concentrated bleaching gels resulted in lower incidences of DH, the bleaching efficacies reported (in terms of ΔE and whitening index (WI)) were considered poor because the outcomes were much less intense and durable as compared to those attained with gels containing high HP concentrations.

Recent approaches focused on the incorporation of metaloxides, such as titanium dioxide (TiO_2, P25 Degussa) and nitrogen-doped titanium dioxide (N_TiO_2) nanoparticles, into the formulation of commercially available bleaching gels containing high HP concentrations [22–24]. In theory, the incorporation of these semiconductors would improve the dissociation of HP into ROS by a photo-physical process where photons are converted into thermal energy. However, despite the theoretical feasibility of the process, experimental bleaching gels containing varying concentrations of metaloxide nanoparticles were demonstrated to be clinically ineffective when compared to unaltered gels containing HP (either 15% or 35%) [22,23]. These unexpected findings are believed to have precipitated from fast and spontaneous dissociation processes that take place when HP is exposed to metaloxides and from other contributing factors such as limited wettability and high viscosity.

A recent study reported on the successful fabrication of N_TiO_2 (6–15 nm) using highly controllable, reproducible, and green solvothermal reactions [25]. In that study, synthesized nanoparticles were incorporated into commercially available dental adhesive resins (OptiBond Solo Plus, Kerr Corp., Orange, CA, USA) with the objective of imparting non-leaching antibacterial and biomimetic functionalities to the parental polymer. According to Huo et al. [26], the synthesis route reported by Esteban Florez et al. [25] results in the attainment of pure and crystalline TiO_2 nanoparticles (anatase phase) that are electron deficient; display high levels of nitrogen doping; have well-defined pore structure, large surface areas, and facilitate the generation of electron-hole pairs; and are capable of efficiently absorbing visible wavelengths (400 to 700 nm) while generating significant amounts of perhydroxyl ($HO_2^{\bullet}$) and hydroxyl ($OH^{\bullet}$) radicals [25], which are long-lived species of oxygen.

Follow-up studies from the same research group demonstrated the successful solvothermal synthesis of TiO_2 nanoparticles that were co-doped with either nitrogen and fluorine (NF_TiO_2) or nitrogen and silver (NAg_TiO_2), functionalized into OptiBond Solo Plus, and tested for antibacterial properties (in dark and light irradiated conditions) against *Streptococcus mutans* using a newly developed and optimized high-throughput bioluminescence assay [27,28]. According to results reported, experimental materials containing 30% of either NF_TiO_2 or NAg_TiO_2 displayed antibacterial behaviors that were comparable to those attained with Clearfil SE Protect (Kuraray Co., Tokyo, Japan; fluoride-releasing material) independently of light irradiation conditions [27]. These findings have not only indicated that the nanotechnology reported has a strong potential to be translated into commercial products capable of sustaining long-term antibacterial properties, but the promising antibacterial effects observed in the absence of light corroborate the findings reported by Huo et al. [26] that nanoparticles synthesized through solvothermal processes are capable of generating long-lived species of oxygen.

Based on that premise and considering that fluorine is one of the most reactive chemical elements known to man, our research group decided to functionalize NF_TiO_2 (NP) into experimental bleaching gels containing HP (6%, 15%, or 35%) and determine the effects of nanoparticles' concentrations (0%, 5%, and 10%) and violet light irradiation on bleaching efficacy, bovine enamel chemical make-up, and surface topography. Additional analyses were focused on revealing how experimental bleaching protocols affect the metabolism and the components of single-species biofilms using a minimally invasive, real-time, and high throughput bioluminescence assay and a concurrent staining technique, along with confocal laser scanning microscopy, respectively. The null hypotheses were that the incorporation of NP would not significantly affect (i) the bleaching efficacy and (ii) the chemical make-up of enamel bleached with the experimental bleaching gels. In addition, it was hypothesized that the incorporation of NP would not (iii) avoid the growth of biofilm on the bleached enamel surface.

2. Materials and Methods

2.1. Experimental Design

The specimens described in Section 2.2 below (n = 200; n = 10/group) were randomly allocated according to the study factors:

Bleaching Agent:

- 0% hydrogen peroxide (0% HP);
- 6% hydrogen peroxide (6% HP);
- 15% hydrogen peroxide (15% HP);
- 35% hydrogen peroxide (35% HP).

NF_TiO_2 Concentration (*v/w*):

- 0% NF_TiO_2;
- 5% NF_TiO_2;
- 10% NF_TiO_2.

Light Activation:

- Dark conditions;
- Visible light (LT).

Analyses of color (ΔE_{00}, ΔWI_D), pH of the experimental gels, mineral composition (carbonate:phosphate ratio), surface topography of the enamel, and microbiological activity evaluation were conducted. Initial surface analyses were performed before bleaching—baseline (T_0). Experimental bleaching protocols consisted of three sessions (T_1 = first bleaching session, T_2 = second bleaching session, T_3 = third bleaching session). Analyses following the bleaching protocols were carried out 14 days after (T_4) the third bleaching session (T_3).

2.2. Specimen Preparation and Experimental Groups

Squared-shaped specimens (enamel-dentin blocks; area = 36.0 mm^2, thickness = 3.0 mm) were obtained from the central buccal area of bovine crowns as described in previous studies [10,29]. The blocks were polished using a rotary polisher (Arotec, São Paulo, SP, Brazil) and abrasive disks (600- and 1200-Grit, Norton Saint-Gobain, Guarulhos, SP, Brazil) and finished using polishing cloths (3M Brazil, Sumaré, SP, Brazil) with diamond suspensions (1 μm, 0.50 μm, and 0.25 μm, Erios, São Paulo, SP, Brazil). Prepared specimens were subjected to Knoop microhardness testing (50.0 g load, 5 s/indentation, 3 indentations/specimen, 100 μm apart; Future Tech FM-ARS, Tokyo, Japan) [30]. Specimens (n = 200; 10/group) with standardized microhardness (296.07 kgf/mm^2 $\pm$ 29.60) were randomly distributed and submitted to bleaching with hydrogen peroxide (HP; 0%, 6%, 15%, and 35%) experimental gels, containing NF_TiO_2 nanoparticles (NP; 0%, 5%, and 10%), and violet light irradiation (LT; with or without):

- G1—No treatment (control group);
- G2—LT;
- G3—HP6;
- G4—HP6 + LT;
- G5—HP15;
- G6—HP15 + LT;
- G7—HP35;
- G8—HP35 + LT;
- G9—HP6 + 5%NP;
- G10—HP6 + 5%NP + LT;
- G11—HP15 + 5%NP;
- G12—HP15 + 5%NP + LT;
- G13—HP35 + 5%NP;
- G14—HP35 + 5%NP + LT;
- G15—HP6 + 10%NP;
- G16—HP6 + 10%NP + LT;
- G17—HP15 + 10%NP;
- G18—HP15 + 10%NP + LT;
- G19—HP35 + 10%NP;
- G20—HP35 + 10%NP + LT.

2.3. Nanoparticles' Synthesis

A detailed description of the synthesis of NF_TiO_2 nanoparticles has been reported in previous publications [25,27,28]. A solution of 1.7 g of $Ti(OBu)_4$ (Aldrich, St. Louis, MO, USA; 97%), 4.6 g of C_2H_5OH (200-proof Decon Labs, King of Prussia, PA, USA), 6.8 g $C_{18}H_{35}NH_2$ (Aldrich; 70%), 7.1 g of $C_{18}H_{34}O_2$ (Aldrich; 90%), and 5% of NH_4F (based on Ti content; crystalline, ACS, Alfa Aesar, Tewksbury, MA, USA) was prepared and mixed with an ethanol–water solution (4%, 18-Milli-Q; total weight = 13.10 g). The prepared solutions were transparent before mixing; however, the final solution clouded instantaneously after mixing due to hydrolysis and some micelle formation. The final solution was placed into a high-pressure reaction vessel (Borosilicate Glass-lined; Paar Series 4593, Bench Top Reactor System, Moline, IL, USA), reacted (180 °C, 24 h, 15 psi), and stirred via an external shaft coupled to a turbine impeller (280 rpm). At the end of the 24-h cycle, the solution was removed from the reaction vessel and transferred to a 50 mL falcon tube with a certain amount of ethanol (200-proof, Decon Labs). The solution was centrifuged for 15 min at 8000 rpm. This procedure was repeated two additional times using 20 mL of ethanol.

2.4. Polymer Synthesis and Incorporation of NPs

Experimental bleaching gels were formulated in our laboratory by mixing a commercially available hydrophilic polymer (12.5 g, Carbomer 940 NF, Spectrum, Gardena, CA, USA) with an aqueous solution (distilled, 400 mL, pH = 11), containing KOH (60%, 20 mL), using a planetary and orbital stand-alone mixer (1 cycle at 2000 rpm for 2 min and 2 additional cycles at 2500 rpm for 3 min each; Speed Mixer, DAC 400.1 FVZ, FlackTek Inc., Laudrum, SC, USA). Immediately after mixing, the resulting polymer (pH~6) was observed to be transparent and free of any undissolved polymer (white agglomerates). The experimental polymer was then stored in a black container for at least 24 h (refrigerator, 8 °C).

Two aliquots (1 mL and 2 mL, respectively) of nitrogen and fluorine co-doped titanium dioxide nanoparticles (NF_TiO_2, ~40 mg/mL) suspended in ethanol (described in Section 2.3 *Nanoparticles Synthesis*) were placed in individual plastic tubes and were centrifuged (8000 rpm, 5 min) in preparation for polymer incorporation procedures. Ethanol-free nanoparticles were then individually mixed into 20 g of the experimental polymer to render gels containing either 5% or 10% NP. Each nanofilled gel was then mixed at 2450 rpm for 20 s (Speed Mixer, DAC 400.1 FVZ, FlackTek Inc., Laudrum, SC, USA). The final gel continued to be transparent and free of visible agglomerates, but its color became pale yellow due to the successful incorporation and dispersion of NP.

2.5. Incorporation of Hydrogen Peroxide (H_2O_2)

Immediately before utilization, experimental gels (either 1 g or 1.5 g, depending on the H_2O_2:polymer ratio) with or without NP (either 5% or 10%) were manually mixed (1:2 (6% or 15% H_2O_2) or 2:3 (35% H_2O_2)) with 1 mL of hydrogen peroxide following previously published protocols (Figure 1) [10,19]. The rationale for the utilization of two distinct H_2O_2:polymer ratios was based on the need to achieve comparable viscosities for all experimental materials investigated.

Figure 1. Appearance of experimental polymers without (**left**) or with (**right**) 10% of NP. Note that the incorporation of NPs in the concentration mentioned rendered experimental materials that were transparent a pale-yellow color and free of large agglomerates.

2.6. Bleaching Protocols

The experimental bleaching protocols investigated consisted of 3 sessions (T_1 = first bleaching session, T_2 = second bleaching session, and T_3 = third bleaching session) 7 days apart. Each 30 min session was based on a single application of the proper experimental gel (with or without nanoparticles), combined or not, with continuously visible light irradiation (20 cycles of 1 min, 30 s intervals between irradiation cycles [19]; 405 ± 15 nm, 1.2 W/cm^2, emission window area = 10.7 cm^2, Bright Max Whitening, MMO, São Carlos, SP, Brazil) according to experimental groups (G1 to G20; see group descriptions in Section 2.2. *Specimen Preparation and Experimental groups*). Figure 2 illustrates specimens subjected to dental bleaching procedures modulated by experimental bleaching gels and visible light irradiation (G14—HP35 + 5%NP + LT). After each session, specimens from all groups were stored (37 °C, dark conditions) in artificial saliva (1.5 mM calcium chloride [$CaCl_2$], 0.9 mM sodium phosphate [NaH_2PO_4], and 0.15 mM potassium chloride [KCl, pH 7.0]). After the third session (T_3), specimens were then stored in artificial saliva for 14 days using the same procedures previously described [10].

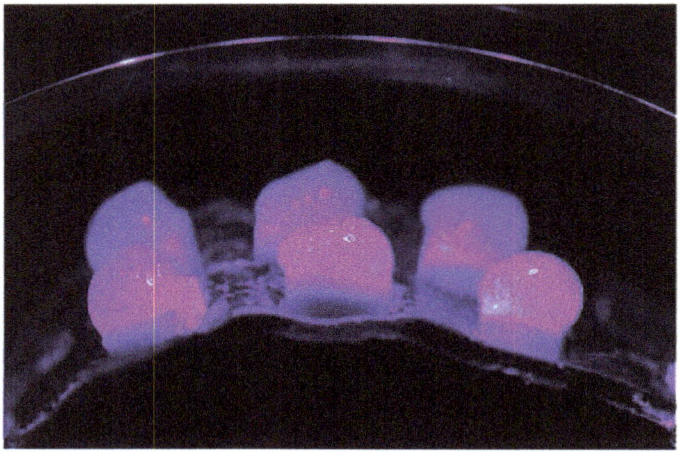

Figure 2. Specimens from G14—HP35 + 5NP + LT being subjected to IPB with visible light irradiation.

2.7. Objective Colorimetric Evaluation

The objective colorimetric evaluation (in terms of L*, a*, and b*) was performed before the first bleaching session (baseline, T_0) and 14 days after the last bleaching session (T_4) using a hand-held digital spectrophotometer (Vita EasyShade, VITA Zahnfabrik H. Rauter GmbH & Co. KG, Bad Sackingen, Germany). Variation of color ($T_4 - T_0$) was determined using the formulae for ΔE_{00} (Equation (1)) [31,32] and ΔWI_D (Equation (2)) [33] as follows:

$$\Delta E_{00} = \sqrt{\left(\frac{\Delta L'}{K_L S_L}\right)^2 + \left(\frac{\Delta C'}{K_C S_C}\right)^2 + \left(\frac{\Delta H'}{K_H S_H}\right)^2 + RT \cdot \left(\frac{\Delta C'}{K_C S_C}\right) \cdot \left(\frac{\Delta H'}{K_H S_H}\right)} \quad (1)$$

$$WI_D = 0.55L^* - 2.32a^* - 1.100b^* \quad (2)$$

2.8. pH Analysis

The temporal evolution (10 min increments, total time = 30 min) of pH was determined for experimental bleaching gels (1 g of each, with or without nanoparticles) irradiated or not with visible light using a calibrated pH meter (AB150, Accumet, Fisher-Scientific, Hampton, NH, USA) to determine the impact of pH on the investigated properties. This analysis was carried out during the last bleaching session (T_3).

2.9. Mineral Content Evaluation

Infrared spectra of bovine enamel at T_0 and T_4 were acquired at three locations per specimen using a Fourier Transform Infrared spectrometer (Nicolet IS50, Thermo Fisher, Madison, WI, USA; scanning parameters: 500–4500 cm^{-1}; resolution 4 cm^{-1}, 10 internal scans per spectrum/location) coupled to a heated attenuated total reflectance (ATR) monolithic diamond crystal (Golden Gate, Specac, Fort Washington, PA, USA). A method previously described [7,34] was utilized to guarantee that the ATR-FTIR measurements were performed exactly at the same locations in each specimen. Enamel spectra (at T_0 and T_4) from each specimen were corrected for the presence of water before being subjected to baseline correction and normalization procedures using the OMNIC software (v7.0, Madison, WI, USA). The areas under the peaks corresponding to CO_3^{2-} υ_2 (886 cm^{-1}), PO_4^{3-} υ_1 (996 cm^{-1}), and PO_4^{3-} υ_2 (1410–1460 cm^{-1}) were calculated after experimental treatments (T_4). The mineral composition of enamel (in terms of the carbonate:phosphate mineral ratio) was determined by integrating the areas under the curves of CO_3^{2-} υ_2 and PO_4^{3-} (υ_1 and υ_2) [7].

2.10. Topography Assessment

An atomic force microscope (MultiMode with Nanoscope V controller, Bruker, Billerica, MA, USA) in ScanAsyst mode coupled with silicon nitride probes (aluminum-coated, triangular, radius = 2 nm, spring constant = 0.4 N/m, Bruker) was used to reveal topographical aspects of specimens (n = 1/group) at T_0 and T_4. Images (A = 625 μm^2; 512 × 512 lines) were acquired (at the same locations at T_0 and T_4) using a scan rate of 0.8 Hz. Images were then flattened before acquiring topographical parameters of interest (R_a (roughness average) and R_q (root mean square roughness)) using the Nanoscope software (v9.0, Bruker).

2.11. Metabolic Status of Non-Disrupted Biofilms

A minimally invasive, real-time, and high throughput bioluminescence assay recently reported by Esteban Florez et al. [27] determined the metabolic status of non-disrupted *Streptococcus mutans* biofilms grown on the surfaces of specimens only after being treated (T_4) and in accordance with experimental groups (n = 18/group) described in Section 2.2 (*Specimen Preparation and Experimental groups*). These specimens were prepared especially for this methodology. In brief, planktonic cultures of *Streptococcus mutans* (JM10) were grown overnight (16 h) in a liquid culture medium (THY) at oral temperature. Cultures having an optical density higher than 0.900 (at 600 nm; corresponding to 6.43 e^{+12} CFU/mL) were used as inoculum to grow biofilms. *S. mutans* biofilms were then grown (24 h, microaerophilic conditions, 37 °C) on the surfaces of sterile specimens (UV-sterilized, 254 nm, 800,000 $\mu J/cm^2$, UVP Crosslinker, model CL-1000, UVP, Fisher Scientific, Hampton, NH, USA) using inoculated biofilm growth media (0.65x THY, 1:50 dilution, 1.0 mL/well) supplemented with sucrose (1%, *w/v*). After 24 h, biofilms were immersed in 1.0 mL of a fresh 1x THY + 1% (*w/v*) glucose recharge medium and were incubated (37 °C, 1 h) before being transferred into the wells of sterile white 24-well plates containing 1.0 mL of a fresh 0.65x THY + 1% (*w/v*) sucrose medium. An aqueous solution (100 mM) of D-Luciferin suspended in a citrate buffer (0.1 M, pH 6.0) was added by a Synergy-HT multimode plate reader (Biotek, Winooski, VE, USA) to the wells containing both the specimens and biofilms in a recharge medium (2:1 ratio [*v/v*] inoculum to D-Luciferin). The metabolic activity of non-disrupted biofilms was assessed (in terms of RLUs) at 590 nm in 2 min increments (a total of 6 min) after the addition of D-Luciferin.

2.12. Staining and Confocal Microscopy

A concurrent staining method previously reported by Khajotia et al. [35] was used to illustrate the impact of experimental bleaching treatments on the 3-dimmensional distribution of nucleic acid, proteins, and extracellular polymeric substance (EPS). To achieve this goal, an additional set of specimens (n = 1/group) were prepared and bleached according to the methods previously described (Sections 2.2 and 2.6). Biofilms were grown on the surfaces of sterile specimens at T_4 using the methods described in Section 2.11. After the 24 h growth period, biofilms were washed with PBS (3x, pH 7.4, 25 °C, 15 s/wash) to remove non-adherent cells. Biofilms were then concurrently stained with Alexa Fluor® 647 conjugate of Concanavalin A (Invitrogen, Waltham, MA, USA; 250 µg/mL), Syto 9 (Molecular Probes, Eugene, OR, USA; 10 µM), and Sypro Red (Invitrogen, USA; 10x). Biofilms were kept hydrated in sterile, ultra-pure water and protected from light until confocal microscopy. Images of biofilms were acquired using a TCS-SP2 MP confocal laser scanning microscope (CLSM, Leica Microsystems, Inc., Buffalo Grove, IL, USA) with Ar (488 nm) and He/Ne (543 and 633 nm) lasers for the excitation of the fluorescent stains within biofilms at three different locations on each specimen's enamel surface. A 63x water immersion microscope objective lens was used. Serial optical sections were recorded from the surface of specimens to the top of biofilms at 0.6 µm intervals on the Z-axis. Three-dimensional images of the biofilms were generated using Volocity software(PerkinElmer, Waltham, MA, USA) to allow the visualization of the distribution of the nucleic acid (green fluorescence), proteins (red fluorescence), and EPS (blue fluorescence) components of biofilms.

2.13. Statistical Analyses

Linear Models (two- and three-way ANOVA) with outcomes including ΔE_{00} and ΔWI_D ($T_4 - T_0$) and the carbonate:phosphate mineral ratio (only at T_4) were fitted. The mineral ratio data were transformed into log. Factors included HP concentration (4 levels: without HP, 6% HP, 15% HP, and 35% HP), NP concentration (3 levels: without NP, 5% NP, and 10% NP), and LT (2 levels: with or without light). A backward model selection strategy was adopted with the full model containing all the main effects, all the two-way interactions (HP*NP, HP*LT, NP*LT), and the three-way interaction (HP*NP*light). A term was removed from the model if its *p*-value was less than 0.05 and the removal process started with the highest order term. In addition, RLU obtained from the metabolic status analysis of the biofilms were submitted to the general linear models test considering the group (G1–G20) and time (0, 2, 4, and 6 min) with post hoc Student–Newman–Keuls tests. All the analyses were conducted using SAS software (version 9.3, SAS Institute, Cary, NC, USA) at a 5% level of significance.

3. Results

3.1. Bleaching Efficacy

The findings reported in Figure 3 have demonstrated that experimental bleaching protocols modulated by gels containing 6%, 15%, and 35% H_2O_2 (without NP or LT) displayed mean values of ΔE_{00} and ΔWI_D that were higher when compared to those of the control groups (no treatment, with, or without LT). The two-way interactions among the factors are displayed in Figure 4. These results have also indicated that the ΔE_{00} and ΔWI_D values varied with HP concentrations ($p < 0.0001$), and bleaching outcomes could be ordered in terms of increasing efficacies where HP6 < HP15 < HP35, respectively. Even though a similar trend was observed when bleaching protocols were modulated by experimental bleaching gels containing 6%, 15%, and 35% HP (without NP) and visible light irradiation (405 nm $\pm$ 15 nm), the ΔE_{00} and ΔWI_D values were higher than those from bleaching protocols with no light irradiation ($p < 0.0001$). The combination of HP and NP further increased the efficacy of experimental bleaching protocols, as denoted by mean ΔE_{00} and ΔWI_D values of HP6 and HP15 incorporated with 5% NPs.

3.2. Analysis of pH

The graphs displayed in Figure 5 illustrate the temporal evolution of pH for experimental gels (6%, 15%, and 35% HP) with or without NP (5% and 10%). It is possible to observe that experimental gels (6%, 15%, and 35% HP), without the incorporation of NP, displayed pH values ($\cong$5.0, 6% HP + LT at 20 min) that were lower when compared to those containing NP. Such behavior was observed to be consistent throughout the observation time (at 0, 10, 20, and 30 min). The results reported have also indicated that such behavior is not influenced by visible light irradiation and that nanofilled bleaching gels displayed comparable pH values after 30 min.

3.3. Mineral Content Evaluation

The linear models showed that only the isolated factors, HP ($p < 0.0001$) and NP ($p < 0.0001$), were significant to the carbonate:phosphate mineral ratio variable. The LT factor and the two-way and three-way interactions were not significant (Figure 6). Figure 7 illustrates the impact of experimental bleaching gels on the mineral ratio CO_3^{2-}/PO_4^{3-} of bovine enamel. It is possible to observe a decrease of T_4 in all groups investigated where values varied from 0.14 ± 0.03 (6HP + LT) to 0.20 ± 0.05 (35HP + NP10% + LT) compared to the control groups (G1 and G2).

Figure 8A–T illustrates the results from the ATR-FTIR analysis of the mineral content of enamel before (baseline (T_0), black curves) and after bleaching (14 days after (T_4), red curves). Figure 8A indicates (in terms of normalized absorbance values) that specimens pertaining to G1 (negative control) displayed spectra characterized by absorbance values that were slightly higher at T_4 (wavenumbers from 800 cm^{-1} to 1150 cm^{-1} and from 1350 cm^{-1} to 1550 cm^{-1}) and by a larger spectral bandwidth at absorbance values between 0.0 and 0.5. In combination, these results indicate that the mineral content of enamel was not altered by storing specimens in artificial saliva for the duration of the experiment. Similar behavior was observed in Figure 8B, which demonstrate that the utilization of violet light irradiation (in the conditions tested) also did not promote any changes to the chemical composition of treated enamel.

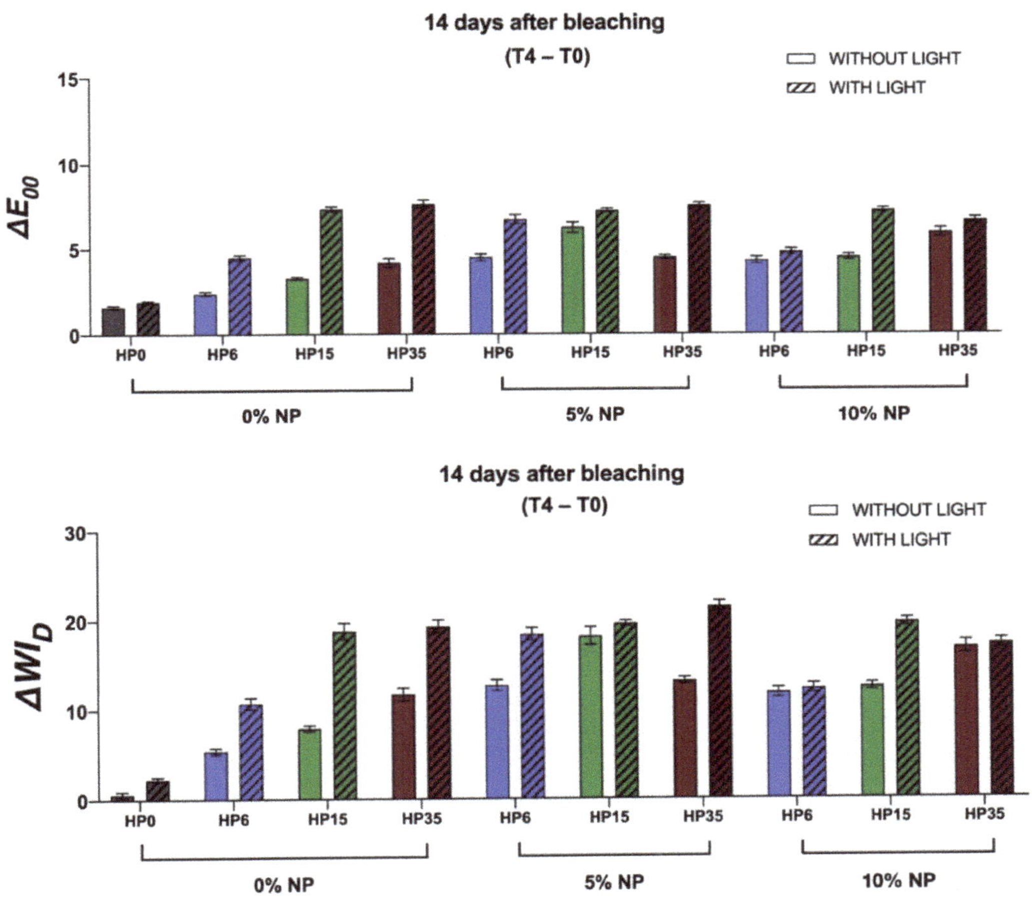

Figure 3. Mean and standard error values of ΔE_{00} and ΔWI_D that were calculated considering the coordinate values collected before (T_0) and 14 days (T_4) after the last bleaching session with 6%, 15%, and 35% HP incorporated or not with NP (either 5 or 10%).

Figure 8 (C, I, and O; HP (6%, 15%, and 35%)), (F, L, and R; HP (6%, 15%, and 35%) + LT), (D, J, and P; HP (6%, 15%, and 35%) + NP (5%)), (G, M, and S; HP (6%, 15%, and 35%) + NP (5%) + LT), (E, K, and O; HP (6%, 15%, and 35%) + NP (10%)), and (H, N, and T; (6%, 15%, and 35%) + NP (10%) + LT) illustrate the results for specimens that were subjected to experimental bleaching protocols. It is possible to observe that specimens treated with HP (either 6%, 15%, and 35%), with or without LT (Figure 8C,F,I,L,O,R), displayed spectra at T_4 that were characterized by lower absorbance values for the CO_3^{2-} υ_2 (886 cm^{-1}) and PO_4^{3-} υ_1 (996 cm^{-1}) peaks. In addition, it was possible to observe that the combination of HP and LT shifted the spectra to the right (wavenumbers between 800 cm^{-1} and 1150 cm^{-1}). Such behavior was more drastic for specimens treated with either 6% or 15% HP and LT (Figure 8F,L). This behavior was not observed in specimens treated with bleaching gels containing NP independent of light irradiation (with or without) or nanoparticles' concentrations (either 5% or 10%). In these instances, spectra were observed to display shapes and absorbance values that were similar to those of the control group (no treatment, stored in artificial saliva) at T_4.

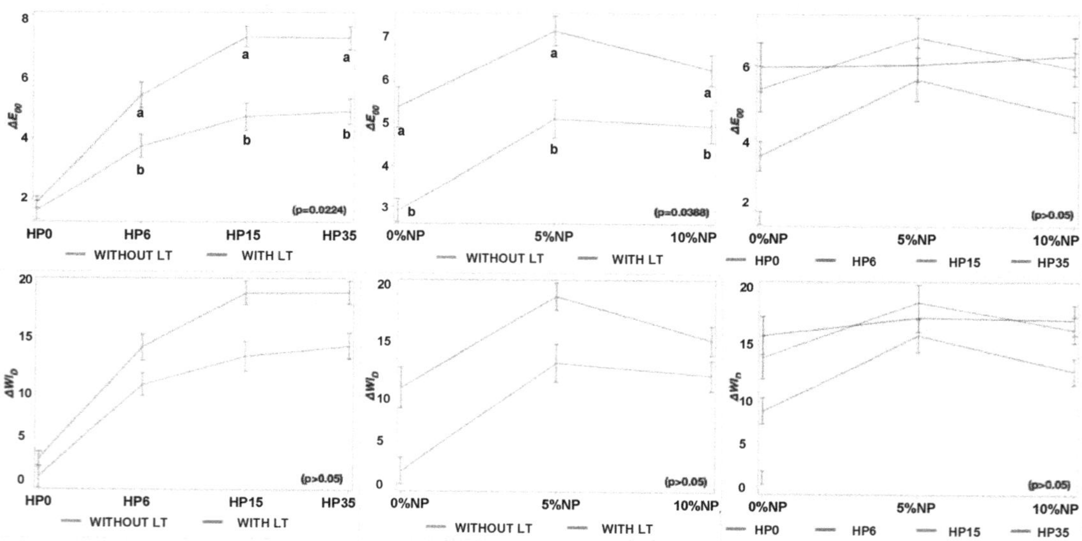

Figure 4. Marginal ΔE_{00} and ΔWI_D means and standard errors within each level of HP (0%, 6%, 15%, 35%), NP (0%, 5%, and 10%), and LT (with or without) according to two-way interactions, HP*LT, NP*LT, and NP*HP, respectively. Distinct letters represent a difference within the same HP or NP concentrations, taking into consideration a 0.05 level of significance.

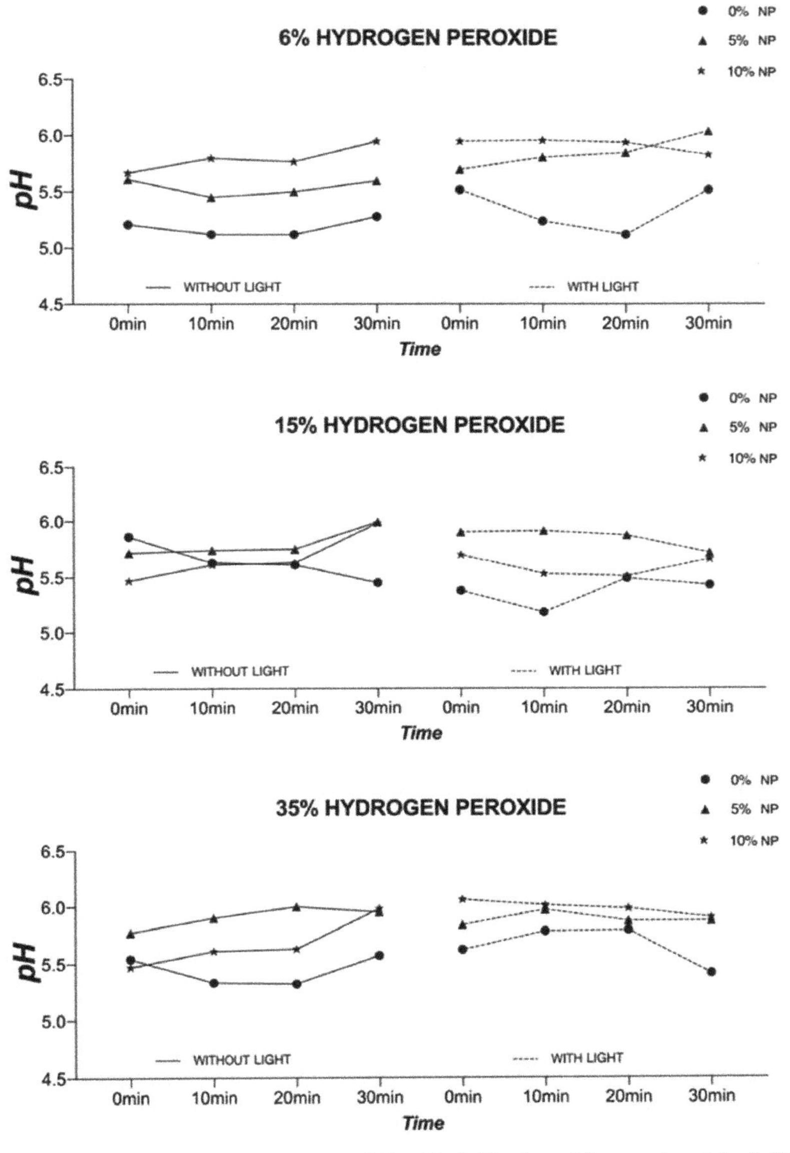

Figure 5. Temporal evolution (0, 10, 20, and 30 min) of pH values of the experimental gels (6%, 15%, and 35% HP) with or without NP and LT.

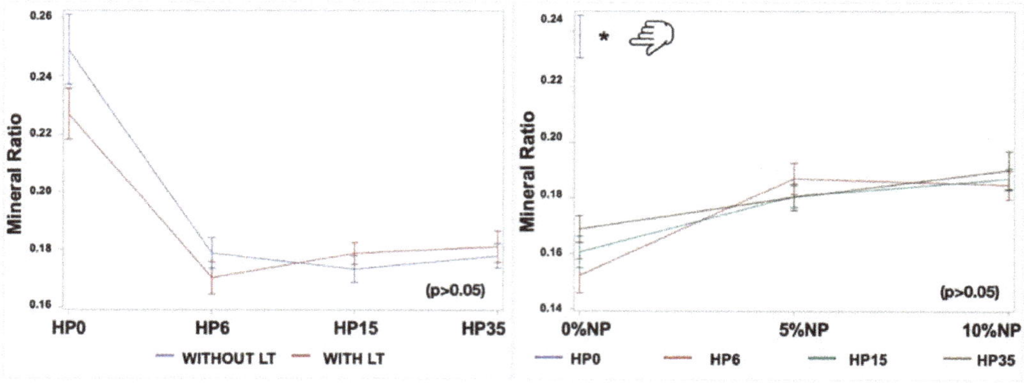

Figure 6. Carbonate:phosphate mineral ratio marginal means and standard errors within each level of HP (0%, 6%, 15%, 35%), NP (0%, 5% and 10%), and LT (with or without) according to the two-way interactions, HP*LT and NP*HP, respectively. The finger icon points to the control group (HP0), and the asterisk represents the difference between the control and the other groups, taking into consideration a 0.05 level of significance.

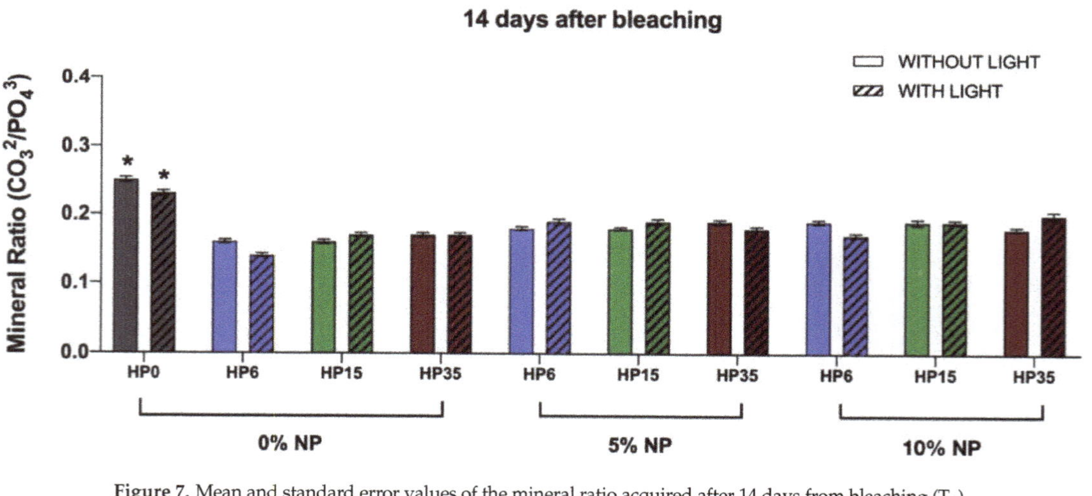

Figure 7. Mean and standard error values of the mineral ratio acquired after 14 days from bleaching (T_4) from the integrated areas of CO_3^{2-} υ_2 to PO_4^{3-} υ_1, υ_2 contours. The asterisks represent the difference between the control groups and the other ones, taking into consideration a 0.05 level of significance.

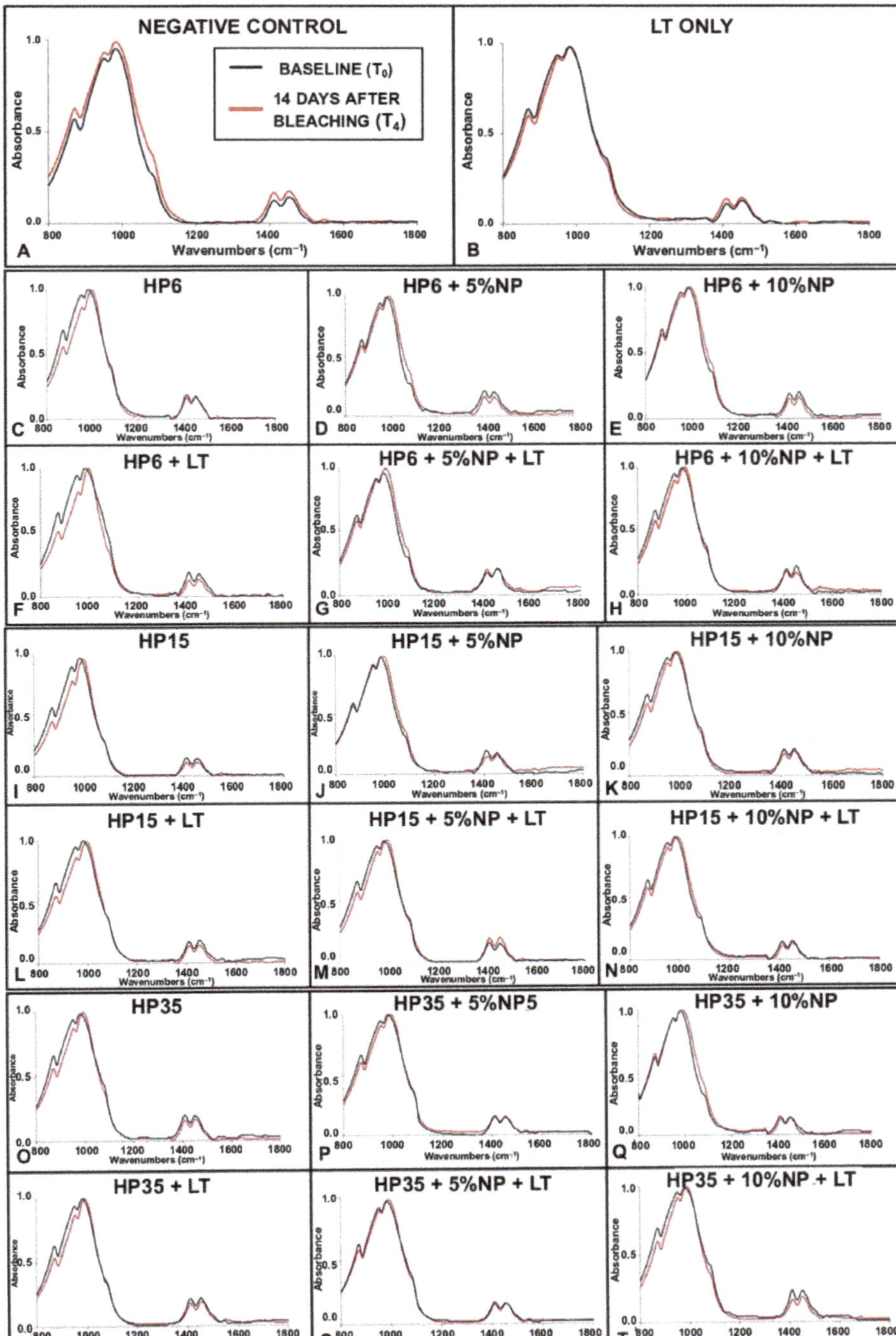

Figure 8. (**A–T**) ATR-FTIR spectra of specimens before (black curves, (T_0)) and 14 days after (red curves, (T_4)) being treated with experimental bleaching gels containing HP (6%, 15%, and 35%), with or without NF_TiO_2, and with or without visible light irradiation (LT).

3.4. Topography Assessment

Illustrative results from the topographical assessment performed with AFM are shown in Figure 9A–NN where it is possible to observe that R_a and R_q values varied from 1.5 nm (HP35 + 5%NP at T_4) to 19.6 nm (HP6 + 10%NP) and from 2.1 nm (HP35 + 5%NP) to 25.2 nm (HP6 + 10%NP), respectively. It is possible to observe through the images that, in most cases, the topography of the surfaces was maintained between T_0 and T_4. Overall, the surfaces were smooth and few of them illustrated the distribution and direction of enamel prisms.

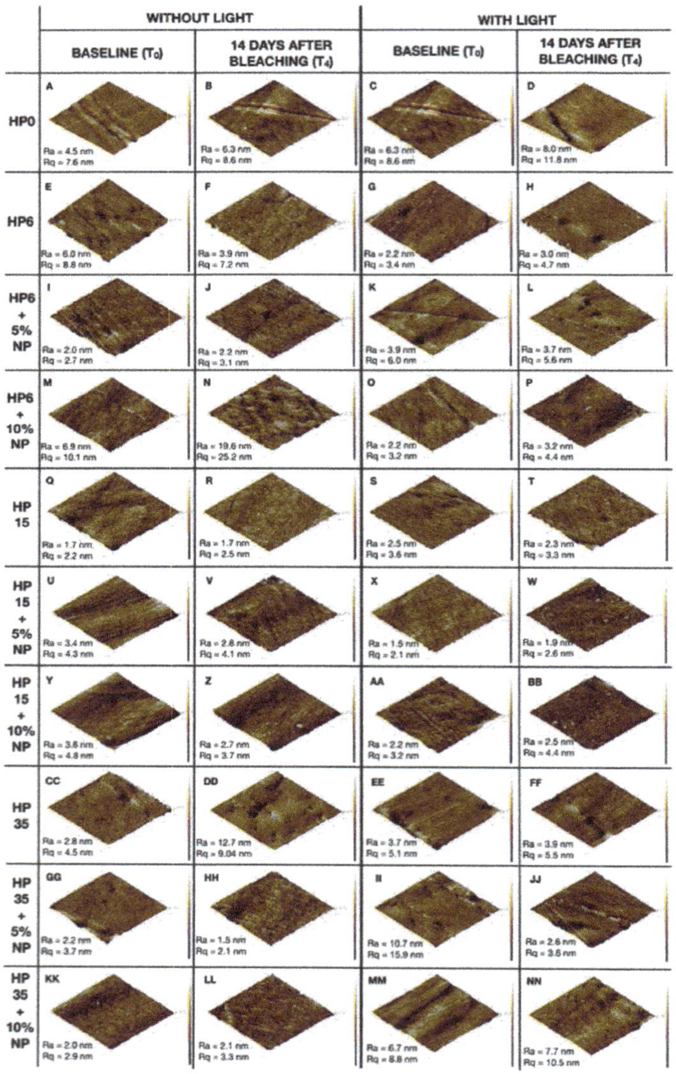

Figure 9. (**A–NN**) Images acquired using atomic force microscopy showing illustrative areas of the enamel before (T_0) and after 14 days (T_4) from bleaching with HP6, HP15, and HP35, with corresponding %NF_TiO_2, and the presence or absence of LT. Control groups are represented by HP0. Ra and Rq displayed in each subfigure indicate roughness average and root mean square roughness, respectively.

3.5. Metabolic Status of Non-Disrupted Biofilms

The general linear model evaluation detected significance for both the isolated factors of group and time ($p < 0.0001$), but no interaction was reported among them (group*time, ($p = 1.000$). Figure 10 illustrates the results, in terms of relative luminescence units (RLU) (mean and standard deviation values 6 min after the addition of D-Luciferin), of the metabolic status of non-disrupted *S. mutans* biofilms grown for 24 h on the surface of the specimens that were previously bleached. It is possible to observe that, except for G9 (HP6 + 5%NP), biofilms grown on specimens treated with experimental bleaching protocols modulated by gels containing HP (6%, 15%, and 35%), with (5% and 10%) or without NP and LT, displayed metabolic statuses that were either comparable to or higher than those observed on G1 (negative control).

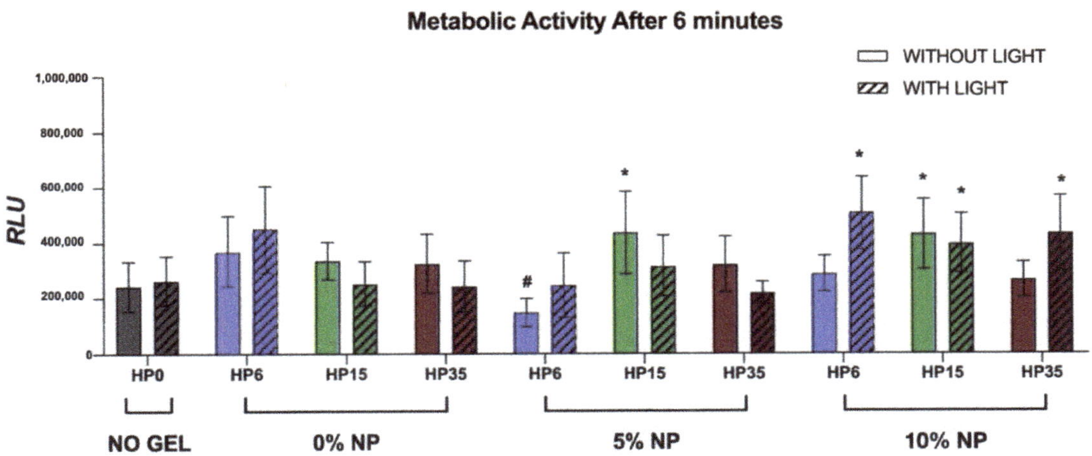

Figure 10. Mean and standard deviations of RLU values 6 min after the addition of D-Luciferin substrate to 24-h *Streptococcus Mutans* (JM10) biofilms. *S. mutans* biofilms were grown on the enamel of individual specimens 14 days after the last bleaching session (T_4). The hashtag sign denotes groups with RLU values that were statistically lower than those from the control group (no treatment-G1), while asterisks represent groups with RLU values that were statistically higher than those from G1.

3.6. Confocal Microscopy

The results from the concurrent staining and confocal laser scanning microscopy analysis are shown in Figure 11A–R as 3D reconstructions of biofilms where it is possible to observe green (nucleic acids), red (proteins), and blue (EPS) fluorescence channels, the impact of experimental bleaching treatments on components of biofilms, and their three-dimensional distributions. It was possible to detect in Figure 11A that biofilms expressed mostly green fluorescence when grown against the surfaces of specimens that were not treated with experimental bleaching gels (G1—negative control). It was possible to observe that biofilms grown on groups that were not light-irradiated surfaces were not only more porous but also displayed fluorescence signals that were mostly red and blue, which indicates that biofilms were expressing proteins and EPS more intensely. A clear trend, in terms of fluorescence signals (either green, red, or blue), could not be observed for specimens treated with experimental bleaching gels containing HP (6%, 15%, or 35%) with (5% and 10%) or without NP and LT.

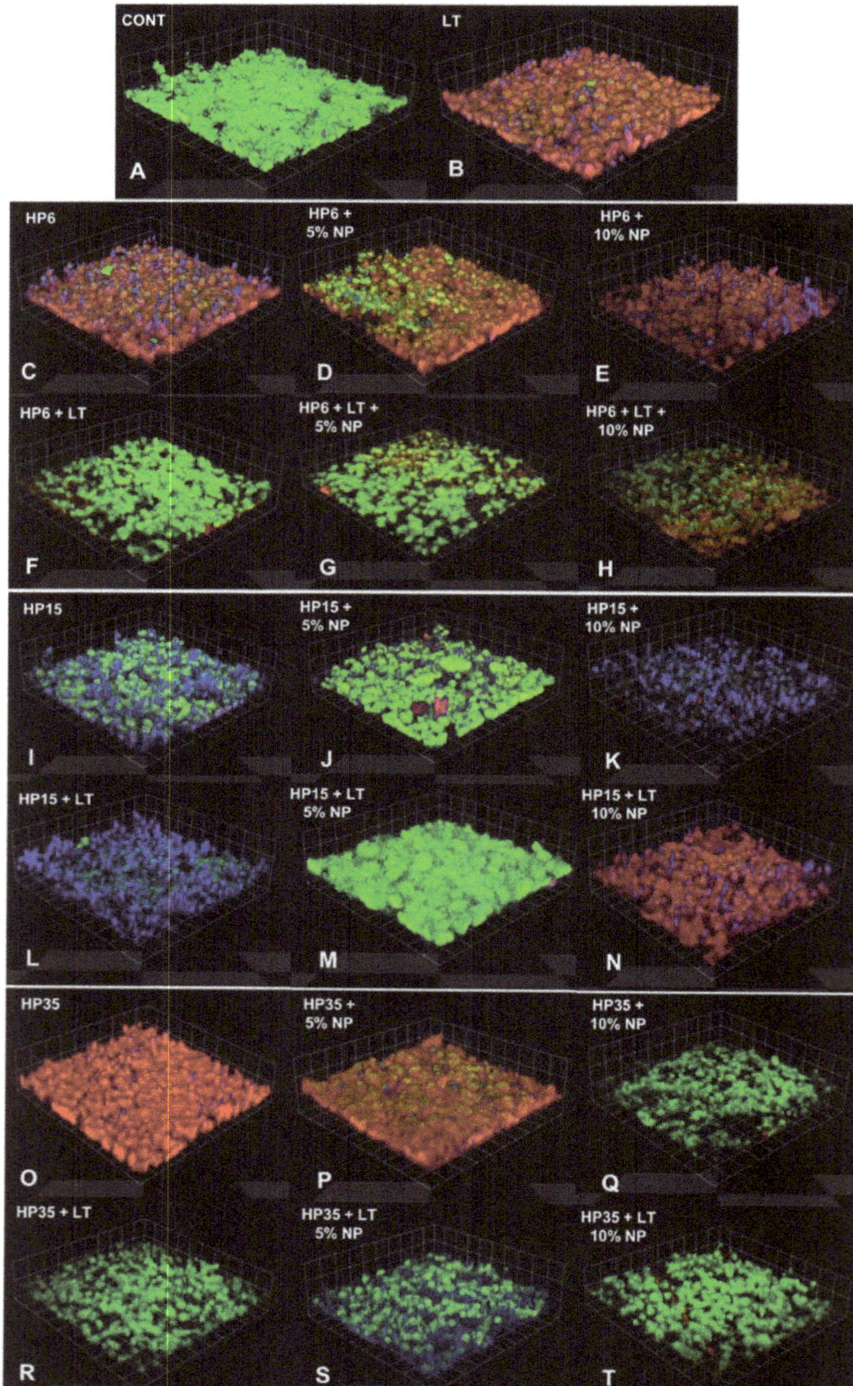

Figure 11. (**A–T**) Three-dimensional reconstructions of the biofilms with concurrent staining observed

using confocal laser scanning microscopy 14 days after (T_4) bleaching with 6%, 15%, and 35% HP, with or without the incorporation of 5% and 10% NP. It is possible to detect the distribution of the nucleic acid (green fluorescence), proteins (red fluorescence), and EPS (blue fluorescence) components of biofilms. Biofilms grown on surfaces bleached without LT were more porous, exhibiting mostly red and blue fluorescence signals. A clear trend, in terms of fluorescence signals (either green, red, or blue), could not be observed for specimens treated with experimental bleaching gels containing HP (6%, 15%, or 35%) with (5% and 10%) or without NP and LT.

4. Discussion

Even though the efficacy of IPB [31] has been previously demonstrated by numerous research groups, post-treatment DH continues to be the most frequently reported adverse effect [15–17]. In fact, a previous study [36] investigating the correlation between the bleaching efficacy and risk/intensity of post-treatment DH has indicated, based on a multi-regression and logistic analysis, that the risk for the occurrence of DH was 120% more likely to precipitate from IPB than from at-home bleaching techniques [36]. In addition, the intensity of painful symptoms was reported to be at least four times stronger for patients treated with IPB than those subjected to at-home treatments.

According to previous studies, the intensity (low, mild, and severe) and duration (short-term or long-term) of DH precipitate directly from the peroxide concentrations and exposure times used [17]. Therefore, the behavior reported [36] is expected because at-home bleaching gels are three-to-six times less concentrated than those used in IPB [1–3], and were demonstrated to be less cytotoxic and to penetrate less into the tooth structure [6,17], thereby diminishing potential risks associated with the vitality of pulpal tissues. Despite these promising results, at-home techniques require long exposure times and result in bleaching outcomes that are only similar to those achieved with IPB [37]. This critical scenario underscores the need for the development of techniques and products that are capable of resolving dental discolorations in a short period of time without causing DH or negatively impacting the properties of teeth (surface, mechanical, and biological).

The present research represents an effort to overcome the limitations cited by developing experimental bleaching gels containing low concentrations of HP, and third-generation titanium dioxide nanoparticles co-doped with nitrogen and fluorine, which have been demonstrated to have non-leaching antibacterial properties [25,27]. Results reported in the present study have demonstrated that the incorporation of NP (5% and 10%, *w/v*) into experimental bleaching gels containing low concentrations of HP (6% and 15%) rendered esthetic outcomes (in terms of ΔE_{00} and ΔWI_D) that were similar to those attained with high-concentrated bleaching gels (HP35), thereby suggesting that the nanotechnology proposed has the potential to resolve mild-to-severe dental discolorations in short periods of time (3 sessions, 30 min/session) and with a lower amount of hydrogen peroxide. Therefore, the first null hypothesis that NP incorporation would not significantly influence the efficacy of experimental bleaching gels was rejected. The utilization of violet light irradiation (LT; 405 nm $\pm$ 15 nm) was shown to improve the efficacy of experimental bleaching gels containing varying concentrations of HP (6%, 15%, and 35%) with or without NP (5% and 10%), as denoted by the statistical outcomes and the mean values of ΔE_{00} and ΔWI_D that were higher than those from experimental bleaching gels (with or without NP) in dark conditions, demonstrating that LT is still fundamentally important to achieving good esthetic outcomes.

Tano et al. [38], while investigating the effects of visible light irradiation (405 nm) emitted from a laser source on the efficacy of HP35 + TiO_2 (0.1% wt/wt) against a methylene blue solution (MB; 1.0 g, 100 ppm in 7.0 g of water), have demonstrated that even small concentrations of HP and TiO_2 can be efficiently used to bleach organic dyes when in the presence of visible light, thereby further supporting the necessity of using light irradiation and the present study's rationale for the selection of concentrations investigated (HP and NP).

Despite these promising in vitro results, findings from randomized clinical trials (RCTs) investigating the efficacy of HP6 with either TiO_2 [39] or N_TiO_2 [22,40] were less encouraging and have shown that experimental bleaching gels tested [22,39,40] produced less DH but were much less effective (in terms of bleaching outcomes) when compared to commercially available products (35% HP). These findings could have precipitated from the combination between the wavelength selected (450 nm ± 15 nm; 2.76 μeV) [22,40] and the bleaching gels containing nanoparticles fabricated by calcination strategies that behave as semiconductors and cannot generate ROS. In combination, the factors cited result in sub-optimal bleaching reactions and poor esthetic outcomes.

The experimental design of the present study was based on the utilization of shorter wavelengths (405 nm ± 15 nm) with higher photon energies (3.06 μeV) and nanoparticles (NF_TiO_2; 6–15 nm) that were synthesized using robust and green solvothermal reactions as previously reported by our group [25,27,28]. Nanoparticles reported herein were shown to have well-defined pore-size distributions, be electron deficient, be capable of producing substantial amounts of ROS [26], and result in experimental materials that are more efficient and less aggressive to the tooth structure. A previous study [25] demonstrated that single-doped nanoparticles (N_TIO_2), fabricated using similar synthetic routes, were capable of absorbing twice as much light (between 200 nm–800 nm) as compared to commercially available nano-TiO_2 (P25, Degussa) [25]. Since LT has been demonstrated by the present study to be fundamentally important for the success of IPB and, taking into consideration that the reported nanoparticles have the ability to intensely absorb visible wavelengths, it can be hypothesized that the investigated experimental protocols could result in good esthetic outcomes.

This hypothesis has been corroborated by findings reported in a recent randomized, controlled, and double-blind clinical trial [41] that investigated the clinical performance (in terms of immediate ΔE_{ab}) of 6% HP with N_TiO_2 when activated by two distinct wavelengths (405 ± 15 nm and 450 ± 15 nm). The results of that study [41] show that the clinical color change of experimental bleaching gels was less pronounced when using blue irradiation (450 ± 15 nm). The authors have also reported that bleaching protocols modulated by violet radiation displayed bleaching efficacies that were comparable to the control group (35% of HP) but were associated with lower levels of DH. Even though the trends observed [41] validate the results of the present study, it is important to underscore that our experimental design was based on the utilization of the CIEDE2000 formula (ΔE_{00}) [32] and the whiteness index (ΔWI_D), which are considered more relevant for dentistry and dental bleaching investigations.

According to Paravina et al. [42], the mean ΔE_{00} values reported for G10 (HP6 + NP5 + LT) are considered excellent and indicate that experimental protocols modulated by LT and 5% NP resulted in bleaching efficacies that were comparable to those from HP15 (G6, G11, G12) and HP35 (G8, G14, G19, G20). These findings could have precipitated from the spontaneous dissociation that hydrogen peroxide undergoes when in the presence of metaloxides such as NF_TiO_2 and suggest that experimental bleaching gels containing low concentrations of NP and HP may result in materials with promising bleaching performances. The higher mean values of ΔE_{00} and ΔWI_D detected in G11 (HP15 + NP5) serve as additional evidence of such a spontaneous dissociation process. This was expected behavior and has been corroborated by previous research from our group that demonstrated that photocatalysts of similar compositions displayed promising antibacterial properties against *Streptococcus mutans* even in the absence of light irradiation [27].

With regard to the temporal evolution of pH, it was possible to observe that the incorporation of NP (5% and 10%) into experimental bleaching gels containing HP (6%, 15%, and 35%) did not adversely impact the values observed. For experimental gels containing 6% HP, the NP incorporation (either 5% or 10%) resulted in pH values that were higher when compared to gels without NP (with or without LT). Gels containing either 15% or 35% of HP, displayed pH values that fluctuated a bit more, but overall, the incorporation of NP seemed to have a stabilizing effect that prevented the acidification of experimental materials during the investigated period of time (30 min). Our findings contradict the results published by Pretel et al. [43], because, even though commercial gels modified by the incorporation of N_TiO_2 (produced by calcination strategies) displayed initial pH values that were high, the materials investigated [43] displayed pH values that were significantly lower after 30 min of observation. On the other hand, Monteiro et al. [44] have shown that the incorporation of 1% TiO_2 nanotubes did not adversely impact the pH values of gels containing either 10% carbamide peroxide (pH $\cong$ 6.5) or 40% HP (pH $\cong$ 7.0). These differences in pH values reported in the literature and by the present study (initial and after 30 min) can be explained by the intrinsic differences in materials' compositions including the type of polymeric matrix and stabilizing agents.

A recent study [43] investigating temporal variations of pH and the electric potential (EP) of three commercially available bleaching gels has demonstrated that there is a strong and positive correlation between pH and EP values, where EP was inversely varied with the evolution of pH. Gentil de Moor et al. [45] have indicated that ROS such as oxygen, hydroperoxyl, sodium hypochlorite, hydrogen peroxide, ozone, and hydroxyl have redox potentials of +1.229 V, +1.510 V, +1.630 V, +1.780 V, +2.075, and +2.800 V, respectively, and therefore, the bleaching gels proposed in the present study are hypothesized to preferentially generate oxygen and hydroperoxyl and to prevent the etching of treated enamel.

In our study, a significant decrease in the mineral ratio was detected for all bleached groups in comparison to the control ones independent of the NP incorporation. Therefore, the second null hypothesis that the NPs would not negatively influence the chemical content of enamel bleached with the experimental gels was rejected. However, the results of the present study suggest a trend of increasing the mineral ratio of enamel bleached with gels containing NP (5% and 10%), which should be further confirmed with additional mechanical and surface testing. Xu et al. [46] already demonstrated the presence of a strong and inverse relationship between the carbonate:phosphate ratio and surface properties of enamel (elastic modulus and hardness). Even though the location of the ATR-FTIR evaluation over time (T_0 and T_4) was standardized, the specimens were stored in saliva for 14 days after bleaching as an additional attempt to mimic, as much as possible, the clinical condition [9,10]. As it has already been demonstrated, the presence of saliva may either uphold or recover the levels of mineralization of enamel after tooth bleaching [47].

Another valuable piece of information provided by the ATR-FTIR assessment was the spectra of enamel, demonstrating that experimental bleaching gels containing HP (6%, 15%, and 35%) but without NP (which are more acidic) have negatively impacted the chemical make-up of treated enamel independently of LT, as denoted by a right-shift of the spectrum between wavenumbers between 800 cm^{-1} and 1150 cm^{-1} and lower absorbance values for $CO_3{}^{2-}$ υ_2 (886 cm^{-1}) and $PO_4{}^{3-}$ υ_1 (996 cm^{-1}). As demonstrated in Figure 5, this behavior was overall not observed in specimens treated with experimental bleaching gels containing NP (either 5% or 10%). In these instances, spectra were observed to have shapes and absorbance values that were similar to those of the control group (no treatment). Although some authors have reported distortions in the peaks mentioned independently of pH values [46,48], Sun et al. [7] have also demonstrated that acidic bleaching agents (30% HP, pH $\cong$ 3.6) not only decreased $CO_3{}^{2-}$ υ_2 (886 cm^{-1}) and $PO_4{}^{3-}$ υ_2 (1410–1460 cm^{-1}) absorbance values but have also right-shifted the spectra (between 800 cm^{-1} and 1150 cm^{-1}) of treated enamel, thereby further corroborating the findings of the present study. In their study [7], the decrease in the mineral content ($CO_3{}^{2-}/PO_4{}^{3-}$) was followed by significant reductions in the enamel's surface microhardness. This concerning trend was not detected when the authors used experimental bleaching agents with a neutral pH [7].

It is important to highlight that the evaluation of the carbonate ($CO_3{}^{2-}$) content provides valuable information to the field because $CO_3{}^{2-}$ represents 2.5% of the weight of the enamel's composition, behaves as a substitute anion for phosphate or hydroxyl groups in hydroxyapatite [45] (HAp $[Ca_{10}(PO_4)_6(OH)_2]$), and is not stoichiometrically distributed in dental enamel [46]. In addition, the carbonate's position (at hydroxyapatite's lattice or surface) not only modifies its shape and size but, more importantly, increases the solubility of carbonated apatites [49,50]. This physico-chemical property of hydroxyapatite could be used to explain the reason why experimental bleaching gels without nanoparticles significantly decreased the mineral ratio of treated enamel. We hypothesize that the trend of the increasing mineral ratio observed when specimens were treated with gels containing NP (5% and 10%) could be explained by the presence of fluorine ions in the crystal lattice of NF_TiO_2 that could potentially alter the composition of hydroxyapatite (into fluorapatite) through an ion-exchange mechanism. Even though the mechanism is not energetically favorable, the formation of fluorapatite could be responsible for the maintenance of the chemical make-up observed once fluorapatite displays solubility levels that are lower than those of carbonated apatites when exposed to acidic pH values [49,50].

Even though large variations in R_a and R_q values could be correlated with intense damage to the enamel, the vast majority of our results have indicated that treatments conducted with or without LT, and modulated by experimental gels (6%, 15%, and 35%) with (5% and 10%) or without NP, were not able to adversely impact the surface topography of treated enamel at T_4, and therefore, the results from the group treated with 6HP+10%NP were considered to not follow the overall trend observed. Additional analyses based on optical profilometry will include a quantitative evaluation of enamel's surface roughness. Contrary to these findings, others [7,12] have shown that enamel surface properties were negatively impacted by bleaching protocols modulated by high-concentrated bleaching gels (either 30% or 40% HP), independently of their pH values (either acidic or neutral), using similar AFM techniques. Therefore, it could be assumed that the compositions tested have the potential to resolve mild-to-severe dental discolorations without negatively impacting the chemical arrangement or surface properties of treated enamel.

These findings are also important from the oral microbiology standpoint because it is well known that surfaces (either biotic or abiotic) with high mean R_a values accumulate more biofilms [51], due to increased surface area and surface energy, and may alter the ecology of biofilms from a state of health into a disease-associated state. The real-time and high throughput bioluminescence assay performed in the present study had the objective of determining (i) if treated enamel surfaces would display a latent antibacterial behavior and (ii) if treated surfaces would support more biofilm growth. As demonstrated by the results shown in Figure 10, experimental bleaching treatments modulated by HP (6%, 15%, and 35%) and NP (5% and 10%) were not capable of rendering antibacterial effects to treated surfaces against non-disrupted biofilms of *Streptococcus mutans* independently of LT, as denoted by RLU values that were either similar or higher to those from the control group (no treatment). Therefore, the last null hypothesis that the incorporation of NP would not avoid the growth of biofilm after bleaching was accepted. In the past, Ittatiruti et al. [51] demonstrated that, by using the viable colony counting assay (CFU/mL), dental bleaching procedures modulated by bleaching gels containing either 25% or 35% of HP did not promote higher accumulations of *S. Mutans* but increased the *S. Sanguinis* biofilm formation.

The only exception to the trend observed in our work was in specimens from G9 (HP6 + NP5) that displayed biofilms with RLU values that were slightly lower than those from the control group (no treatment). These findings suggest that enamel surfaces treated with HP6 + 5%NP could potentially become antibacterial by the utilization of the nanotechnology proposed, but additional studies are necessary to confirm these findings and elucidate potential mechanisms of action associated. The qualitative results from the confocal microscopy analysis (Figure 11) have indicated that enamel surfaces treated with experimental bleaching protocols modulated by HP (6%, 15%, and 35%) and NP (5% and 10%) with or without LT, promoted the growth of biofilms displaying different three-dimensional distributions of nucleic acids, proteins, and EPS, thereby suggesting that bleached enamel may indeed impact the accumulation and growth of oral biofilms. However, these results should be interpreted with caution because the qualitative data reported cannot be considered representative due to the small number of specimens analyzed.

Subsequent studies from our group will investigate the mechanisms of action by which the proposed nanoparticles (i) improve the efficacy of experimental bleaching gels, (ii) maintain the mineral content of treated enamel, and (iii) modulate the three-dimensional distribution of biofilms' components. Our group is also planning the execution of a controlled, randomized, and double-blind clinical trial to determine the clinical efficacy of experimental bleaching gels proposed herein.

5. Conclusions

The present study has successfully demonstrated the synthesis of experimental bleaching gels using hydrophilic polymers and functionalized nanoparticles. The nanotechnology reported was demonstrated to significantly improve the bleaching efficacy of experimental materials independent of hydrogen peroxide or light irradiation and did not adversely impact the surface properties or chemical make-up of treated enamel. The results of the present study have shown that experimental materials were not capable of rendering antibacterial effects to treated enamel but were observed to alter the three-dimensional distribution of components within *S. mutans* biofilms. Subsequent studies should not only investigate the mechanisms of action by which the proposed nanotechnology improves bleaching reactions but also demonstrate the clinical efficacy of the proposed materials.

Author Contributions: Conceptualization, M.K., F.L.E.F., and V.C.; methodology, M.K., R.D.H., and M.Z.D.P.; formal analysis, Y.D.Z. and S.S.K.; investigation, M.K.; resources, J.H., S.S.K., F.L.E.F., and V.C.; data curation, M.K.; writing—original draft preparation, M.K.; writing—review and editing, F.L.E.F. and V.C.; visualization, F.L.E.F. and V.C.; supervision, F.L.E.F. and V.C.; project administration, F.L.E.F. and V.C.; funding acquisition, J.H., F.L.E.F., and V.C. All authors have read and agreed to the published version of the manuscript.

Funding: This research was funded by São Paulo State Research Foundation (FAPESP) [#2019/02393-6 (M.K.) and #2020/06782-4 (V.C.)], by the Oklahoma Center for the Advancement of Science and Technology (HR20-121; F.L.E.F.), the OUCOD Faculty Research Pilot Program (J.H.), and the National Institutes of Health (NIGMS, #1U54GM104938). The authors gratefully acknowledge Fulbright Brazil, Fulbright Scholarship Board, and The Bureau of the Educational and Cultural Affairs of the United States Department of State for a scholarship granted to the first author (M.K.) who participated in the program Doctoral Dissertation Research Award. This study was also supported in part by Coordenação de Aperfeiçoamento de Pessoal do Nível Superior (CAPES)-001.

Institutional Review Board Statement: Not applicable.

Informed Consent Statement: Not applicable.

Data Availability Statement: Not applicable.

Conflicts of Interest: The authors declare no conflict of interest.

References

1. Rodríguez-Martínez, J.; Valiente, M.; Sánchez-Martín, M. Tooth whitening: From the established treatments to novel approaches to prevent side effects. *J. Esthet. Restor. Dent.* **2019**, *31*, 431–440. [CrossRef] [PubMed]
2. Kury, M.; Lins, R.B.E.; Resende, B.D.A.; Picolo, M.Z.D.; André, C.B.; Cavalli, V. The influence of the renewal or the single application of the peroxide gel on the efficacy and tooth sensitivity outcomes of in-office bleaching—A systematic review and meta-analysis. *J. Esthet. Restor. Dent.* **2021**, *34*, 490–502. [CrossRef] [PubMed]
3. Kwon, S.R.; Wertz, P.W. Review of the Mechanism of Tooth Whitening. *J. Esthet. Restor. Dent.* **2015**, *27*, 240–257. [CrossRef] [PubMed]
4. Basting, R.T.; Amaral, F.L.; França, F.M.; Flório, F.M. Clinical comparative study of the effectiveness of and tooth sensitivity to 10% and 20% carbamide peroxide home-use and 35% and 38% hydrogen peroxide in-office bleaching materials containing desensitizing agents. *Oper. Dent.* **2012**, *37*, 464–473. [CrossRef]
5. Pinto, A.; Bridi, E.C.; Amaral, F.; França, F.; Turssi, C.P.; Pérez, C.A.; Martinez, E.F.; Flório, F.M.; Basting, R.T. Enamel Mineral Content Changes After Bleaching With High and Low Hydrogen Peroxide Concentrations: Colorimetric Spectrophotometry and Total Reflection X-ray Fluorescence Analyses. *Oper. Dent.* **2017**, *42*, 308–318. [CrossRef]
6. Cavalli, V.; da Silva, B.G.; Berger, S.B.; Marson, F.C.; Tabchoury, C.P.M.; Giannini, M. Decomposition Rate, pH, and Enamel Color Alteration of At-Home and In-Office Bleaching Agents. *Braz. Dent. J.* **2019**, *30*, 385–396. [CrossRef]
7. Sun, L.; Liang, S.; Sa, Y.; Wang, Z.; Ma, X.; Jiang, T.; Wang, Y. Surface alteration of human tooth enamel subjected to acidic and neutral 30% hydrogen peroxide. *J. Dent.* **2011**, *39*, 686–692. [CrossRef]
8. Kemaloğlu, H.; Tezel, H.; Ergücü, Z. Does post-bleaching fluoridation affect the further demineralization of bleached enamel? An in vitro study. *BMC Oral Health* **2014**, *14*, 113. [CrossRef]

9. Cavalli, V.; Da Rosa, D.A.; Da Silva, D.P.; Kury, M.; Liporoni, P.C.S.; Soares, L.E.S.; Martins, A.A. Effects of experimental bleaching agents on the mineral content of sound and demineralized enamels. *J. Appl. Oral Sci.* **2018**, *26*, e20170589. [CrossRef]
10. Kury, M.; Antonialli, F.M.; Soares, L.E.S.; Tabchoury, C.P.M.; Giannini, M.; Florez, F.L.E.; Cavalli, V. Effects of violet radiation and nonthermal atmospheric plasma on the mineral contents of enamel during in-office dental bleaching. *Photodiagnosis Photodyn. Ther.* **2020**, *31*, 101848. [CrossRef]
11. Grazioli, G.; Valente, L.L.; Isolan, C.P.; Pinheiro, H.A.; Duarte, C.G.; Münchow, E.A. Bleaching and enamel surface interactions resulting from the use of highly-concentrated bleaching gels. *Arch. Oral Biol.* **2018**, *87*, 157–162. [CrossRef] [PubMed]
12. Bilge, K.; Kılıç, V. Effects of different remineralizing agents on color stability and surface characteristics of the teeth following vital bleaching. *Microsc. Res. Tech.* **2021**, *84*, 2206–2218. [CrossRef] [PubMed]
13. Olmedo, D.E.R.P.; Kury, M.; Resende, B.A.; Cavalli, V. Use of antioxidants to restore bond strength after tooth bleaching with peroxides. *Eur. J. Oral Sci.* **2021**, *129*, e12773. [CrossRef] [PubMed]
14. Da Silva, A.P.; de Oliveira, R.; Cavalli, V.; Arrais, C.A.; Giannini, M.; de Carvalho, R.M. Effect of peroxide-based bleaching agents on enamel ultimate tensile strength. *Oper. Dent.* **2005**, *30*, 318–324. [PubMed]
15. Ortega-Moncayo, M.G.; Aliaga-Sancho, P.; Pulido, C.; Gutierrez, M.F.; Rodriguez-Salazar, E.; Burey, A.; León, K.; Román-Oñate, Y.; Arrais, C.A.G.; Loguercio, A.D.; et al. Is the use of a potassium nitrate dentifrice effective in reducing tooth sensitivity related to in-office bleaching? A randomized triple-blind clinical trial. *J. Esthet. Restor. Dent.* **2021**. [CrossRef] [PubMed]
16. Piknjač, A.; Soldo, M.; Illeš, D.; Zlatarić, D.K. Patients' Assessments of Tooth Sensitivity Increase One Day Following Different Whitening Treatments. *Acta Stomatol. Croat.* **2021**, *55*, 280–290. [CrossRef] [PubMed]
17. Soares, D.G.; Basso, F.G.; Hebling, J.; Costa, C.A.D.S. Concentrations of and application protocols for hydrogen peroxide bleaching gels: Effects on pulp cell viability and whitening efficacy. *J. Dent.* **2014**, *42*, 185–198. [CrossRef]
18. Chen, C.; Huang, X.; Zhu, W.; Ding, C.; Huang, P.; Li, R. H_2O_2 gel bleaching induces cytotoxicity and pain conduction in dental pulp stem cells via intracellular reactive oxygen species on enamel/dentin disc. *PLoS ONE* **2021**, *16*, e0257221. [CrossRef]
19. Kury, M.; Wada, E.E.; Palandi, S.D.S.; Picolo, M.Z.D.; Giannini, M.; Cavalli, V. Colorimetric evaluation after in-office tooth bleaching with violet LED: 6- and 12-month follow-ups of a randomized clinical trial. *Clin. Oral Investig.* **2021**, *26*, 837–847. [CrossRef]
20. Gallinari, M.D.O.; Cintra, L.T.A.; Barboza, A.C.S.; da Silva, L.M.A.V.; de Alcantara, S.; dos Santos, P.H.; Fagundes, T.C.; Briso, A.L.F. Evaluation of the color change and tooth sensitivity in treatments that associate violet LED with carbamide peroxide 10%: A randomized clinical trial of a split-mouth design. *Photodiagnosis Photodyn. Ther.* **2020**, *30*, 101679. [CrossRef]
21. Kobayashi, R.S.; Picolo, M.Z.D.; Kury, M.; Resende, B.D.A.; Florez, F.L.E.; Cavalli, V. Effects of dental bleaching protocols with violet radiation on the color and chemical composition of stained bovine enamel. *Photodiagnosis Photodyn. Ther.* **2021**, *34*, 102194. [CrossRef] [PubMed]
22. Bortolatto, J.F.; Trevisan, T.C.; Bernardi, P.S.; Fernandez, E.; Dovigo, L.N.; Loguercio, A.D.; Junior, O.B.d.; Pretel, H. A novel approach for in-office tooth bleaching with 6% H_2O_2/TiO_N and LED/laser system-a controlled, triple-blinded, randomized clinical trial. *Lasers Med. Sci.* **2016**, *31*, 437–444. [CrossRef] [PubMed]
23. Bortolatto, J.F.; Pretel, H.; Floros, M.C.; Luizzi, A.C.; Dantas, A.A.; Fernandez, E.; Moncada, G.; de Oliveira, O.B., Jr. Low Concentration H_2O_2/TiO_N in Office Bleaching: A Randomized Clinical Trial. *J. Dent. Res.* **2014**, *93* (Suppl. 7), 66s–71s. [CrossRef] [PubMed]
24. Cuppini, M.; Leitune, V.C.B.; de Souza, M.; Alves, A.K.; Samuel, S.M.W.; Collares, F.M. In vitro evaluation of visible light-activated titanium dioxide photocatalysis for in-office dental bleaching. *Dent. Mater. J.* **2019**, *38*, 68–74. [CrossRef] [PubMed]
25. Florez, F.L.E.; Hiers, R.D.; Larson, P.; Johnson, M.; O'Rear, E.; Rondinone, A.J.; Khajotia, S.S. Antibacterial dental adhesive resins containing nitrogen-doped titanium dioxide nanoparticles. *Mater. Sci. Eng. C* **2018**, *93*, 931–943. [CrossRef] [PubMed]
26. Huo, Y.; Bian, Z.; Zhang, X.; Jin, Y.; Zhu, J.; Li, H. Highly Active $TiO_{2-x}N_x$ Visible Photocatalyst Prepared by N-Doping in Et3N/EtOH Fluid under Supercritical Conditions. *J. Phys. Chem. C* **2008**, *112*, 6546–6550. [CrossRef]
27. Florez, F.L.E.; Hiers, R.D.; Zhao, Y.; Merritt, J.; Rondinone, A.J.; Khajotia, S.S. Optimization of a real-time high-throughput assay for assessment of Streptococcus mutans metabolism and screening of antibacterial dental adhesives. *Dent. Mater.* **2020**, *36*, 353–365. [CrossRef]
28. Florez, F.L.E.; Trofimov, A.; Ievlev, A.; Qian, S.; Rondinone, A.J.; Khajotia, S.S. Advanced characterization of surface-modified nanoparticles and nanofilled antibacterial dental adhesive resins. *Sci. Rep.* **2020**, *10*, 9811. [CrossRef]
29. Palandi, S.; Kury, M.; Picolo, M.Z.D.; Coelho, C.S.S.; Cavalli, V. Effects of activated charcoal powder combined with toothpastes on enamel color change and surface properties. *J. Esthet. Restor. Dent.* **2020**, *32*, 783–790. [CrossRef]
30. Eskelsen, E.; Catelan, A.; Hernades, N.M.A.P.; Soares, L.E.S.; Cavalcanti, A.N.; Aguiar, F.H.B.; Liporoni, P.C.S. Physicochemical changes in enamel submitted to pH cycling and bleaching treatment. *Clin. Cosmet. Investig. Dent.* **2018**, *10*, 281–286. [CrossRef]
31. Joiner, A.; Luo, W. Tooth colour and whiteness: A review. *J. Dent.* **2017**, *67*, S3–S10. [CrossRef] [PubMed]
32. Sharma, G.; Wu, W.; Dalal, E.N. The CIEDE2000 color-difference formula: Implementation notes, supplementary test data, and mathematical observations. *Color Res. Appl.* **2004**, *30*, 21–30. [CrossRef]
33. Pérez, M.D.M.; Ghinea, R.; Rivas, M.J.; Yebra, A.; Ionescu, A.M.; Paravina, R.D.; Herrera, L.J. Development of a customized whiteness index for dentistry based on CIELAB color space. *Dent. Mater.* **2016**, *32*, 461–467. [CrossRef]
34. Bistey, T.; Nagy, I.P.; Simó, A.; Hegedűs, C. In vitro FT-IR study of the effects of hydrogen peroxide on superficial tooth enamel. *J. Dent.* **2007**, *35*, 325–330. [CrossRef]

35. Khajotia, S.S.; Smart, K.H.; Pilula, M.; Thompson, D.M. Concurrent quantification of cellular and extracellular components of biofilms. *J. Vis. Exp.* **2013**, *82*, e50639. [CrossRef] [PubMed]
36. Rezende, M.; Loguercio, A.D.; Kossatz, S.; Reis, A. Predictive factors on the efficacy and risk/intensity of tooth sensitivity of dental bleaching: A multi regression and logistic analysis. *J. Dent.* **2015**, *45*, 1–6. [CrossRef]
37. De Geus, J.L.; Wambier, L.M.; Kossatz, S.; Loguercio, A.D.; Reis, A. At-home vs In-office Bleaching: A Systematic Review and Meta-analysis. *Oper. Dent.* **2016**, *41*, 341–356. [CrossRef]
38. Tano, E.; Otsuki, M.; Kato, J.; Sadr, A.; Ikeda, M.; Tagami, J. Effects of 405 nm Diode Laser on Titanium Oxide Bleaching Activation. *Photomed. Laser Surg.* **2012**, *30*, 648–654. [CrossRef]
39. Vildósola, P.; Bottner, J.; Avalos, F.; Godoy, I.; Martín, J.; Fernández, E. Teeth bleaching with low concentrations of hydrogen peroxide (6%) and catalyzed by LED blue (450 ± 10 nm) and laser infrared (808 ± 10 nm) light for in-office treatment: Randomized clinical trial 1-year follow-up. *J. Esthet. Restor. Dent.* **2017**, *29*, 339–345. [CrossRef]
40. Martín, J.; Vildósola, P.; Bersezio, C.; Herrera, A.; Bortolatto, J.; Saad, J.R.; Oliveira, O.B., Jr.; Fernández, E. Effectiveness of 6% hydrogen peroxide concentration for tooth bleaching—A double-blind, randomized clinical trial. *J. Dent.* **2015**, *43*, 572–965. [CrossRef]
41. Trevisan, T.C.; Bortolatto, J.F.; Rizzi, G.; Meloto, B.T.; Dantas, A.A.R.; Junior, O.B.D.O. Clinical performance of 6% hydrogen peroxide containing TiO_2N nanoparticles activated by LED in varying wavelengths—A randomized clinical trial. *Lasers Med. Sci.* **2021**, *37*, 2017–2024. [CrossRef] [PubMed]
42. Paravina, R.D.; Pérez, M.M.; Ghinea, R. Acceptability and perceptibility thresholds in dentistry: A comprehensive review of clinical and research applications. *J. Esthet. Restor. Dent.* **2018**, *31*, 103–112. [CrossRef] [PubMed]
43. Pretel, H.; Costa, J.L.D.S.G.; Florez, F.L.E.; Nogueira, B.R.; Junior, O.B.D.O. Assessment of the temporal variation of electrical potential and pH of different bleaching agents. *Heliyon* **2021**, *7*, e08452. [CrossRef]
44. Monteiro, N.R.; Basting, R.T.; Amaral, F.L.B.D.; França, F.M.G.; Turssi, C.P.; Gomes, O.P.; Filho, P.N.L.; Kantovitz, K.R.; Basting, R.T. Titanium dioxide nanotubes incorporated into bleaching agents: Physicochemical characterization and enamel color change. *J. Appl. Oral Sci.* **2020**, *28*, e20190771. [CrossRef]
45. De Moor, R.J.G.; Verheyen, J.; Diachuk, A.; Verheyen, P.; Meire, M.A.; De Coster, P.J.; Keulemans, F.; De Bruyne, M.; Walsh, L.J. Insight in the Chemistry of Laser-Activated Dental Bleaching. *Sci. World J.* **2015**, *2015*, 650492. [CrossRef] [PubMed]
46. Xu, C.; Reed, R.J.; Gorski, J.P.; Wang, Y.; Walker, M.P. The distribution of carbonate in enamel and its correlation with structure and mechanical properties. *J. Mater. Sci.* **2012**, *47*, 8035–8043. [CrossRef]
47. Sa, Y.; Sun, L.; Wang, Z.; Ma, X.; Liang, S.; Xing, W.; Jiang, T.; Wang, Y. Effects of Two In-Office Bleaching Agents with Different pH on the Structure of Human Enamel: An In Situ and In Vitro Study. *Oper. Dent.* **2013**, *38*, 100–110. [CrossRef]
48. Orilisi, G.; Tosco, V.; Monterubbianesi, R.; Notarstefano, V.; Özcan, M.; Putignano, A.; Orsini, G. ATR-FTIR, EDS and SEM evaluations of enamel structure after treatment with hydrogen peroxide bleaching agents loaded with nano-hydroxyapatite particles. *PeerJ* **2021**, *9*, e10606. [CrossRef] [PubMed]
49. Von Euw, S.; Wang, Y.; Laurent, G.; Drouet, C.; Babonneau, F.; Nassif, N.; Azaïs, T. Bone mineral: New insights into its chemical composition. *Sci. Rep.* **2019**, *9*, 8456. [CrossRef]
50. Cacciotti, I. Cationic and Anionic Substitutions in Hydroxyapatite. In *Handbook of Bioceramics and Biocomposites*; Antoniac, I.V., Ed.; Springer International Publishing: Cham, Switzerland, 2014; pp. 1–68.
51. Ittatirut, S.; Matangkasombut, O.; Thanyasrisung, P. In-office bleaching gel with 35% hydrogen peroxide enhanced biofilm formation of early colonizing streptococci on human enamel. *J. Dent.* **2014**, *42*, 1480–1486. [CrossRef]

nanomaterials

MDPI

Article

Characterization of Experimental Nanoparticulated Dental Adhesive Resins with Long-Term Antibacterial Properties

Rochelle Denise Hiers [1], Pedro Huebner [2], Sharukh Soli Khajotia [1] and Fernando Luis Esteban Florez [1,*]

1 Division of Dental Biomaterials, Department of Restorative Sciences, College of Dentistry, University of Oklahoma Health Sciences Center, Oklahoma City, OK 73117, USA
2 Department of Mechanical Engineering, University of Utah, Salt Lake City, UT 84112, USA
* Correspondence: fernando-esteban-florez@ouhsc.edu

Abstract: Experimental adhesives with functional nitrogen-doped titanium dioxide nanoparticles (N_TiO_2) have been shown to display improved properties. However, these materials have not been characterized regarding their degree of conversion (DC), biaxial flexure strength (BFS), surface roughness (SR), elastic modulus (EM), and long-term antibacterial functionalities. Experimental adhesives were synthesized by dispersing N_TiO_2 (10%, 20%, or 30%, v/v%) into OptiBond Solo Plus (OPTB, Kerr Corp., USA). Unpolymerized adhesives (volume = 50 µL/drop, n = 3/group) were individually placed onto a heated (37 °C) attenuated total reflectance (ATR) monolithic diamond crystal (Golden Gate, Specac). The spectra of composites were obtained with a Fourier-transform infrared (FTIR) spectrometer (Nicolet IS50; 500–4500 cm^{-1}; resolution = 4 cm^{-1}, 10 internal scans/spectrum) before and after polymerization. Disk-shaped specimens (diameter = 6.0 mm, thickness = 0.5 mm) for BFS (n = 12/group), SR and EM (n = 3/group), and for antibacterial testing (n = 18/group/time-point) were fabricated and photopolymerized (1 min each; 385–515 nm, 1000 mW/cm^2; VALO). DC values (%) were calculated from pre- and post-polymerization spectra using the two-frequency method and tangent-baseline technique. BFS was assessed using a universal testing machine (Instron 68TM-5, crosshead speed = 1.27 mm/min, 25 °C). SR and EM were investigated using an atomic force microscope (Multimode 8) with aluminum-coated silicon probes (8 nm pyramidal tip, spring constant 40 N/m, Bruker). Antibacterial testing was performed by growing *Streptococcus mutans* biofilms (UA159-*ldh*, 37 °C, microaerophilic) on the surfaces of specimens for 24 h and then measuring the relative luminescence units (RLU) with a Biotek Synergy HT multi-well plate reader. Results demonstrate that experimental materials containing 10%, 20%, and 30% of N_TiO_2 displayed higher levels of DC, had better mechanical properties, and were able to exert strong and durable antibacterial properties without visible light irradiation and after extended periods of simulated shelf-life and aging in PBS. The reported experimental materials are expected to increase the service lives of polymer-based bonded restorations by decreasing the incidence of secondary caries.

Keywords: metal nanoparticles; anti-bacterial agents; dental materials; *Streptococcus mutans*

1. Introduction

Dental caries continues to pose a major health burden in numerous countries. This biofilm-originated disease [1] is estimated to affect 3.5 billion people [2] and to represent around 4.6% of the total global expenditures in healthcare [3]. Clinical manifestations include the progressive and irreversible dissolution [4,5] of dental hard tissues (e.g., enamel, dentin, and cementum), cavitation, pain, and tooth loss. The treatment of dental caries revolves around the mechanical removal of disorganized and infected tissues using hand-held instruments (e.g., either static or rotary) and the placement of a dental biomaterial (e.g., metal, polymer, or ceramic) to restore the esthetics and masticatory function of affected tissues.

Over the years, composite resins became the first choice of restorative materials amongst patients and clinicians due to its mercury-free compositions [6] and superior properties (e.g., handling and esthetic) [7–9]. In fact, composite resin restorations are the most prevalent biomedical intervention in human beings with more than 800 million placed every year [10]. Despite such widespread acceptance and utilization, previous studies indicated that these materials are associated with limited-service lives (between 5–7 years), [11] experience polymerization shrinkage and accumulate more biofilms, when compared to other restorative materials [9,12,13]. When combined, these factors may shift the ecology of the oral cavity from a healthy state into a disease-associated state [14].

Streptococcus mutans, a Gram-positive and facultatively anaerobic bacteria, has been widely accepted as a major contributor to the development of dental caries (primary and secondary) due to their (i) ability to adhere and accumulate onto the surfaces of teeth through a disaccharide-dependent mechanism, (ii) ability to deposit an extracellular matrix that protects cells from external aggressors, (iii) ability to metabolize a wide variety of complex carbohydrates into organic acids, and (iv) the ability to thrive in acidic environments [15]. Even though *S. mutans* is not solely responsible for the occurrence of dental caries, these undisputable features have made *S. mutans* an important model organism in oral antibacterial research [16].

Secondary caries develop between dental adhesive resins and the tooth structure, and are considered the primary reason for the failure of polymer-based bonded restorations [17]. According to previous studies, current dental adhesive resins are formulated using a combination of hydrophilic and hydrophobic components [18] that phase-separate when applied onto water-rich tissues [19]. This chemical instability leads to incomplete envelopment of exposed collagen fibrils and the establishment of porous hybrid layers that are prone to failure by biodegradation, hydrolysis, esterases, and biofilms [20]. This significant problem has precipitated the execution of several studies to improve the physical, chemical, and biological properties of current polymer formulations. Ideally, dental adhesive resins should be able to establish interfaces that are hermetically sealed, are dimensionally stable, prevent the formation of cariogenic biofilms, and precipitate highly organized crystalline structures to fill the gaps from the incomplete envelopment of collagen fibrils and polymerization shrinkage [21–23].

Despite significant investments by the manufacturing and scientific communities, newly developed materials containing antibacterial agents, quaternary ammonium compounds, [13] or metaloxide nanoparticles (e.g., zinc and titanium) [24,25] failed to sustain long-term antibacterial properties and did not extend the service lives of composite restorations, thereby underscoring the need for the development of novel materials with long-term biological properties. Pérez-Mondragón et al., [26] while investigating the shelf-life stability in terms of the degree of conversion, ultimate tensile strength, and color of dental adhesive resins (commercially available and experimental) at different periods of simulated shelf-life (37 °C; 6, 18, and 24 months), have demonstrated that experimental materials displayed higher shelf-life stability (in terms of the degree of conversion), when compared to commercially available materials. However, experimental materials displayed mechanical and optical properties that were similar ($p > 0.05$) to those of the control group, independently of the time-point considered [26].

Recent advances in the field of material sciences and nanotechnology enabled the synthesis and incorporation of metaloxide nanoparticles (e.g., zinc oxide, silver, and titanium) into dental polymers (denture base, dental adhesives, and composite resins). These nanoparticulated systems have become prevalent in many areas of dental research, including orthodontics, dental materials, bleaching, implants, and prosthodontics [27], because of their intrinsic optical, physical, biological, and chemical properties [28,29]. Recent studies investigated the utility of silver, zinc, copper, titanium, calcium fluoride, and magnesium nanoparticles [30,31] in the prevention of secondary caries because they have been shown to disrupt bacterial metabolism and biofilm formation [32].

Titanium dioxide (TiO_2, anatase, rutile, or brookite) is known for its relevant physical, chemical, antimicrobial, and biocompatibility properties [33]. Nanoparticles of TiO_2 (TiO_2-np) are typically white, have diameters around 25 nm, have high refractive index, are corrosion resistant, display high microhardness values [20,34], and were demonstrated to be effective against numerous microorganisms, including *Candida albicans*, *Staphylococcus aureus*, *Pseudomonas aeruginosa*, *Escherichia coli*, and *Lactobacillus acidophilus* [35]. However, despite these relevant characteristics, TiO_2-np have large bandgaps (3.2204 eV, for anatase) and require the utilization of UV irradiation to generate different types of reactive oxygen species (ROS) [34]. Even though the photocatalysis of TiO_2-np is feasible from the electronic standpoint, the UV dose of energy required to promote surface disinfection has been demonstrated to be harmful to eukaryotic cells and tissues, [36] which significantly restricts its widespread utilization in dental applications.

Doping the crystal lattice of TiO_2-np with metals and non-metals has been previously shown to decrease the bandgap of Titania (2.47 eV) [37] and allow the utilization of visible light irradiation, which is widely used in dentistry, for the generation of ROS. The synthesis, incorporation, and covalent functionalization of visible light-responsive nitrogen-doped titanium dioxide nanoparticles (N_TiO_2) into a commercially available dental adhesive resin (OptiBond Solo Plus, Kerr Corp., USA; OPTB) has been recently reported by Esteban Florez et al. [34,38,39] Experimental adhesive resins displayed strong antibacterial and biomimetic properties when irradiated with visible light [34] and were less soluble and more biocompatible [20], when compared to commercially available materials, which suggests that nanoparticulated materials may hold the promise to decrease the incidence of secondary caries and to extend the service lives of polymer-based adhesive restorations.

Therefore, the objective of the present study was to characterize the (i) degree of conversion at the time of polymer synthesis and after two years of simulated shelf-life, (ii) biaxial flexure strength, (iii) flexural modulus, (iv) surface roughness, (v) elastic modulus, and (vi) long-term antibacterial properties of experimental dental adhesive resins, containing varying concentrations of N_TiO_2 (10%, 20%, and 30%, *v*/*v*%).

2. Materials and Methods

2.1. Synthesis of N_TiO_2

Synthesis of N_TiO_2 nanoparticles has been described in detail in previous publications [17,29,35,36] from our group and will be summarized here. Nitrogen-doped TiO_2 nanoparticles (N_TiO_2) were synthesized via a 2-step process. The first step involved a solvothermal synthesis of pure TiO_2. In a typical reaction, a solution comprised of 1.7 g of Ti (IV)-butoxide (Aldrich, St. Louis, MO, USA, 97%), 4.6 g ethanol (Decon Labs, 200 proof), 6.8 g oleylamine (Aldrich, 70%), and 7.1 g oleic acid (Aldrich, St. Louis, MO, USA, 90%) was prepared, then mixed with 20 mL of 4% H_2O in ethanol (18-MΩ Milli-Q; Decon Labs, King of Prussia, PA, USA). The solution was reacted using a high-pressure reaction vessel (Paar Series 5000 Multiple Reactor System) and continuous stirring at 180 °C for 24 h. Solutions went through a series of washing steps with anhydrous ethanol to remove extraneous surfactants. The synthesized TiO_2 nanoparticles were stored in ethanol. To produce nitrogen-doped TiO_2 nanoparticles, aliquots were then reacted with an equal volume of triethylamine (Aldrich, St. Louis, MO, USA, 99.5%), and underwent a second period in a high-pressure reaction vessel, at 140 °C for 12 h. The N_TiO_2 particles (size distribution= 6–15 nm, anatase) were rinsed 3 times with anhydrous ethanol. The final N_TiO_2 nanoparticle solution was stored in ethanol, and the concentration of particles was determined gravimetrically and was approximately 35 mg/mL.

2.2. Synthesis of Experimental Adhesive Resins and Specimen Fabrication

Experimental dental adhesive resins were synthesized by dispersing (Q700 sonicator, QSonica, LLC, USA) 10%, 20%, or 30% of N_TiO_2 (*v*/*v* %, suspended in ethanol) into OptiBond Solo Plus (Kerr Corp., OPTB, Composition: Bis-GMA, HEMA, GDMA, GPDM, ethanol, CQ, ODMAB, BHT, filler particles, and coupling factor A174). The rationale for selecting these concentrations of nanoparticles was based on a previous study where strong initial antibacterial properties were demonstrated in both dark and light-irradiated conditions [34]. Disk-shaped specimens (n = 18/group; diameter = 6.00 mm, thickness = 0.50 mm) of unaltered OPTB and experimental dental adhesive resins were fabricated using a custom stainless-steel mold. Glass coverslips (No. 2, VWR International, Radnor, PA, USA, LLC) were used to give specimens a smooth surface finish. Specimens were polymerized using blue light irradiation (1000 mW/cm^2, 1 min) emitted from a broadband LED light-curing unit (VALO, Ultradent Products, Inc., South Jordan, UT, USA). Specimens of both unaltered and experimental adhesive resins were then UV-sterilized (254 nm, 800,000 μJ/cm^2, UVP Crosslinker, model CL-1000, UVP, Upland, CA, USA).

2.3. *Degree of Conversion*

Adhesives described in Section 2.2. were assessed for degree of conversion at the time of polymer synthesis (NEW) and after two years of simulated shelf-life (OLD; dark conditions, 25 °C) using an attenuated total reflectance (ATR) diamond crystal (KR-5 lens, Golden Gate model GS10542-K; Specac, Inc. Fort Washington, PA, USA) coupled to a Fourier transform infrared spectrometer (500–4000 cm^{-1}; resolution 4 cm^{-1}, 10 internal scans per spectrum; Nicolet IS50, Thermo Fisher Scientific, Waltham, MA, USA). Uncured drops (volume = 50 μL/drop, n = 3/group) of each material were individually dispensed onto the ATR crystal (at 37 °C). Spectra of materials in the unpolymerized state were then recorded. Materials were individually photopolymerized using a broad-band LED light curing unit (385–515 nm, 1000 mW/cm^2, 20 sec., VALO, Ultradent Products, Inc., South Jordan, UT, USA) before obtaining the spectra of materials in the polymerized state. Values of DC reported were calculated using the two-frequency method [40] and tangent-baseline technique [41].

2.4. *Biaxial Flexure Strength*

The biaxial flexure strength (BFS) at the time of polymer synthesis was investigated to test the bulk mechanical properties of experimental adhesive resins. This method was selected because it eliminates spurious edge failures that are typically associated with three-point bending testing and results are independent of flaw direction [42]. Specimens (n = 12/group) were fabricated, as described in Section 2.2., and were subjected to BFS testing using a universal testing machine (Instron, model 33R4468, cross-head speed = 1.27 mm/min, 25 °C, Norwood, MA, USA) until failure.

2.5. *Nanoscale Surface and Mechanical Characterization*

The concurrent characterization of surface roughness and elastic modulus of specimens (n = 3/group) fabricated, as described in Section 2.2., was conducted using an atomic force microscope (AFM, Multimode 8, Bruker Corporation, Billerica, MA, USA) with aluminum-coated silicon probes (RTESPA-300, Bruker, 8 nm pyramidal tip, spring constant, 40 N/m). In brief, following AFM calibration, as directed by the manufacturer, three discrete 20 × 20 μm areas on the surface of each specimen were randomly selected for characterization. The quantitative nanomechanical (QNM) test was configured to scan the surfaces of samples at a rate of 0.2 Hz and 8 μm/s. Peak forces were limited to 80–120 nN, depending on the specimen, so as to achieve the necessary amount of surface deformation for mechanical data modelling and to protect the tip of the probe from excessive wear. Within each area, square matrices of 256 × 256 equally spaced data points were captured and used to (i) characterize the average surface roughness (Ra) measured in nm, and (ii) estimate the average elastic modulus of samples (E), measured in Pa, according to the Derjagin, Muller, Toropov (DMT) model of elastic contact, which is recommended in the case of small tips and stiff samples with small adhesion [43]. All testing was performed in air at ambient conditions (25 °C).

2.6. Bacterial Strain and Growth of Biofilms

Streptococcus mutans strain UA159 (JM10:pJM1-ldh, luc+, Spc^R, *luc* under the control of the *ldh* promoter) was utilized as the model organism in the present study. [34,35,42] The selection of antibiotic-resistant colonies was performed on two passages of TH plates (Todd-Hewitt, BD Difco, New Jersey, NJ, USA), supplemented with 0.3% yeast extract (EMD Millipore Sigma, Burlington, MA, USA) and 800 μg/mL of spectinomycin (MP Biomedicals, Santa Ana, CA, USA). The plates were incubated under anaerobic conditions at 37 °C for 48 h. Planktonic cultures of *S. mutans* (JM10) were grown in THY culture medium at 37 °C for 16 h. Cultures having optical density (OD_{600}) levels equal to or higher than 0.900 (corresponding to 6.43 e^{+12} CFU/mL) were used as inoculum to grow biofilms. Optimal biofilm growth parameters identified during a previous study from our group [1:50 dilution, 0.65× THY + 1% (*w*/*v*) sucrose, 1000 μL] [6] were then used to grow the biofilms. Aliquots (1.0 mL) of inoculated biofilm growth media were dispensed into the wells of sterile 24-well microtiter plates (Falcon, Corning, NY, USA), containing sterile specimens. Biofilms were grown for 24 h (static cultures, microaerophilic conditions, 37 °C). An additional set of specimens fabricated with OPTB was treated with 2% chlorhexidine gluconate (CHX) for 2 min and served as the control group.

2.7. High Throughput Bioluminescence Assay

After the growth period, biofilms were replenished with 1.0 mL of fresh 1^x THY + 1% (*w*/*v*) glucose recharge medium and were incubated at 37 °C for 1 h. Replenished biofilms were transferred into the wells of sterile white 24-well plates, containing 1.0 mL of fresh 0.65× THY + 1% (*w*/*v*) sucrose medium. D-Luciferin aqueous solution (100 mM) suspended in 0.1 M citrate buffer (pH 6.0) was added by a computer-controlled system in a Synergy HT Multi-mode microplate reader (Agilent Biotek, Winooski, VT, USA) to the wells containing both the specimens with biofilms and recharge medium in 2:1 ratio (*v*/*v*) of inoculum:D-Luciferin. Luciferase metabolic activity in non-disrupted *S. mutans* biofilms was evaluated at 590 nm in 2 min increments (6 min total) after the addition of D-Luciferin substrate, in terms of relative luminescence units (RLUs). After completion of the assay, the timepoint with the lowest coefficient of variation was selected for results analysis. High throughput bioluminescence procedures reported in the present study were conducted, following a previously published and validated protocol [38].

2.8. Long-Term Antibacterial Properties

In the present study, the testing of long-term antibacterial properties was divided into two parts. In part one (simulated shelf-life), unpolymerized adhesive resins identified in Section 2.2. (unaltered [OPTB] and experimental [OPTB + N_TiO_2) were synthesized and stored (dark conditions, 25 °C) in the original containers provided by the manufacturer (5 mL, black bottles, Kerr Corp., Orange, CA, USA) for the duration of the study (24 months). At four specific time-points (T1 = 0, T2 = 6, T3 = 12 and T4 = 24 months), specimens (n = 18/group/time-point; total number of specimens = 90) were fabricated, sterilized, and monomer-extracted, following the procedures and methods described in Section 2.2. Twenty four-hour biofilms were then grown onto the surfaces of specimens, according to protocols described in Section 2.6. and bioluminescence procedures were performed, as described in Section 2.7. In part two (longevity of antibacterial properties), an additional set of specimens (n = 18/group/time-point; total number of specimens = 450) were fabricated immediately after the synthesis of experimental adhesive resins. Fabricated specimens were then UV-sterilized, as described in Section 2.2., before being individually stored in 50 mL of sterile PBS (pH 7.4, 37 °C, dark conditions, sealed Falcon tubes) for the duration of the study (12 months). At specific time-points (T1 = 0, T2 = 1, T3 = 3, T4 = 6 and T5 = 12 months) specimens were then subjected to biofilm growth and bioluminescence procedures, as described before.

2.9. Statistical Analysis

Values of degree of conversion were statistically analyzed using t-Tests and post hoc Student–Newman–Keuls tests (SAS software, version 9.3; SAS Institute, Cary, NC, USA). Values of biaxial flexure strength were statistically analyzed using general linear models and Student–Newman–Keuls post hoc tests (SAS software, version 9.3; SAS Institute, Cary, NC, USA). Values of elastic modulus and roughness were statistically analyzed using one-way ANOVA and Tukey post hoc tests (JASP, v. 0.15, University of Amsterdam, The Netherlands). RLU values indicating the viability of non-disrupted biofilms of *S. mutans* grown against the surfaces of both unaltered and experimental dental adhesive resins were statistically analyzed using two-factor general linear models (GLM) and post hoc Student–Newman–Keuls tests (SAS software, version 9.3; SAS Institute, Cary, NC, USA). All tests were conducted with a level of significance of 95%.

3. Results

Figure 1A,B demonstrate that N_TiO_2 was produced using solvothermal reactions and, as described in Section 2.1., resulted in nanoparticles that had an approximate spherical shape, had smooth surfaces, and mostly exhibiting some faceting.

(A) (B)

Figure 1. (**A**) SEM (magnification 100 k×) and (**B**) TEM (magnification 500×) micrographs of non-immobilized N_TiO_2.

Figure 2 illustrates the mean and standard deviation values of DC for the investigated adhesives (unaltered and experimental). Results reported have clearly indicated that DC values decreased after two years of simulated shelf-life independently of material (unaltered or experimental) or N_TiO_2 concentration (10%, 20% and 30%) considered. Values of DC varied from 82.17% (NEW, 10% N_TiO_2) to 85.23% (NEW, 30% N_TiO_2) and from 79.73% (OLD, 10% N_TiO_2) to 82.47% (OLD, 20% N_TiO_2). The smallest reductions in DC values after two years of simulated shelf-life were detected on experimental materials containing 20% N_TiO_2 (1.86%), which suggests that the incorporation and functionalization of N_TiO_2 into OPTB does not adversely impact the shelf-life stability of the parental polymer over the course of 24 months. It is well known that restorative materials with low DC values are associated with inadequate properties (physical, mechanical, optical, and biological) and display short service lives. Therefore, these results are critically important, as the experimental materials' DC has been in congruence with controls within the commercial manufacturer shelf-life specification of two years.

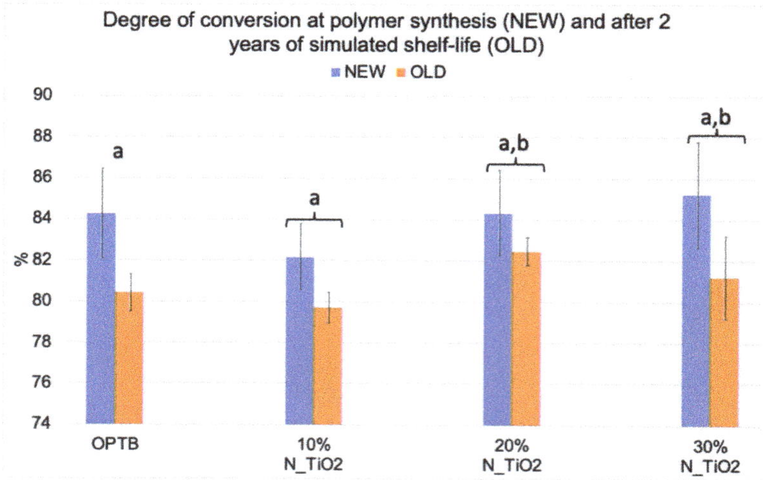

Figure 2. Mean and standard deviation values of degree of conversion (DC) of unaltered (OPTB) and experimental adhesives (OPTB + N_TiO$_2$ [10%, 20% or 30%, *v*/*v*]). Connecting braces denote no statistical differences ($p > 0.05$) between NEW and OLD within each group. Letters above bars illustrate the SNK rankings and denote the presence of significant ($p < 0.05$) intergroup differences.

Figure 3A,B illustrate the results of the biaxial flexure strength test. It can be observed that the incorporation of N_TiO$_2$ resulted in experimental materials with superior mechanical properties, as denoted by values of BFS and flexural modulus that were numerically higher and statistically significant ($p < 0.05$; flexural modulus), when compared to those of OPTB. This confirms our hypothesis that experimental materials containing varying concentrations of N_TiO$_2$ would display better bulk mechanical properties, when compared to the parental polymer.

Figure 4A–F presents the results of the quantitative nanoscale mechanical characterization of adhesives investigated in terms of surface roughness (4A), elastic modulus (4B), and surface deformation for (C) OPTB, (D) OPTB+10% N_TiO$_2$, (E) OPTB+20% N_TiO$_2$, and (F) OPTB+30% N_TiO$_2$. It is evident that the incorporation of nanoparticles into OPTB resulted in experimental materials capable of producing specimens with surface roughness values that were comparable ($p = 0.999$) to those produced with the parental polymer. The elastic modulus of experimental materials containing 20% and 30% of N_TiO$_2$ were demonstrated to be numerically higher and statistically different ($p = 0.001$), when compared to that of the parental polymer, which indicates that experimental materials produced specimens that were much stiffer than those produced with OPTB (control group).

Figure 5 shows the mean and standard error values of luciferase metabolic activity in terms of relative light units (RLUs) for *S. mutans* biofilms grown on the surfaces of specimens fabricated during the simulated shelf-life portion of the experiment. It can be observed that biofilms grown against the surfaces of OPTB (control group) displayed the highest levels of luciferase metabolic activity amongst all groups investigated and independent of the time-point considered. Twelve-month findings reported for specimens pertaining to the control group (OPTB) should be interpreted with caution because these could have resulted from the intrinsic limitations of the study that include the utilization of a semi-defined biofilm growth medium and the leaching of unreacted hydrophilic monomers [44]. Results reported suggest that experimental materials' antibacterial properties varied in a concentration-dependent manner, and experimental materials could be rank-ordered in terms of increasing antibacterial properties (10% < 20% < 30%), as denoted by lower RLU values and distinct SNK rankings. The lowest levels of metabolic activity were detected in biofilms topically treated with CHX for 2 min. It can be noted when observing the results from time-points T2, T3, and T4 (6, 12 and 24 months, respectively), that storage time has negatively affected all materials investigated (unaltered and experimental), as denoted by metabolic activities that were progressively higher (in terms of RLU values). It was also possible to observe that experimental materials containing 30% of N_TiO_2 rendered biofilms with metabolic activities that were comparable to those treated with CHX (at time-points T1 and T4).

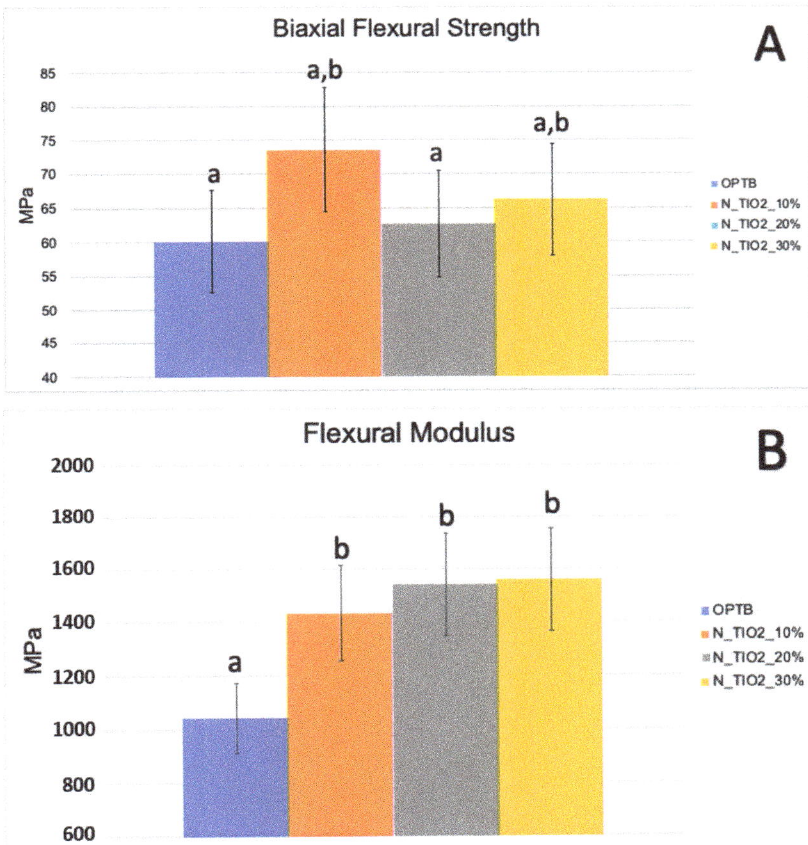

Figure 3. Mean and standard deviation values of (**A**) biaxial flexure strength (BFS) and (**B**) flexural modulus of unaltered (OPTB) and experimental adhesives (OPTB + N_TiO_2 [10%, 20%, or 30%, *v*/*v*]). Letters above bars illustrate the SNK rankings and denote the presence of significant ($p < 0.05$) intergroup differences.

Figure 6 illustrates the percentage reduction in metabolic activity for experimental groups investigated, where it becomes obvious that experimental adhesive resins containing functionalized N_TiO_2 could sustain long-term antibacterial properties, as denoted by metabolic reductions that ranged from 40% (T1, 10% of N_TiO_2) to 87% (T4, 30% of N_TiO_2). The highest reductions in metabolic activity were observed in biofilms treated with CHX (reductions ranging from 88–98%). This was an expected result, because fresh CHX was topically applied onto biofilms at each specific time-point.

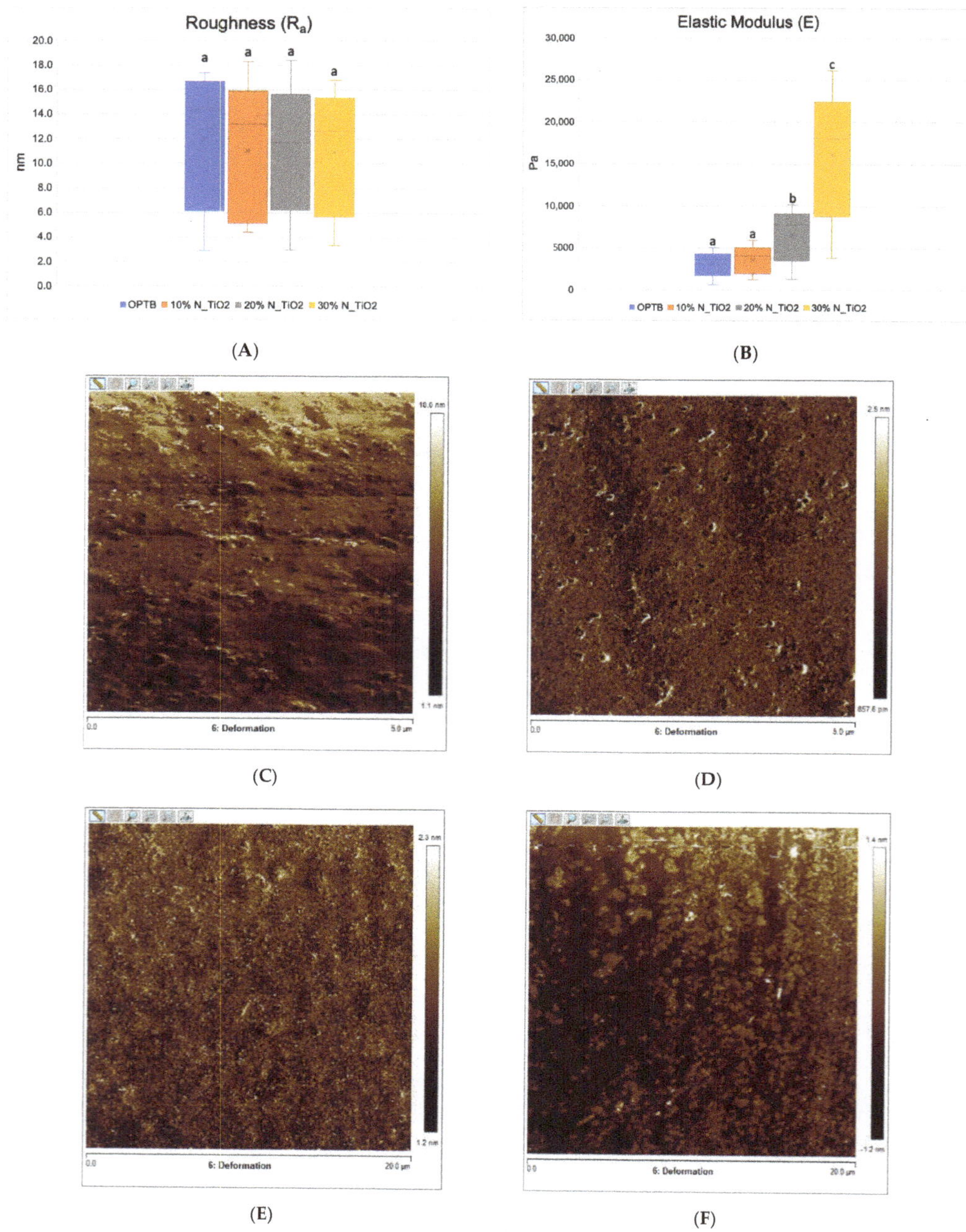

Figure 4. Mean and standard deviation values of (**A**) surface roughness and (**B**) elastic modulus of adhesives containing N_TiO_2. (**C**–**F**) Raw two-dimensional atomic force images, preceding filters, and post-processing for data analysis, illustrating the mapping of samples deformation during QNM analysis for (**C**) OPTB, (**D**) OPTB+10% N_TiO_2, (**E**) OPTB+20% N_TiO_2, and (**F**) OPTB+30% N_TiO_2. Side bars on individual images illustrate the range of deformation experienced by each material. Letters above bars in images (**A**,**B**) illustrate the SNK rankings and denote the presence of differences that are statistically significant ($p < 0.05$).

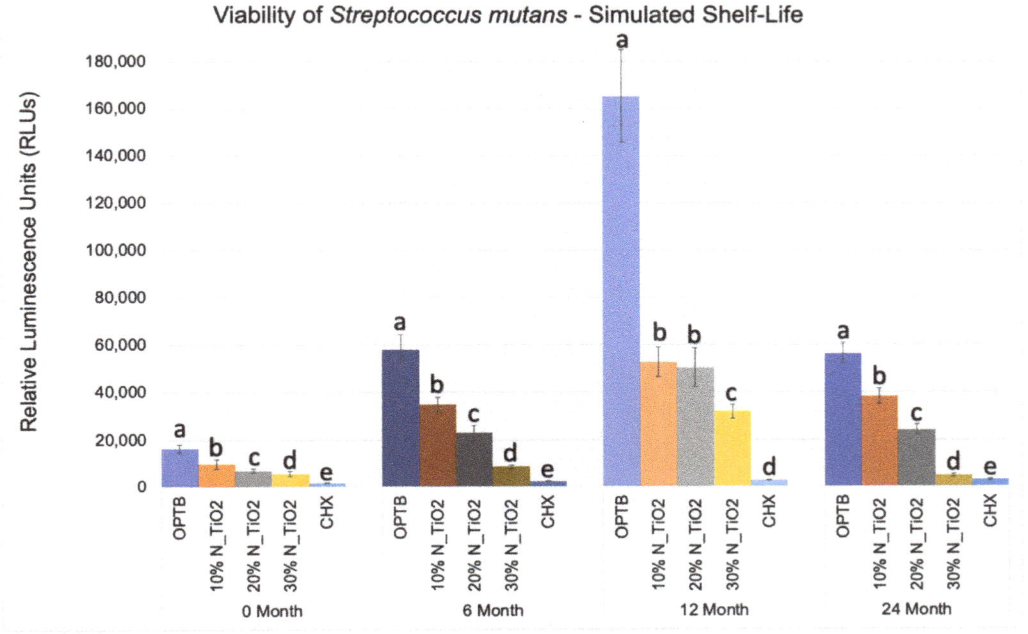

Figure 5. Expression of luciferase activity ($n = 18$/group) was quantified in terms of bioluminescence. Biofilms on specimens that were fabricated from dental adhesives that were aged for 0, 6, 12, and 24 months had sustained lower bioluminescent values when exposed to higher nanoparticle concentrations. CHX is the positive control group. Letters above bars illustrate the SNK rankings and denote mean values that are statistically different ($p < 0.001$) within timepoints of 0, 6, 12, and 24 months.

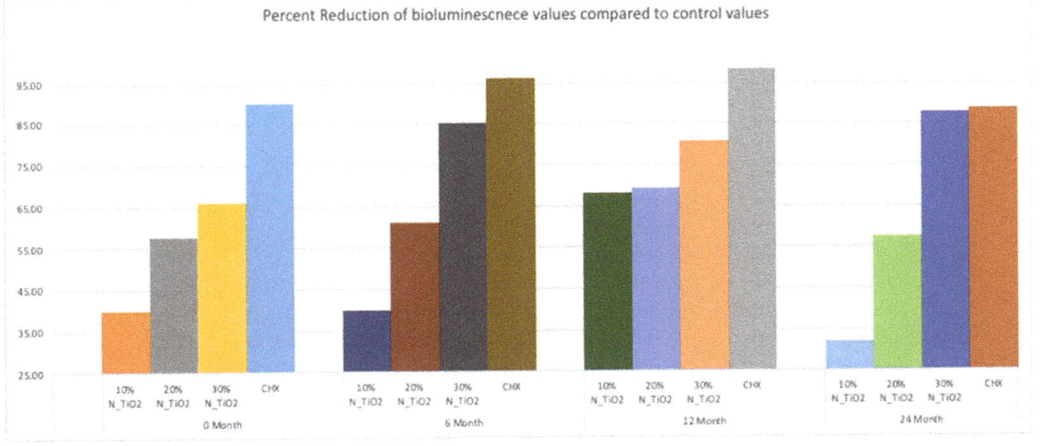

Figure 6. Percent reduction of bioluminescence values of experimental adhesive resins that were fabricated with aged adhesive resins, when compared to unaltered OPTB. Biofilms grown on specimens with higher concentrations of nanoparticles showed greater decrease in bioluminescence. CHX serves as a well-established positive control.

The results of the second part of the experiment (longevity of antibacterial properties) are shown in Figure 7 in terms of mean and standard error values of luciferase metabolic activity (RLUs). In this portion of the study, 24 h *S. mutans* biofilms were grown after specific time-points: T1 = 0, T2 = 1, T3 = 3, T4 = 6, and T5 = 12 months; on the surfaces of specimens (in PBS, 37 °C, duration of study); fabricated with unaltered (OPTB) or experimental adhesive resins containing varying concentrations of N_TiO_2 (10%, 20% or 30%, *v*/*v*%).

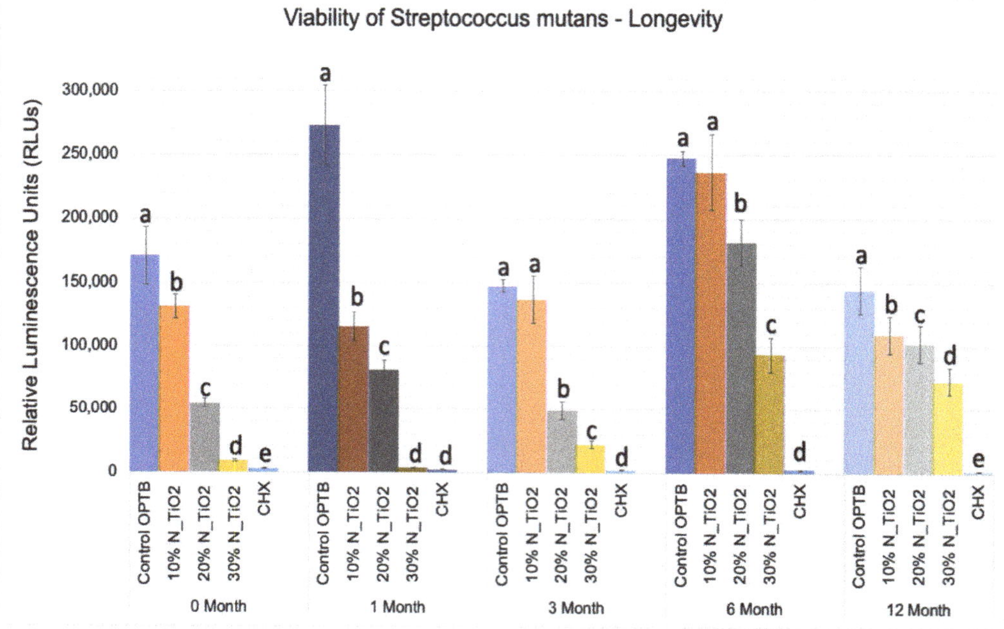

Figure 7. Expression of luciferase activity (n = 18/group) was quantified in terms of bioluminescence. Biofilms grown on specimens of dental adhesives that were then aged for 0, 1, 3, 6, and 12 moths in PBS exhibited lower bioluminescent values when exposed to higher nanoparticle concentrations. Letters above bars illuastrate the SNK rankings and denote mean values that are statistically different ($p < 0.001$) within timepoints of 0, 1, 3, 6, and 12 months.

Results reported in Figure 7 agree with those from Figures 5 and 6 and suggest that experimental materials' antibacterial properties varied in a concentration-dependent manner, where materials containing higher N_TiO_2 concentrations displayed antibacterial properties (10% < 20% < 30%) that were stronger, as denoted by RLU values that were consistently smaller and statistically different ($p < 0.001$) than those from control group specimens (OPTB). The antibacterial properties of experimental adhesive resins containing 30% of N_TiO_2 were comparable to those from CHX (topical treatment, 2 min) at time-points T1 and T2 (0 and 1 month) but were less pronounced in time-points T3, T4, and T5 (3, 6, and 12 months, respectively).

This less-than-ideal behavior indicates that long-term storage in PBS adversely impacts the properties of the materials investigated. However, when comparing these results to those from the control group, it is obvious that experimental materials proposed were able to sustain relevant antibacterial properties against non-disrupted *S. mutans* biofilms. Figure 8 helps to further illustrate that point by displaying the results of metabolic reduction (in %) among experimental groups, when compared to the control (OPTB). The group treated with CHX displayed the most consistent and effective antibacterial behavior amongst all groups investigated and displayed viability reductions that ranged between 98 and 99%. Biofilms grown against the surfaces of specimens fabricated with experimental adhesive resins displayed metabolic reductions that ranged from 4.36% (T1, 10% of N_TiO_2) to 98.63% (T2, 30% of N_TiO_2). In combination, reported results indicate that experimental materials were indeed capable of sustaining relevant long-term (up to 24 months) antibacterial properties against non-disrupted *S. mutans* biofilms without the need for visible light irradiation.

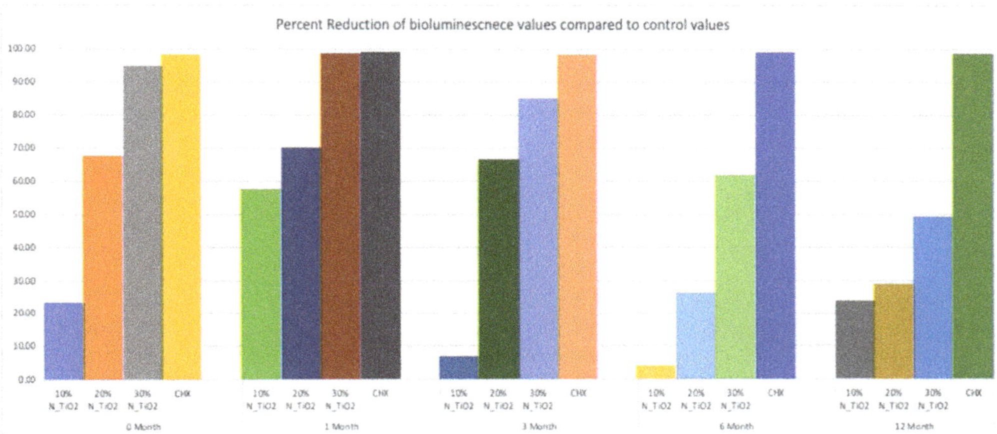

Figure 8. Percent reduction of bioluminescence values of experimental adhesive resins that were fabricated and aged in PBS, when compared to unaltered OPTB. Biofilms grown on fabricated specimens that exhibited higher concentrations of nanoparticles showed greater decrease in bioluminescence, however the percent reduction decreases over time. CHX is to serve as a well-established positive control.

4. Discussion

Several approaches have been tested to impart long-term antibacterial properties to dental adhesive resins. These include the utilization of fluorinated graphene [45], eugenyl methacrylate (EgMA) [46], chlorhexidine (0.2% diacetate or 2% digluconate) [47], tt-farnesol [48], sodium hypochlorite (6%) [49], benzalkonium chloride [50], and epigallocatechin-3-gallate [51], among others. However, even though experimental materials were able to initially decrease the viability of oral microorganisms, reports have indicated that newly developed materials display a low degree of polymerization, reduced mechanical properties, and leaching of uncured monomers [52] and were not able to display significant long-term antibacterial properties, or were capable of extending the service lives of polymer-based adhesive restorations. The photocatalysis of metaloxide nanoparticles has been considered as an alternative approach [53,54] because of their proven effectivity against microorganisms relevant to public health [34]. However, UV-irradiation requirements have restricted its widespread use in dental applications and resulted in materials displaying rough surfaces and degraded polymer matrixes. [25,55] Subsequent studies investigated the antibacterial efficacy of heterogeneous photocatalysis and doped nanoparticles against *S. mutans* biofilms [34]. According to results reported, experimental adhesives containing N_TiO_2 (50%, 67% and 80%) displayed strong antibacterial and biomimetic properties when irradiated with visible light (410 ± 10 nm, 310.07 J/cm^2). Specimens fabricated with experimental adhesives reported were shown (SEM/EDX data) to have smooth surfaces and polymer matrixes that were not degraded [34].

Follow-up studies investigated if relevant antibacterial properties could be achieved in the absence of visible light and to determine the impact of N_TiO_2 incorporation on the sorption, solubility, and biocompatibility of experimental adhesives [20]. Results indicated that experimental adhesives were less soluble, absorbed less water, were more biocompatible and more antibacterial, when compared to the parental polymer (OPTB). When comparing experimental materials' antibacterial properties to those of Clearfil SE Protect (Kuraray, Co.), which is a commercially available, MDPB-containing, and fluoride-releasing dental adhesive resin, experimental adhesives containing N_TiO_2 displayed comparable ($p > 0.05$) efficacies against non-disrupted *S. mutans* biofilms (either 24 or 48 h) in dark conditions, [38] thereby demonstrating that the nanotechnology proposed could impart relevant toxicity against *S. mutans* without visible light irradiation. Despite these promising results, previous studies failed to investigate other relevant properties of experimental adhesives containing N_TiO_2 (10%, 20%, and 30%), including the degree of conversion at the time of polymer synthesis and after two years of simulated shelf-life, the biaxial flexure strength, flexural modulus, surface roughness, elastic modulus, and long-term antibacterial properties (12 and 24 months).

The rationale for fabricating N_TiO_2 using solvothermal reactions under supercritical conditions of temperature and pressure precipitates from the fact that this synthetic route has been previously shown [56] to be highly reproducible and to yield pure and crystalline nanoparticles (6–15 nm, anatase phase) with high levels of nitrogen doping (N/Ti molar ratio = 3.4%), when compared to traditional calcination strategies (N/Ti molar ratio = 1.3%). Nanoparticles fabricated by this process are electron deficient, display smooth surfaces, have high specific surface area, well-defined pore architecture, can generate substantial amounts of ROS [56], and absorb twice as much visible light, when compared to their calcinated and undoped counterparts [34]. The synthesis of experimental adhesives was conducted using a previously published protocol because such a route allows for the functionalization of non-agglomerated N_TiO_2 and results in experimental materials with proven antibacterial properties [56]. Figure 2 illustrates that the functionalization of N_TiO_2 into OPTB, which is a self-etch and 5th generation dental adhesive resin, resulted in experimental materials displaying DC values that were either comparable or higher than those from the parental polymer. In addition, the reported results have indicated that experimental adhesives containing 20% of N_TiO_2 displayed the smallest decrease in DC values after two years of simulated shelf-life, thereby indicating that the nanotechnology proposed does not adversely impact the stability of the parental polymer and has the ability to maintain DC values at levels that are considered clinically relevant.

Mohammed and Riad, [57] while investigating the effects of silver nanoparticles (6.25, 12.5, 25, 50 and 100 μg/mL) on the antibacterial properties and degree of conversion of a self-etch adhesive resin (Universal Bond, 3M ESPE, St. Paul, USA), have demonstrated that the incorporation of commercially available nanoparticles (in ethanol suspension or powder) resulted in experimental materials with DC values (26.14% ± 4.47% [in suspension] and 47.72% ± 4.47% [powder]) that were numerically lower and statistically different ($p < 0.001$, in suspension) than the DC values from the control group (50.31% ± 4.04%, no nanoparticles). These findings have highlighted the importance of the present study' results because the authors [57] have clearly demonstrated that simple incorporation of metaloxide nanoparticles into existing polymer formulations does not necessarily lead to the development of materials with improved properties, such as the ones reported herein. The biaxial flexure strength and flexural modulus data presented in Figure 3A,B indicate that the incorporation of N_TiO_2 into OPTB resulted in experimental materials displaying improved mechanical properties. The findings of the present study have been corroborated by Giannini et al. [58] while investigating the effects of filler particles on the mechanical properties of commercially available (Adper Single Bond [3M ESPE] and Prime and Bond NT [Dentsply]) dental adhesive resins. According to results reported [58], materials containing filler particles in their compositions displayed values of biaxial flexural strength and modulus that were higher to those from unfilled adhesive resins.

The results from the nanoscale mechanical characterization (Figure 4A–F) of adhesives investigated have indicated that the functionalization of N_TiO_2 into OPTB resulted in experimental materials with surface properties that were comparable to those of the parental polymer. These results are fundamentally important from the mechanical and biological standpoints because it is well known that increased surface roughness results in larger surface areas, higher aggregation, and growth of oral biofilms, and to premature mechanical failures. Results reported have also demonstrated that experimental materials containing 20% and 30% of N_TiO_2 had elastic moduli that were significantly ($p < 0.001$) higher than that of OPTB, which indicates that experimental materials could potentially withstand the harsh conditions of the oral cavity. The results of the present study have been corroborated by the findings recently reported by Azmy et al. [59] who demonstrated that the incorporation of nanoparticles (ZrO_2, TiO_2, and SiO_2) into dental polymers resulted in experimental materials displaying superior wear resistance and flexural strength, thereby further supporting the experimental design and the nanotechnology proposed here.

Figure 5 illustrated that experimental adhesives investigated in the present study decreased, in a concentration-dependent manner, the viability of non-disrupted *S. mutans* biofilms (24 h) grown against the surfaces of specimens fabricated after specific periods of simulated shelf-life (T1 = 0, T2 = 6, T3 = 12 and T4 = 24 month). Results in Figure 6 illustrates that the attained reductions in metabolic activity (30% of N_TiO_2) were similar to those observed from biofilms topically treated with CHX (at T1 and T4).

The second part of the experiment was designed to demonstrate if specimens fabricated (immediately after the process described in Section 2.2) and stored in PBS (37 °C, duration of study) would be able to sustain strong and long-term (longevity) antibacterial properties. It is possible to observe in Figure 7 that experimental adhesives decreased, in a concentration-dependent manner, the metabolic activity of non-disrupted *S. mutans* biofilms independently of the time-point considered (T1 = 0, T2 = 1, T3 = 3, T4 = 6, and T5 = 12 months). Figure 8 shows that adhesives with 30% of N_TiO_2 displayed long-term antibacterial properties that were comparable to CHX (at time-points T1 and T2). Even though PBS storage was observed to decrease the antibacterial efficacy of all materials investigated, metabolic viability reductions (at T3, T4, and T5) attained with 30% of N_TiO_2 ranged from 49.30% (at T5) to 85.11% (at T3), which is still considered significant. Careful consideration must be exercised by the reader when comparing the antibacterial efficacies attained with either CHX (reductions ranged from 88% to 98%) or experimental adhesives investigated (4.36% [T1, 10% of N_TiO_2] to 98.63% [T2, 30% of N_TiO_2]) because fresh CHX (2%) was topically applied (for 2 min) onto biofilms immediately before bioluminescence procedures.

Taken together, the results of the present study have shown for the first time that synthetic strategies adopted (for nanoparticles and experimental adhesives) translated into experimental adhesives displaying promising non-leaching and long-term antibacterial properties (up to 24 months) without the need for visible light irradiation. Furthermore, specimens fabricated using experimental materials reported in the present study displayed surface characteristics that were similar to those fabricated with the parental polymer (OPTB) and did not exhibit any evidence of phase separation between the polymer matrix and nanoparticles during the periods of simulated shelf-life investigated. The absence of phase separation is a strong indication of the successful incorporation and functionalization of nanoparticles into commercially available dental polymers and may lead to materials displaying superior biological, mechanical, and physical properties [39].

The results of the present study have been corroborated by Melo et al. [60] while investigating the long-term antibacterial properties of experimental dental adhesive resins containing silver nanoparticles (0.05%) and cationic quaternary ammonium monomers (0–40%). According to the reported results [60], specimens (n = 6/group; diameter = 9.0 mm) were able to significantly ($p < 0.05$) decrease CFU/mL counts of *S. mutans*, total *Streptococci*, and total microorganisms in a dental plaque biofilm model after 24 h and after 6 months. Zhang et al. [61] investigated the long-term antibacterial properties of a novel bonding agent containing dimethylaminohexadecyl methacrylate (DMAHDM) after aging in water (37 °C; 1, 30, 90 and 180 days) and reported [61] that experimental materials were able to sustain significant ($p < 0.05$) antibacterial properties after 6 months. Machado et al. [62] recently investigated the long-term efficacy (up to 6 months, water storage, 37 °C) of experimental adhesive resins (66.66% Bis-GMA and 33.33% HEMA) containing triclosan-loaded chitosan particles (TLP) against *S. mutans* biofilms, and reported that experimental materials containing higher concentrations of TLP displayed significantly ($p < 0.05$) lower CFU/mL values (for planktonic cultures and biofilms), when compared to the parental polymer (without TLP) [62]. These results further corroborate the present study's rationale for the utilization of functionalized and non-agglomerated nanoparticles displaying non-leaching and long-term antibacterial properties (up to 24 months).

Future studies from our group will investigate (i) the long-term antibacterial and biomimetic properties of experimental adhesive resins containing third-generation co-doped metaloxide nanoparticles, (ii) the mechanisms of action by which immobilized nanoparticles downregulate the viability of cells and (iii) what genes and metabolic pathways are adversely affected by experimental materials containing single- or co-doped nanoparticle.

5. Conclusions

The present study has reported the synthesis of nitrogen-doped titanium dioxide nanoparticles and their successful incorporation into a commercially available dental adhesive resin. Experimental adhesives were demonstrated to display higher values of degree of conversion after polymer synthesis and after two years of simulated-shelf life and to display better mechanical properties. The antibacterial efficacies of experimental materials reported were comparable to CHX at specific time-points and sustained strong antibacterial properties during extended periods of simulated shelf-life (24 months) and aging in PBS (12 months). The results of the present study suggest that experimental materials reported may extend the service lives of polymer-based bonded restorations by decreasing the viability of pathogenic oral microorganisms, typically associated with the occurrence of secondary caries.

Author Contributions: Conceptualization, F.L.E.F., P.H. and R.D.H.; methodology, F.L.E.F., P.H. and R.D.H.; software, S.S.K.; statistical analysis, S.S.K. and P.H.; investigation, R.D.H. and P.H.; resources, F.L.E.F.; data curation, R.D.H. and P.H.; writing—original draft preparation, R.D.H.; writing—review and editing, R.D.H., F.L.E.F., P.H. and S.S.K.; supervision, F.L.E.F.; project administration, F.L.E.F.; funding acquisition, F.L.E.F. All authors have read and agreed to the published version of the manuscript.

Funding: This research was funded by Oklahoma Center for the Advancement of Science and Technology (OCAST), grant number HR20-121 to FLEF.

Data Availability Statement: Datasets generated and analyzed in the present study are available from the corresponding author on reasonable request.

Conflicts of Interest: The authors declare no conflict of interest. The funders had no role in the design of the study; in the collection, analyses, or interpretation of data; in the writing of the manuscript; or in the decision to publish the results.

References

1. Ferracane, J.L. Resin-based composite performance: Are there some things we can't predict? *Dent. Mater.* **2013**, *29*, 51–58. [CrossRef] [PubMed]
2. Listl, S.; Galloway, J.; Mossey, P.; Marcenes, W. Global economic impact of dental diseases. *J. Dent. Res.* **2015**, *94*, 1355–1361. [CrossRef] [PubMed]
3. Demarco, F.F.; Collares, K.; Correa, M.B.; Cenci, M.S.; de Moraes, R.R.; Opdam, N.J. Should my composite restorations last forever? Why are they failing? *Braz. Oral Res.* **2017**, *31*, 92–99. [CrossRef] [PubMed]
4. Featherstone, J. The continuum of dental caries—Evidence for a dynamic disease process. *J. Dent. Res.* **2004**, *83*, 39–42. [CrossRef]
5. Featherstone, J.D. The science and practice of caries prevention. *J. Am. Dent. Assoc.* **2000**, *131*, 887–899. [CrossRef]
6. Brunthaler, A.; Konig, F.; Lucas, T.; Sperr, W.; Schedle, A. Longevity of direct resin composite restorations in posterior teeth. *Clin. Oral Investig.* **2003**, *7*, 63–70. [CrossRef]
7. Watts, D.; Marouf, A.; Al-Hindi, A. Photo-polymerization shrinkage-stress kinetics in resin-composites: Methods development. *Dent. Mater.* **2003**, *19*, 1–11. [CrossRef]

8. Drummond, J.L. Degradation, Fatigue, and Failure of Resin Dental Composite Materials. *J. Dent. Res.* **2008**, *87*, 710–719. [CrossRef]
9. Beyth, N.; Domb, A.J.; Weiss, E.I. An in vitro quantitative antibacterial analysis of amalgam and composite resins. *J. Dent.* **2007**, *35*, 201–206. [CrossRef]
10. Heintze, S.D.; Loguercio, A.D.; Hanzen, T.A.; Reis, A.; Rousson, V. Clinical efficacy of resin-based direct posterior restorations and glass-ionomer restorations—An updated meta-analysis of clinical outcome parameters. *Dent. Mater.* **2022**, *38*, e109–e135. [CrossRef]
11. Burke, F.J.; Wilson, N.H.; Cheung, S.W.; Mjor, I.A. Influence of patient factors on age of restorations at failure and reasons for their placement and replacement. *J. Dent.* **2001**, *29*, 317–324. [CrossRef]
12. Rajan, G.; Raju, R.; Jinachandran, S.; Farrar, P.; Xi, J.; Prusty, B.G. Polymerisation shrinkage profiling of dental composites using optical fibre sensing and their correlation with degree of conversion and curing rate. *Sci. Rep.* **2019**, *9*, 3162. [CrossRef] [PubMed]
13. Beyth, N.; Bahir, R.; Matalon, S.; Domb, A.J.; Weiss, E.I. *Streptococcus mutans* biofilm changes surface-topography of resin composites. *Dent. Mater.* **2008**, *24*, 732–736. [CrossRef] [PubMed]
14. Spencer, P.; Ye, Q.; Misra, A.; Goncalves, S.E.P.; Laurence, J.S. Proteins, Pathogens, and Failure at the Composite-Tooth Interface. *J. Dent. Res.* **2014**, *93*, 1243–1249. [CrossRef] [PubMed]
15. Lemos, J.A.; Burne, R.A. A model of efficiency: Stress tolerance by *Streptococcus mutans*. *Microbiology* **2008**, *154 Pt 11*, 3247. [CrossRef] [PubMed]
16. Lemos, J.; Palmer, S.; Zeng, L.; Wen, Z.; Kajfasz, J.; Freires, I.; Abranches, J.; Brady, L. The biology of *Streptococcus mutans*. *Microbiol. Spectr.* **2019**, *7*, 1–26. [CrossRef] [PubMed]
17. Mjör, I.A.; Toffentti, F. Secondary caries: A literature review with case reports. *Quintessence Int.* **2000**, *31*, 165–179.
18. Malacarne, J.; Carvalho, R.M.; de Goes, M.F.; Svizero, N.; Pashley, D.H.; Tay, F.R.; Yiu, C.K.; Carrilho, M.R.d.O. Water sorption/solubility of dental adhesive resins. *Dent. Mater.* **2006**, *22*, 973–980. [CrossRef]
19. Lee, Y.; An, S.Y.; Park, Y.J.; Yu, F.H.; Park, J.C.; Seo, D.G. Cytotoxic effects of one-step self-etching adhesives on an odontoblast cell line. *Scanning* **2016**, *38*, 36–42. [CrossRef]
20. Esteban Florez, F.L.; Kraemer, H.; Hiers, R.D.; Sacramento, C.M.; Rondinone, A.J.; Silvério, K.G.; Khajotia, S.S. Sorption, solubility and cytotoxicity of novel antibacterial nanofilled dental adhesive resins. *Sci. Rep.* **2020**, *10*, 13503. [CrossRef]
21. Song, L.; Ye, Q.; Ge, X.; Misra, A.; Laurence, J.S.; Berrie, C.L.; Spencer, P. Synthesis and evaluation of novel dental monomer with branched carboxyl acid group. *J. Biomed. Mater. Res. B Appl. Biomater.* **2014**, *102*, 1473–1484. [CrossRef] [PubMed]
22. Moszner, N.; Hirt, T. New polymer-chemical developments in clinical dental polymer materials: Enamel–dentin adhesives and restorative composites. *J. Polym. Sci. Pol. Chem.* **2012**, *50*, 4369–4402. [CrossRef]
23. Ergun, G.; Egilmez, F.; Uctasli, M.B.; Yilmaz, S. Effect of light curing type on cytotoxicity of dentine-bonding agents. *Int. Endod. J.* **2007**, *40*, 216–223. [CrossRef] [PubMed]
24. Cai, Y.; Strömme, M.; Zhang, P.; Engqvist, H.; Welch, K. Photocatalysis induces bioactivity of an organic polymer based material. *RSC Adv.* **2014**, *4*, 57715–57723. [CrossRef]
25. Welch, K.; Cai, Y.; Engqvist, H.; Stromme, M. Dental adhesives with bioactive and on-demand bactericidal properties. *Dent. Mater.* **2010**, *26*, 491–499. [CrossRef]
26. Pérez-Mondragón, A.A.; Cuevas-Suárez, C.E.; García-Serrano, J.; Trejo-Carbajal, N.; Lobo-Guerrero, A.; Herrera-González, A.M. Adhesive Resins with High Shelf-Life Stability Based on Tetra Unsaturated Monomers with Tertiary Amines Moieties. *Polymers* **2021**, *13*, 1944. [CrossRef]
27. Moraes, G.; Zambom, C.; Siqueira, W.L. Nanoparticles in Dentistry: A Comprehensive Review. *Pharmaceuticals* **2021**, *14*, 752. [CrossRef]
28. Jeevanandam, J.; Barhoum, A.; Chan, Y.S.; Dufresne, A.; Danquah, M.K. Review on nanoparticles and nanostructured materials: History, sources, toxicity and regulations. *Beilstein J. Nanotechnol.* **2018**, *9*, 1050–1074. [CrossRef]
29. Corrie, S.R.; Thurecht, K.J. *Nano-Bio Interactions: Guiding the Development of Nanoparticle Therapeutics, Diagnostics, and Imaging Agents*; Springer: Berlin/Heidelberg, Germany, 2016; Volume 33, pp. 2311–2313.
30. Ali, A.; Ismail, H.; Amin, K. Effect of nanosilver mouthwash on prevention of white spot lesions in patients undergoing fixed orthodontic treatment-a randomized double-blind clinical trial. *J. Dent. Sci.* **2022**, *17*, 249–255. [CrossRef]
31. Sun, L.; Chow, L.C. Preparation and properties of nano-sized calcium fluoride for dental applications. *Dent. Mater.* **2008**, *24*, 111–116. [CrossRef]
32. Nizami, M.Z.I.; Xu, V.W.; Yin, I.X.; Yu, O.Y.; Chu, C.-H. Metal and Metal Oxide Nanoparticles in Caries Prevention: A Review. *Nanomaterials* **2021**, *11*, 3446. [CrossRef] [PubMed]
33. Bahadur, J.; Agrawal, S.; Panwar, V.; Parveen, A.; Pal, K. Antibacterial properties of silver doped TiO2 nanoparticles synthesized via sol-gel technique. *Macromol. Res.* **2016**, *24*, 488–493. [CrossRef]
34. Esteban Florez, F.L.; Hiers, R.D.; Larson, P.; Johnson, M.; O'Rear, E.; Rondinone, A.J.; Khajotia, S.S. Antibacterial dental adhesive resins containing nitrogen-doped titanium dioxide nanoparticles. *Mater. Sci. Eng. C* **2018**, *93*, 931–943. [CrossRef]
35. Matsunaga, T.; Tomoda, R.; Nakajima, T.; Wake, H. Photoelectrochemical sterilization of microbial cells by semiconductor powders. *FEMS Microbiol. Lett.* **1985**, *29*, 211–214. [CrossRef]
36. International Commission on Non-Ionizing Radiation Protection. Guidelines on limits of exposure to ultraviolet radiation of wavelenghts between 180 nm and 400 nm (incoherent optical radiation). *Health Phys.* **2004**, *87*, 171–186. [CrossRef]

37. Livraghi, S.; Paganini, M.C.; Giamello, E.; Selloni, A.; Di Valentin, C.; Pacchioni, G. Origin of Photoactivity of Nitrogen-Doped Titanium Dioxide under Visible Light. *J. Am. Chem. Soc.* **2006**, *128*, 15666–15671. [CrossRef]
38. Esteban Florez, F.L.; Hiers, R.D.; Zhao, Y.; Merritt, J.; Rondinone, A.J.; Khajotia, S.S. Optimization of a real-time high-throughput assay for assessment of *Streptococcus mutans* metabolism and screening of antibacterial dental adhesives. *Dent. Mater.* **2020**, *36*, 353–365. [CrossRef]
39. Esteban, F.L.; Trofimov, A.A.; Ievlev, A.; Qian, S.; Rondinone, A.J.; Khajotia, S.S. Advanced characterization of surface-modified nanoparticles and nanofilled antibacterial dental adhesive resins. *Sci. Rep.* **2020**, *10*, 9811. [CrossRef]
40. Rueggeberg, F.A.; Hashinger, D.T.; Fairhurst, C.W. Calibration of FTIR conversion analysis of contemporary dental resin composites. *Dent. Mater.* **1990**, *6*, 241–249. [CrossRef]
41. Soyland, R.; Currier, G.F.; Kadioglu, O.; Li, J.; Esteban Florez, F.L.; Rueggeberg, F.A.; Khajotia, S.S. Pairing orthodontic adhesive resins and light-curing units for optimal degree of conversion. *J. World Fed. Orthod.* **2020**, *9*, 68–74. [CrossRef]
42. Palin, W.M.; Fleming, G.J.P.; Marquis, P.M. The reliability of standardized flexure strength testing procedures for a light-activated resin-based composite. *Dent. Mater.* **2005**, *21*, 911–919. [CrossRef] [PubMed]
43. Butt, H.-J.; Cappella, B.; Kappl, M. Force measurements with the atomic force microscope: Technique, interpretation and applications. *Surf. Sci. Rep.* **2005**, *59*, 1–152. [CrossRef]
44. Bourbia, M.; Ma, D.; Cvitkovitch, D.G.; Santerre, J.P.; Finer, Y. Cariogenic bacteria degrade dental resin composites and adhesives. *J. Dent. Res.* **2013**, *92*, 989–994. [CrossRef]
45. Maryoosh, R.M.; Al-Shamma, A.M.W. Shear bond strength of fluorinated graphene nanoparticles modified dental adhesives. *Ann. Trop. Med. Public Health* **2020**, *23*, 1–6. [CrossRef]
46. Almaroof, A.; Niazi, S.A.; Rojo, L.; Mannocci, F.; Deb, S. Evaluation of dental adhesive systems incorporating an antibacterial monomer eugenyl methacrylate (EgMA) for endodontic restorations. *Dent. Mater.* **2017**, *33*, e239–e254. [CrossRef] [PubMed]
47. André, C.B.; Chan, D.C.; Giannini, M. Antibacterial-containing dental adhesives' effects on oral pathogens and on *Streptococcus mutans* biofilm: Current perspectives. *Am. J. Dent.* **2018**, *31*, 37b–41b.
48. Leyva Del Rio, D.; Sartori, N.; Tomblin, N.B.; Phark, J.H.; Pardi, V.; Murata, R.M.; Duarte, S., Jr. Bioactive Dental Adhesive System With tt-Farnesol: Effects on Dental Biofilm and Bonding Properties. *Front. Bioeng. Biotechnol.* **2020**, *8*, 865. [CrossRef]
49. Cha, H.S.; Shin, D.H. Antibacterial capacity of cavity disinfectants against *Streptococcus mutans* and their effects on shear bond strength of a self-etch adhesive. *Dent. Mater. J.* **2016**, *35*, 147–152. [CrossRef]
50. Comba, A.; Maravic, T.; Valente, L.; Girlando, M.; Cunha, S.R.; Checchi, V.; Salgarello, S.; Tay, F.R.; Scotti, N.; Breschi, L.; et al. Effect of benzalkonium chloride on dentin bond strength and endogenous enzymatic activity. *J. Dent.* **2019**, *85*, 25–32. [CrossRef]
51. Barros Silva, P.G.; Terto, C.N.N.; da Cunha Andrade, R.; Fernandes, A.R.; Rolim, J.P.M.L.; Neri, J.R. Antimicrobial evaluation, degree of solubility, and water sorption of universal dental adhesive incorporated with epigallocatechin-3-gallate: An in-vitro study. *Eur. J. Dent.* **2020**, *9*, 79–83. [CrossRef]
52. Beyth, N.; Yudovin-Farber, I.; Bahir, R.; Domb, A.J.; Weiss, E.I. Antibacterial activity of dental composites containing quaternary ammonium polyethylenimine nanoparticles against *Streptococcus mutans*. *Biomaterials* **2006**, *27*, 3995–4002. [CrossRef]
53. Lee, M.-J.; Kim, M.-J.; Kwon, J.-S.; Lee, S.-B.; Kim, K.-M. Cytotoxicity of Light-Cured Dental Materials according to Different Sample Preparation Methods. *Materials* **2017**, *10*, 288. [CrossRef] [PubMed]
54. Gotti, V.B.; Correr, A.B.; Lewis, S.H.; Feitosa, V.P.; Correr-Sobrinho, L.; Stansbury, J.W. Influence of nanogel additive hydrophilicity on dental adhesive mechanical performance and dentin bonding. *Dent. Mater.* **2016**, *32*, 1406–1413. [CrossRef] [PubMed]
55. Cai, Y.; Stromme, M.; Melhus, A.; Engqvist, H.; Welch, K. Photocatalytic inactivation of biofilms on bioactive dental adhesives. *J. Biomed. Mater. Res. B Appl. Biomater.* **2014**, *102*, 62–67. [CrossRef]
56. Huo, Y.; Bian, Z.; Zhang, X.; Jin, Y.; Zhu, J.; Li, H. Highly Active TiO2-xNx Visible Photocatalyst Prepared by N-Doping in Et3N/EtOH Fluid under Supercritical Conditions. *J. Phys. Chem. C* **2008**, *112*, 6546–6550. [CrossRef]
57. Mohammed, H.F.; Riad, M.I. The effect of silver nanoparticles incorporation in the self-etch adhesive system on its antibacterial activity and degree of conversion: An in-vitro study. *F1000Research* **2019**, *8*, 244. [CrossRef]
58. Giannini, M.; Mettenburg, D.; Arrais, C.A.; Rueggeberg, F.A. The effect of filler addition on biaxial flexure strength and modulus of commercial dentin bonding systems. *Quintessence Int.* **2011**, *42*, e39–e43. [PubMed]
59. Azmy, E.; Al-Kholy, M.R.Z.; Fattouh, M.; Kenawi, L.M.M.; Helal, M.A. Impact of Nanoparticles Additions on the Strength of Dental Composite Resin. *Int. J. Biomater.* **2022**, *2022*, 1165431. [CrossRef] [PubMed]
60. Melo, M.A.; Weir, M.D.; Cheng, L.; Zhang, K.; Xu, H.H.K. Broad-spectrum, long-term antibiofilm features of metallic nanoparticles and antibacterial monomers on dental adhesive and resin composite surfaces. In Proceedings of the 8th ASM—American Society for Microbiology, Washington, DC, USA, 7–11 October 2018.
61. Zhang, N.; Zhang, K.; Weir, M.D.; Xu, D.J.; Reynolds, M.A.; Bai, Y.; Xu, H.H.K. Effects of water-aging for 6 months on the durability of a novel antimicrobial and protein-repellent dental bonding agent. *Int. J. Oral Sci.* **2018**, *10*, 18. [CrossRef]
62. Machado, A.H.S.; Garcia, I.M.; Motta, A.d.S.d.; Leitune, V.C.B.; Collares, F.M. Triclosan-loaded chitosan as antibacterial agent for adhesive resin. *J. Dent.* **2019**, *83*, 33–39. [CrossRef]

nanomaterials

Article

Shear Bond Strength and Color Stability of Novel Antibacterial Nanofilled Dental Adhesive Resins

Qing Hong [1], Alexandra C. Pierre-Bez [1], Matheus Kury [2], Mark E. Curtis [3], Rochelle D. Hiers [4], Fernando L. Esteban Florez [4,*] and John C. Mitchell [1,*]

1 College of Dental Medicine, Midwestern University, Glendale, AZ 85308, USA
2 Division of Operative Dentistry, Department of Restorative Dentistry, Piracicaba School of Dentistry, University of Campinas, Piracicaba 13414-903, Brazil
3 Mewbourne School of Petroleum and Geological Engineering, University of Oklahoma, Norman, OK 73019, USA
4 Division of Dental Biomaterials, Department of Restorative Sciences, College of Dentistry, The University of Oklahoma Health Sciences Center, Oklahoma City, OK 73117, USA
* Correspondence: fernando-esteban-florez@ouhsc.edu (F.L.E.F.); jmitch@midwestern.edu (J.C.M.)

Abstract: Experimental adhesives containing co-doped metaloxide nanoparticles were demonstrated to display strong and long-term antibacterial properties against *Streptococcus mutans* biofilms. The present study represents an effort to characterize the shear-bond strength (SBS) and color stability (CS) of these novel biomaterials. Experimental adhesives were obtained by dispersing nitrogen and fluorine co-doped titanium dioxide nanoparticles (NF_TiO_2, 10%, 20% or 30%, v/v%) into OptiBond Solo Plus (OPTB). Dentin surfaces were wet-polished (600-Grit). Specimens (n = 5/group) of Tetric EvoCeram were fabricated and bonded using either OPTB or experimental (OPTB + NF_TiO_2) adhesives. Specimens were stored in water (37 °C) for twenty-four hours (T1), three months (T2), and six months (T3). At T1, T2, or T3, specimens were removed from water storage and were tested for SBS. Disc-shaped specimens (n = 10/group; d = 6.0 mm, t = 0.5 mm) of adhesives investigated were fabricated and subjected to thermocycling (10,000 cycles, 5–55 °C, 15 s dwell time). Specimens' colors were determined with a VITA Easyshade® V spectrophotometer (after every 1000 cycles). SBS data was analyzed using two-way ANOVA and post-hoc Tukey tests, while CS data was analyzed using one-way ANOVA and post-hoc Tukey tests (α = 0.05). Mean values of SBS ranged from 16.39 ± 4.20 MPa (OPTB + 30%NF_TiO_2) to 19.11 ± 1.11 MPa (OPTB), from 12.99 ± 2.53 MPa (OPTB + 30% NF_TiO_2) to 14.87 ± 2.02 (OPTB) and from 11.37 ± 1.89 (OPTB + 20% NF_TiO_2) to 14.19 ± 2.24 (OPTB) after twenty-four hours, three months, and six months of water storage, respectively. Experimental materials had SBS values that were comparable ($p > 0.05$) to those from OPTB independently of nanoparticle concentration or time-point considered. Experimental materials with higher NF_TiO_2 concentrations had less intense color variations and were more color stable than OPTB even after 10,000 thermocycles. In combination, the results reported have demonstrated that experimental adhesives can establish strong and durable bonds to human dentin while displaying colors that are more stable, thereby suggesting that the antibacterial nanotechnology investigated can withstand the harsh conditions within the oral cavity without compromising the esthetic component of dental restorations.

Keywords: metaloxide nanoparticles; antibacterial; shear-bond strength; dental materials; human dentin

1. Introduction

Composite resins and dental adhesive resins are typically used to repair and augment the function and esthetics of mineralized dental tissues. These mercury-free restorative materials display outstanding handling and esthetic properties, have good mechanical and physical properties, and are associated with minimally invasive [1] and ultraconservative restorative techniques [2]. However, despite their widespread acceptance and utilization, polymer-based adhesive restorations were demonstrated to have limited-service lives (5.7 years) [3] and to primarily fail by secondary caries [4]. Incomplete envelopment of collagen fibrils [3], polymerization shrinkage [5], hydrolysis [6], biodegradation (salivary esterases and biofilms) [7], and upregulation of pathogenic biofilms [8] are some of the typical limitations associated with current dental adhesive resins. Previous studies have indicated that polymer degradation byproducts accumulate at the tooth–adhesive interface and increase the virulence of caries-producing bacteria [9], thereby shifting the ecology of biofilms from a state of health into a disease-associated state [10,11]. The long-term accumulation of degradation byproducts at the adhesive interface has been suggested to result in a degradative positive feedback loop that is responsible for the catastrophic failure of the hybrid layer and the short longevity of polymer-based adhesive restorations [12–14].

Several approaches have been investigated to improve the clinical performance and longevity of dental adhesive resins. These include the incorporation of fluoride, quaternary ammonium dimethacrylates (QADM), silver (Ag) [15], zinc oxide (ZnO), and titanium dioxide nanoparticles (TiO_2) [16,17] into commercially available materials [18–22]. Nitrogen-doped TiO_2 produced via sol-gel (N-TiO_2) was demonstrated to enhance the strength and in vitro antibacterial properties of composite resins against *Escherichia coli* when used as a filler [23]. Salehi et al. demonstrated that orthodontic brackets coated with N-TiO_2 prevented the growth of *Streptococcus mutans* over a period of three months, which was considered to be an effective strategy in preventing enamel demineralization during orthodontic therapy [24]. Sodagar et al., while studying the antibacterial effect of TiO_2 in orthodontic resins, found that experimental materials containing TiO_2 (1%, 5%, or 10%, wt/wt) promoted significant in vitro microbial reductions against *S. mutans*, *Streptococcus sanguinis*, and *Lactobacillus acidophilus* [25]. The antibacterial efficacy of silver-doped TiO_2 (Ag_TiO_2) has also been evaluated [26]. Results reported have indicated that, independent of presentation (in suspension or immobilized), Ag_TiO_2 was capable of eradicating *S. mutans* planktonic cultures (strain NG8) when exposed to visible light irradiation (1500 lux) [26]. Despite these promising results, strategies previously reported were not capable of reducing the incidence of secondary caries or to extend the service lives of polymer-based bonded restorations, thereby underscoring a critical need for the development and characterization of novel materials displaying long-term and non-leaching antibacterial and biomimetic properties.

Nitrogen-doped titanium dioxide nanoparticles obtained using two-step solvothermal reactions (N_TiO_2, 6–15 nm) [27] have been recently characterized and incorporated into a fifth generation and commercially available dental adhesive resin (OptiBond Solo Plus, Kerr Corp., Orange, CA, USA; OPTB). Results indicated that N_TiO_2 were spherical, had smooth surfaces, and were capable of absorbing two times more visible light when compared to undoped TiO_2 (P25, Degussa, Germany) [27]. Experimental adhesives containing N_TiO_2 (50, 67, and 80%, v/v) were demonstrated to have strong antibacterial and biomimetic properties when irradiated with visible light (410 $\pm$ 10 nm, 3 h irradiation = 38.75 J/cm^2, 24 h irradiation = 310.07 J/cm^2) [27]. These promising findings precipitated the execution of a follow-up study that focused on the characterization of the water sorption, solubility, and cytotoxicity of experimental adhesives containing N_TiO_2 (25 and 30%, v/v) [28]. Materials tested were demonstrated to be less soluble, absorb less water, and to have cytotoxicity properties that were comparable ($p > 0.05$) to those of three commercially available and FDA-approved materials (OptiBond Solo Plus (Kerr Corp., Orange, CA, USA), Clearfil SE Protect (Kuraray Noritake Dental, Tokyo, Japan), and Adper Scotchbond (3M ESPE, St. Paul, MN, USA)) [28].

In combination, previous results cited indicate that N_TiO_2-containing materials could withstand the harsh conditions in the oral cavity and could express strong antibacterial properties at the adhesive interface without adversely impacting the viability of human pulp cells. Esteban Florez et al. [29], while investigating the utilization of simple methods to synthesize, surface-modify, and functionalize co-doped nanoparticles (nitrogen and fluorine (NF_TiO_2) or nitrogen and silver (NAg_TiO_2)) into OPTB, have demonstrated that strategies used resulted in the covalent functionalization of nanoparticles, the establishment of smooth interfaces between nanoparticles and polymer chains, and experimental materials displaying polymeric networks with preserved morphology, structure and three-dimensional lamellar distribution, which indicates that experimental materials tested could potentially display good mechanical properties [29]. Despite these encouraging findings, previous studies [30,31] indicated that the incorporation of nanoparticles (NAg = 0.1% and NACP = 10%, 20%, and 30%) into dental adhesive resins results in experimental materials displaying high viscosity and limited ability to flow into dentinal tubules, as denoted by the formation of resin tags that were shorter than those attained with the parental polymer (Scotchbond Multi-Purpose, 3M, St. Paul, MN, USA).

These results are critical because micromechanical interlocking is the primary mechanism of adhesion to enamel and dentin [32], and the longevity of polymer-based dental restorations has been shown to depend on the quality and durability of the hybrid layer. Since masticatory forces are described as a shearing phenomenon [33], previous studies indicated [34] that bond strength should be tested in shear mode to yield results that are acceptable and relevant from the clinical standpoint [33]. Sirisha and Tanikonda, while reviewing the validity of bond strength tests [35,36], indicated, based on previous scientific evidence [37], that it is important to measure the immediate and aged (three months in water storage, 37 °C) bond strength to predict the clinical effectiveness of dental adhesive resins. Studies conducted by Hidari et al. [38] and Takamizawa et al. [39] have found that in vitro bond strength of commercial self-etch adhesive resins to dentin surfaces was unchanged by storage in water (3, 6, and 12 months). Other studies have reported unchanged or increased bond strengths of self-etch adhesives to dentin at six months and one year. A study in primates found that tensile bond strength of a self-etching primer to teeth extracted 24 h, six months, and one year after restoration placement was not significantly different [40]. Additional studies reported decreased dentin bond strength of etch-and-rinse and self-etch adhesives following long-term water storage [41,42]. Conflicting evidence on long-term performance of adhesive resins warrants additional studies to understand bond durability and color stability of novel dental adhesive resins. Therefore, the objectives of the present study were to characterize the shear bond strength (SBS) and the color stability (CS) of novel experimental adhesive resins containing varying concentrations (10%, 20%, and 30%, *v*/*v*%) of antibacterial NF_TiO_2. Null hypotheses tested were that experimental materials would display values of shear bond strength and color stability (in terms of ΔE_{ab} and ΔE_{00}) that were similar ($p > 0.05$) to those attained with the commercially available dental adhesive resin.

2. Materials and Methods

2.1. Synthesis of Nanoparticles

NF_TiO_2 was synthesized in the laboratories of the Division of Dental Biomaterials at the University of Oklahoma Health Sciences Center College of Dentistry using protocols recently published [29,43–45]. In a typical reaction, a solution of 1.7 g of titanium butoxide (Aldrich, St. Louis, MO, USA, 97%), 6.8 g of oleylamine (Aldrich, St. Louis, MO, USA, 70%), 7.10 g of oleic acid (Aldrich, St. Louis, MO, USA, 90%), 0.065 g of ammonium fluoride (Alfa Aesar, Tewksbury, MA, USA), and 0.01250 g of tetramethylammonium hydroxide (Aldrich, St. Louis, MO, USA) was prepared and then mixed with a solution containing 13.10 g of ethyl alcohol (Decon Labs, King of Prussia, PA, USA, 200 proof) and 0.40 g of ultrapure water. Each solution was transparent before mixing, but the final solution became clouded due to micelle formation and some hydrolysis. The final solution was placed in a high-pressure reaction vessel (Parr reactor series 4593, Parr Instrument Company, Moline, IL, USA) lined with a boron-silicate glass liner (Parr Instrument Company, Moline, IL, USA), reacted (180 °C, 24 h, 15 psi), and stirred using an external shaft coupled to a propeller blade (280 rpm). Upon cooling, the solution was decanted and washed with anhydrous ethanol (3 × 1 min/wash) to remove extraneous surfactant. NF_TiO_2 nanoparticles were then stored in ethanol (concentration ≅ 40 mg/mL).

2.2. Synthesis of Experimental Adhesives

Experimental dental adhesive resins were formulated by ultrasonically dispersing 10%, 20%, and 30% (v/v, suspended in ethyl alcohol) of NF_TiO_2 into OptiBond Solo Plus (OPTB, Kerr Corp., Orange, CA, USA) using a Q700 sonicator (QSonica, Newtown, CT, USA). The rationale for selecting these concentrations was based on previous publications demonstrating that single-doped or co-doped nanoparticles displayed strong initial antibacterial and biomimetic properties with or without visible light irradiation at similar concentrations [27,45–47]. Experimental adhesives were then stored (dark conditions, 25 °C) in the original containers provided by the manufacturer until use (5 mL, black bottles, Kerr Corp., Orange, USA).

2.3. Fabrication of Specimens

2.3.1. For Shear Bond Strength Test

De-identified and caries-free extracted human molars (n = 110) were longitudinally sectioned (diamond saw, Model 650, South Bay Technology Inc., San Clemente, CA, USA) under copious water irrigation to expose dentin surfaces. Each half tooth section was embedded in an acrylic block and wet-polished through 600-grit SiC paper (Buehler, Lake Bluff, IL, USA) before being randomly distributed to groups investigated. Specimens (n = 15/group, height = 2.38 mm, diameter = 2.00 mm) of Tetric EvoCeram (Ivoclar Vivadent Inc., Amherst, MA, USA) were fabricated using a mounting jig (Ultradent Products, South Jordan, UT, USA) and were bonded (37% H_3PO_4, 15 s; light cured with Bluephase Style, 20 s top irradiation and 10 s/each side after removal of jig) using the proper adhesive in each group. Figure 1 shows bonded specimens prior to shear bond strength testing. Specimens were then individually stored in distilled water (dark conditions, at 37 °C) for twenty-four hours (T1), three months (T2), and six months (T3). An additional set of specimens (n = 15/group/time-point) were fabricated using the same procedures previously described but were bonded (following manufacturer instructions) using unaltered and commercially-available OptiBond Solo Plus (Kerr Corp., Orange, CA, USA), and served as the control group in the present study. At specific time-points (T1, T2 or T3), specimens pertaining to each group investigated were removed from water storage and were lightly dried in preparation for SBS testing procedures.

Figure 1. A representative specimen of Tetric EvoCeram composite resin that was bonded using either OPTB, or experimental adhesives containing 10%, 20%, or 30% of NF_TiO_2. Red circle highlights the location where specimens were typically fabricated.

2.3.2. For Color Stability Test

Disc-shaped specimens (n = 10/group; d = 6.0 mm, t = 0.5 mm) were manually fabricated in a custom-made metallic mold using the adhesives investigated (unaltered (OPTB) or experimental (OPTB+ NF_TiO_2)). All specimens were photopolymerized (385–515 nm, 1000 mW/cm^2, 60 s, VALO, Ultradent Products, Inc.) against glass.

2.4. Shear Bond Strength Test

After the completion of each water storage period (T1, T2 and T3, respectively) specimens from each group were individually tested for SBS. Individual specimens were secured using a metal clamp coupled to an Ultradent shear bond strength testing jig (Figure 2) coupled to an ElectroPuls (model E-3000) universal mechanical testing machine (Instron, Norwood, MA, USA). A notched-edge blade was applied parallel to the face of the specimen within 0.25 mm. SBS tests were performed at a crosshead speed of 0.5 mm/min until fracture or when the applied load dropped by 40% or more, which was then identified as the point of failure. Values of SBS (in MPa) were calculated by dividing the peak load at failure by specimens' surface areas.

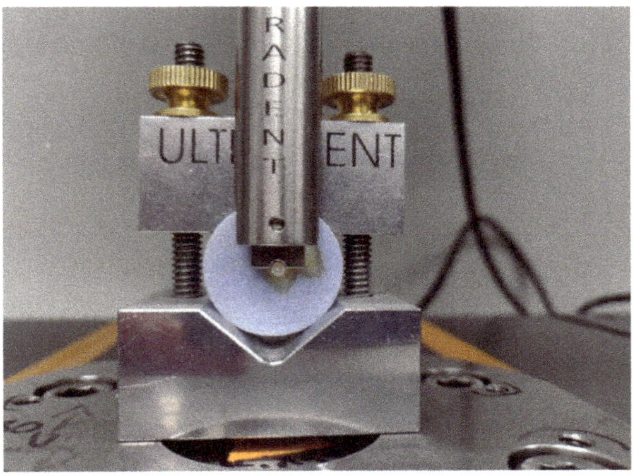

Figure 2. Specimen loaded in the Ultradent shear bond strength testing jig coupled to the E-3000 Instron Universal Testing Machine prior to testing with the notched-edge blade mounted as close to the dentin surface as possible (within 0.25 mm).

2.5. Scanning Electron Microscopy

Samples were allowed to dry for approximately 18 hrs at room temperature and then mounted to aluminum SEM stubs using a conductive carbon adhesive tape. A grounding strip of copper tape was applied to connect the top surface of the sample to the Al stub. Samples were coated with Au/Pd in a Denton Desk V sputter coater system to ensure conductivity. Imaging was performed in a Helios Nanolab™ 600 Dualbeam FIB/SEM (FEI, Hillsboro, OR, USA) at 2 kV accelerating voltage and 0.34 nA beam current. The detector used was a solid-state annular backscatter (ABS) detector. FEI MAPS 2.0 software was used to acquire and stitch together individual SEM images, given that the ROI was not completely within the field of view at the lowest magnification of the SEM.

2.6. Color Stability Test

Specimens fabricated as described in Section 2.3.2. were tested for color stability using a hand-held reflectance spectrophotometer (VITA Easyshade® V, VITA, Bad Säckingen, Germany). Specimens' colors were determined immediately after fabrication and after every increment of 1000 thermocycles (5–55 °C, 15 s dwell time) to a total of 10,000 thermocycles. Raw experimental data was then tabulated, and specimens' color stability was calculated (in terms of ΔE_{ab} and ΔE_{00}) using the CIELAB and CIEDE2000 [48] equations below and following a previously published protocol [49]. The rationale for the selection of thermocycling as the method to test the color stability of materials investigated was based on a previous study [50] that indicated that thermocycling best simulates the typical physical conditions found in the oral environment (in terms of temperature and humidity).

$$\Delta E_{ab} = [(L_2 - L_1)^2 + (a_2 - a_1)^2 + (b_2 - b_1)^2]^{1/2}$$

$$\Delta E_{00} = [(\Delta L'/K_L S_L)^2 + (\Delta C'/K_C S_C)^2 + (\Delta H'/K_H S_H)^2 + RT \times (\Delta C'/K_C S_C) \times (\Delta H'/K_H S_H)]^{1/2}$$

where H stands for hue and C for chroma.

2.7. Statistical Analysis

Mean and standard deviation values of SBS were calculated and were used to determine statistically significant differences among groups investigated using two-way ANOVA and post-hoc Tukey tests ($\alpha = 0.05$). Color stability data were assessed for normality and homoscedasticity. Mean values that met statistical assumptions were analyzed using one-way ANOVA and post hoc Tukey tests. Statistical analyses in the present study were performed ($\alpha = 0.05$) using GraphPad Prism version 9.0.0 for Windows (GraphPad Software, San Diego, CA, USA; available at www.graphpad.com accessed on 15 November 2022) and SPSS 23 (IBM).

3. Results

Figure 3A–D illustrate the macroscopic aspects of failed interfaces attained using either OPTB (3A), or experimental adhesives containing 10% (3B), 20% (3C) or 30% (3D) of NF_TiO_2 after six months of water storage (37 °C, dark storage conditions).

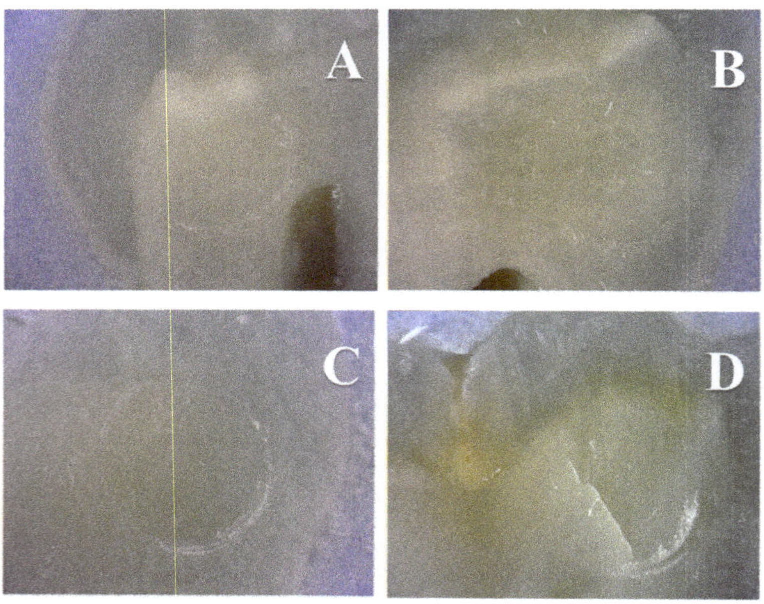

Figure 3. Macroscopic aspects of failed interfaces established either with OPTB (**A**) or experimental adhesives containing 10% (**B**), 20% (**C**), or 30% (**D**) of NF_TiO_2 and after six months of water storage (37 °C, dark storage conditions). Representative figures indicate that failures were cohesive in resin composite independently of the type of adhesive (unaltered or experimental) or concentration of nanoparticles (10%, 20%, or 30%) considered.

Figure 4A–D reveals the microscopic aspect of failed interfaces attained using either OPTB (4A) or experimental adhesives containing 10% (4B), 20% (4C), or 30% (4D) of NF_TiO_2 after six months of water storage (37 °C, dark storage conditions).

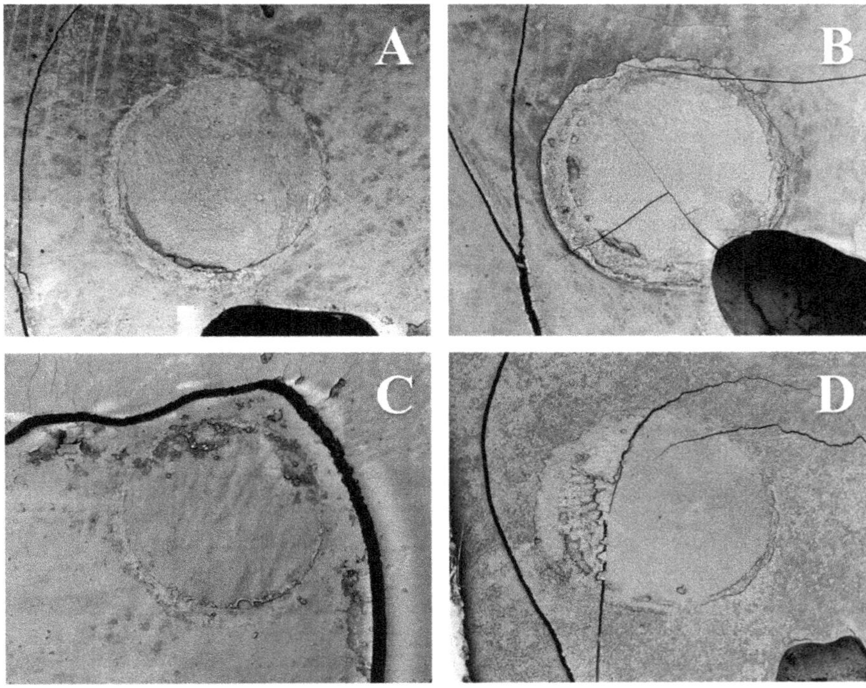

Figure 4. Microscopic aspects of failed interfaces established either with OPTB (**A**) or experimental adhesives containing 10% (**B**), 20% (**C**), or 30% (**D**) of NF_TiO_2 and after six months of water storage (37 °C, dark storage conditions). Images confirm the findings of Figure 3 and indicate that failures were cohesive in resin composite independently of the type of adhesive (unaltered or experimental) or concentration of nanoparticles (10%, 20%, or 30%) considered. Cracks on images are the result of specimen dehydration, sputter coating and extended periods under vacuum to allow the collection of 25 equally-sized images for the illustration of the entire region of interest at the lowest magnification.

Mean and standard deviation values of SBS from each group investigated were calculated as described in Section 2.4 and are shown in Figure 5. Two-way ANOVA and post-hoc Tukey tests ($\alpha = 0.05$) were used to determine the significance level of inter- and intra-group differences at each time point (T1, T2 and T3). Mean values of SBS ranged from 16.39 ± 4.20 MPa (OPTB + 30% NF_TiO_2) to 19.11 ± 1.11 MPa (OPTB), from 12.99 ± 2.53 MPa (OPTB + 30% NF_TiO_2) to 14.87 ± 2.02 (OPTB), and from 11.37 ± 1.89 (OPTB + 20% NF_TiO_2) to 14.19 ± 2.24 (OPTB) after twenty-four hours, three months, and six months of water storage, respectively. Even though parameters "time" ($p < 0.0001$) and "adhesive" ($p = 0.0005$) were observed to be significant predictors of response when analyzed individually, the interaction between parameters of interest (time*adhesive) was shown to not be significant ($p = 0.8536$), which indicates that the functionalization of NF_TiO_2 into OPTB did not adversely impacted OPTB's ability to establish adhesive interfaces that were strong and durable.

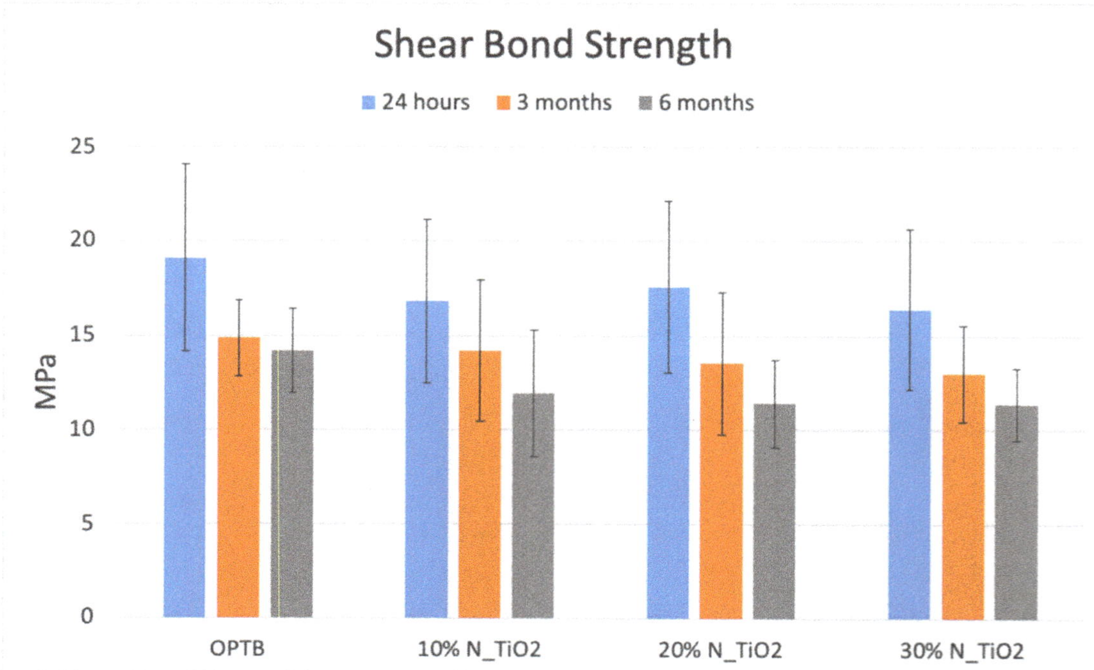

Figure 5. Shear bond strength of unaltered (OPTB) or experimental adhesive resins (OPTB + NF_TiO_2). Blue, orange, and grey bars indicate 24 h, three months, and six months of water storage (37 °C), respectively.

Table 1 illustrates, in terms of intra-group percent change, the temporal variation in SBS mean values experienced by specimens after each water storage period (T2 and T3) relative to the baseline (T1), where it is possible to observe that SBS values significantly ($p < 0.001$) decreased overtime and independently of the adhesive (control or experimental) and nanoparticles' concentrations (10%, 20%, or 30%) considered.

Table 1. Intra-group percent change in shear bond strength over time.

	24 h–3 Months	3 Months–6 Months
OPTB	−22.1%	−4.6%
10% N_TiO_2	−15.4%	−16.0%
20% N_TiO_2	−22.8%	−15.9%
30% N_TiO_2	−20.7%	−12.5%

Table 2 demonstrates the temporal percent change in SBS values relative to the parental polymer (OPTB), where it is possible to observe that after twenty-four hours of water storage, experimental groups containing either 20% or 30% of NF_TiO_2 displayed variations in SBS values that were either positive (4.5%) or negative (−2.4%), respectively. After three months of water storage, and except for the group containing 10% of NF_TiO_2 that had a positive variation in SBS values (+3.3%), all other groups had SBS variations that were negative in nature and ranged from −5.5% (OPTB + 30% NF_TiO_2) to −1.4% (OPTB + 20% NF_TiO_2). This trend could not be observed after six months of water storage, and all experimental groups were associated with positive variations in SBS values that ranged from 0.4% (OPTB + 30% NF_TiO_2) to 5.5% (OPTB + 10% NF_TiO_2).

Table 2. Percent Change—Experimental adhesives from OptiBond Solo Plus.

	24 h	3 Months	6 Months
10% N_TiO_2	0.0%	3.3%	5.5%
20% N_TiO_2	4.5%	−1.4%	0.8%
30% N_TiO_2	−2.4%	−5.5%	0.4%

In combination, these results suggest that the functionalization of NF_TiO_2 into OPTB resulted in experimental materials that can establish strong and durable bonds to human dentin. In addition, the statistical analysis performed indicated that materials investigated (control or experimental) had similar (time*adhesive; $p = 0.8131$) SBS values and would be appropriate for clinical use, thereby supporting the utilization of the antibacterial nanotechnology investigated in the present study. Figure 6A–C and Tables 3 and 4 illustrate the results from the color stability analysis (in terms of ΔE_{00} and ΔE_{ab}, respectively), where it is possible to observe that after 1000 thermocycles, experimental adhesives investigated displayed color variations that were smaller than that of the parental polymer and were below the perceptibility threshold, which indicates that experimental materials were more color stable than OPTB independently of the concentration of nanoparticles considered. From 3000 to 10,000 thermocycles, all materials investigated (unaltered or experimental) were observed to display color variations that were higher than the thresholds of perceptibility (PT) and acceptability (AT). The analysis of ΔE_{00} obtained from 5000 to 10,000 thermocycles indicated that experimental adhesives containing 30% of NF_TiO_2 displayed color variation values that were consistently lower when compared to the control group (OPTB). These results have clearly demonstrated that the incorporation of NF_TiO_2 into OPTB did not alter the optical behavior of OPTB and seemed to downregulate the color variations observed, which further supports the clinical utilization of the antibacterial nanotechnology investigated here.

Table 3. Mean and standard deviation values of color variation (in terms of ΔE_{00}) after each specific thermocycle investigated.

	ΔE_{00}									
GROUP	NUMBER OF CYCLES									
	1000	2000	3000	4000	5000	6000	7000	8000	9000	10,000
OPTB	1.3 (0.2)	1.6 (0.4)	2.0 (0.5)	1.8 (0.5)	2.4 (0.7)	3.0 (0.6)	3.0 (0.6)	3.3 (0.6)	4.1 (0.6)	4.1 (0.6)
10% NF_TiO_2	0.9 (0.3)	1.6 (0.4)	2.2 (0.5)	2.8 (0.6)	3.5 (0.8)	3.7 (0.5)	4.0 (0.7)	4.0 (0.7)	4.6 (0.8)	4.9 (0.8)
20% NF_TiO_2	0.7 (0.3)	2.7 (1.2)	2.2 (0.5)	2.9 (0.9)	2.8 (0.4)	3.7 (0.3)	3.6 (0.3)	4.0 (0.3)	4.2 (0.5)	4.3 (0.4)
30% NF_TiO_2	0.7 (0.2)	1.5 (0.4)	2.0 (0.5)	2.0 (0.3)	2.1 (0.5)	2.6 (0.6)	2.5 (0.6)	2.9 (0.5)	3.3 (0.7)	3.8 (0.7)

Table 4. Mean and standard deviation values of color variation (in terms of ΔE_{ab}) after each specific thermocycle investigated.

	ΔE_{ab}									
GROUP	NUMBER OF CYCLES									
	1000	2000	3000	4000	5000	6000	7000	8000	9000	10,000
OPTB	1.0 (0.2)	1.7 (0.6)	2.3 (0.6)	1.8 (0.6)	2.5 (0.7)	3.2 (0.7)	3.2 (0.7)	3.6 (0.8)	4.5 (0.7)	4.4 (1.1)
10% NF_TiO_2	1.1 (0.4)	1.7 (0.6)	2.5 (0.8)	3.2 (0.9)	3.9 (0.9)	4.2 (0.6)	4.6 (1.1)	4.8 (1.1)	5.1 (0.9)	5.3 (1.0)
20% NF_TiO_2	0.9 (0.5)	3.8 (1.8)	2.5 (0.7)	3.5 (1.4)	2.8 (0.5)	4.1 (0.7)	3.9 (0.5)	4.4 (0.5)	4.7 (0.7)	4.6 (0.5)
30% NF_TiO_2	0.9 (0.5)	1.7 (0.5)	2.2 (1.0)	2.3 (0.5)	2.2 (0.6)	3.0 (0.7)	2.6 (0.7)	3.1 (0.7)	3.5 (0.7)	4.2 (0.6)

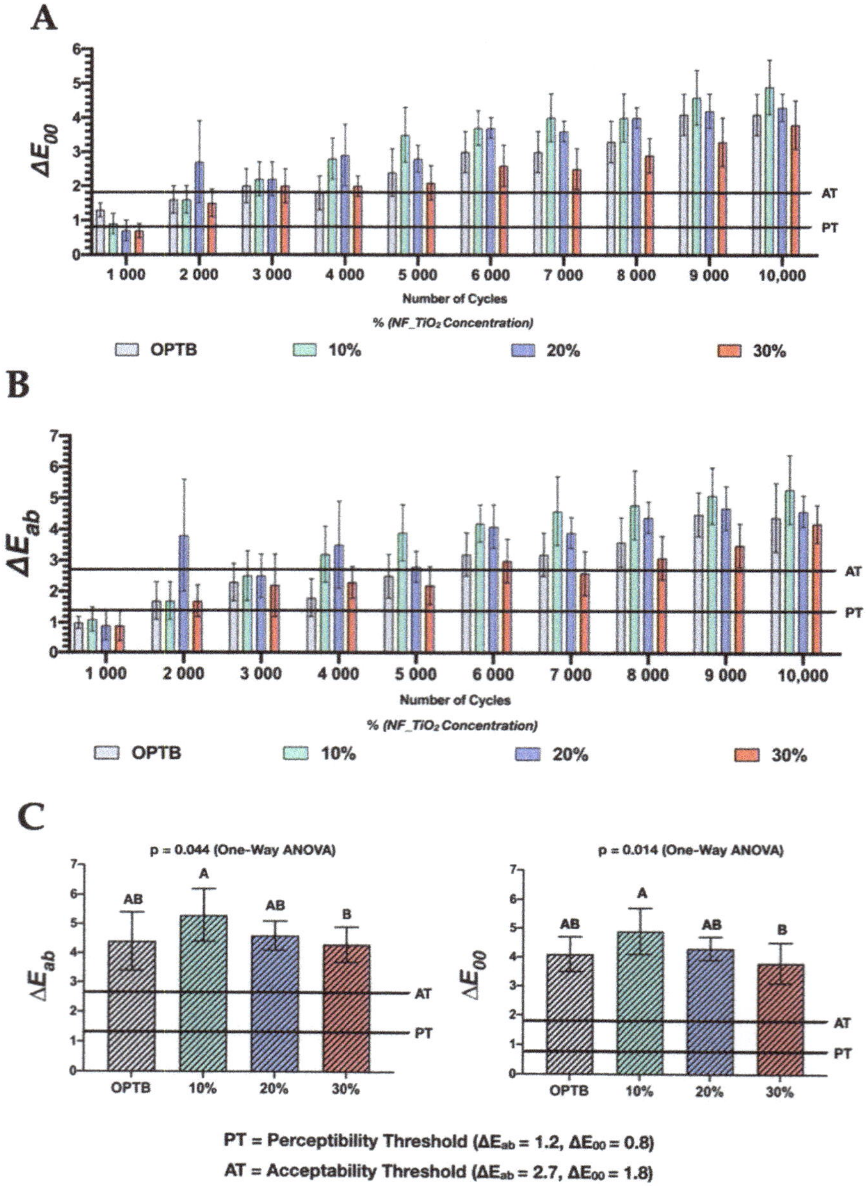

Figure 6. Color stability in terms of (**A**) ΔE_{00} and (**B**) ΔE_{ab} of unaltered and experimental dental adhesive resins. (**C**) the color stability of materials investigated after 10,000 thermocycles. Dissimilar letters above individual bars denote the presence of statistically significant differences ($p < 0.05$).

4. Discussion

Recent advancements in the field of nanotechnology allowed the utilization of nanoparticles and nanostructured materials in many health-related fields, including biosensing, cancer treatment, and drug delivery [51]. Metaloxide nanoparticles (TiO_2, ZnO, Ag_TiO_2, N_TiO_2 and NF_TiO_2) have gained the attention of researchers due to their ability to generate reactive oxygen species (ROS) and their proven strong antimicrobial and biomimetic effects [52]. Nanoparticles have been utilized in orthodontics, prosthodontics, and restorative dentistry with the objective to improve materials' mechanical properties and reduce the microbial load adjacent or attached to different types of restorations [53]. Previous studies confirmed that the incorporation of metaloxide nanoparticles into dental polymers resulted in experimental materials with improved physical, mechanical, biological, and biocompatibility properties, thereby supporting the continued development and characterization of nanofilled materials [29,54,55]. However, despite these encouraging findings, recent reports have raised concerns regarding the incorporation of nanoparticles into dental adhesive resins, because nanofilled materials displayed high viscosity and limited ability to flow into dentinal tubules [30,31].

The present study represents an effort to characterize the SBS and the CS of experimental adhesive resins containing varying concentrations of NF_TiO_2. Results reported in Figure 5 have demonstrated that mean values of SBS peaked after 24 h of water immersion and decreased with the evolution of time (at T2 and T3) independently of adhesive (commercial or experimental) and concentration of nanoparticles considered (10%, 20%, or 30%). According to Esteban Florez et al. [28], the functionalization of hydrolysis-resistant nanoparticles (N_TiO_2) into dental adhesive resins results in materials that are less soluble, absorb less water, and have higher specific gravity, which in theory could decrease the detrimental effects of water on bond strength. Al-Saleh et al. [56], while investigating the influence of metaloxide nanoparticles on the bond strength and viscosity of dental adhesive resins, have demonstrated that materials containing 5% of either TiO_2 or ZrO_2 displayed non-Newtonian rheological behaviors, pseudo-plasticity properties, and improved flow, which suggests that nanofilled adhesive resins could establish strong interfaces with the tooth structure. Results reported [56] have also indicated that aging (thermocycling, 10,000 cycles, 5 °C and 55 °C, 5 s dwell time) led to significant reductions in SBS values independently of the material considered. The authors reported that adhesives containing TiO_2 nanoparticles were shown to produce enhanced hybrid layers when compared to those attained with ZrO_2-containing adhesives, as denoted by the formation of resin tags that were longer [56]. In combination, the findings cited corroborate the results of the present study and the rationale for the selection of the nanoparticles investigated.

Sun et al. [57], while investigating the effect of varying concentrations of TiO_2 (in terms of mass fraction, 0.08%, 0.10%, 0.12%, 0.02%, and 0.5%) in adhesive resins demonstrated that experimental materials displayed SBS values that were comparable ($p > 0.05$) to the parental polymer and indicated the presence of concentration thresholds (minimum and maximum) in which TiO_2 would yield a positive effect on bond strength. In contrast, the results of the present study revealed that all materials investigated displayed comparable ($p > 0.05$) mean values of SBS at twenty-four hours, three months, and six months independently of concentration of nanoparticles, thereby further corroborating that the functionalization of nanoparticles did not adversely impact the SBS of OPTB. Prior research comparing several self-etch adhesives found that OPTB displayed SBS values that were higher and statistically different ($p < 0.05$) when compared to those from AdheSE, Adper Prompt Self-Etch Adhesive, Clearfil SE Bond, and One-Up Bond F [58]. Even though our results revealed that OPTB displayed SBS mean values that were numerically higher than those of experimental materials at each time point (T1, T2 and T3), differences reported were not statistically significant ($p = 0.8131$).

The results of the present study indicated that no differences in SBS were detected between experimental adhesives containing NF_TiO_2. This was an unexpected behavior, because previous studies [30,31] have indicated that the incorporation of nanoparticles into dental polymers results in materials with higher viscosity and limited ability to establish adequate hybrid layers. According to Ashraf et al. [31], the agglomeration of nanoparticles reduces the potential enhancement of mechanical properties in nanocomposites due to restriction of interfacial area and inadequate dispersion of agglomerates. Esteban Florez et al. [29], while characterizing experimental adhesives containing as-synthesized and surface-modified nanoparticles using cutting-edge scientific technologies including small-angle neutron scattering (SANS) and time-of-flight secondary ion mass spectroscopy (ToF-SIMS), have demonstrated that the functionalization of non-agglomerated nanoparticles did not affect the composition or the structure of the polymeric network as denoted by materials displaying preserved morphology, radius of gyration, and three-dimensional lamellar distribution [29], which further supports the utilization of the antibacterial nanotechnology investigated in the present study.

Color stability results are shown in Figure 6A–C, where it is possible to observe that thermocycling (5–55 °C, 15 s dwell time) adversely impacted the color of all materials investigated. After 1000 thermocycles, variations in color were observed to be below PT and AT for experimental materials containing either 20% or 30% of NF_TiO_2. From 4000 to 10,000 thermocycles, all materials investigated displayed variations in color that consistently increased above PT and AT. When comparing (Tables 3 and 4) the color variation of OPTB and OPTB + 30% NF_TiO_2, it becomes obvious that the functionalization of NF_TiO_2 into OPTB resulted in materials that displayed the numerical values of color variation (in terms of ΔE_{ab} and ΔE_{00}) that were consistently lower than that of OPTB, thereby suggesting that NF_TiO_2 functionalization reduced the impact of thermocycling on the color of OPTB. Figure 6C highlights this trend by demonstrating that color variations detected (OPTB + 30% of NF_TiO_2) were numerically smaller and statistically significant (ΔE_{ab}: $p = 0.044$, ΔE_{00}: $p = 0.014$) compared with those from OPTB after 10,000 thermocycles. Šimunović et al. [59], while investigating the color stability of orthodontic dental adhesive resins, indicated that degree of conversion, water sorption, polymer composition, temperature, and humidity impact the color stability of dental biomaterials through a complex and multifactorial process.

Previous studies [60,61] have indicated that continuous temperature challenges, such as the one used in the present study, typically lead to internal stresses and increased water sorption, and suggested, based on previous scientific evidence [62], that 10,000 cycles correspond to one year of clinical service. El-Rashidy et al. [63], while evaluating the effects of two aging protocols (thermocycling (37–57 °C) for 10,000 cycles and storage in either tea or red wine) on the color stability (in terms of ΔE_{00}) of commercially available composite resins, have demonstrated that all materials investigated (Omnichroma (Tokuyama Dental, Tokyo, Japan) and Filtek Z350 XT (3M ESPE, St. Paul, MN, USA)) displayed color variations that were statistically significant ($p < 0.001$) and considered unacceptable (above PT and AT) after 10,000 cycles. The results published by Štruncová et al. [50] have corroborated the findings of the present study, when the authors demonstrated that the incorporation of metaloxide nanoparticles (NAg, 0.005–0.025 wt%) resulted in experimental materials displaying improved color stability even after 10,000 thermocycles. According to the authors, color changes observed have precipitated from the composition of the polymer composition, the type of initiator-activator system, and the level of filler particles' silanization. The incorporation of NAg in the concentrations reported [50] was shown to not change the color of materials investigated, which further supports the utilization of metaloxide nanoparticles in dental polymers for restorative applications.

Results reported in the present study have allowed us to fully accept the first part of the null hypothesis tested, that experimental materials with varying concentrations (10%, 20%, and 30%) of NF_TiO_2 would display SBS values that were comparable ($p > 0.05$) to those of commercially available adhesive resins. The second part of the null hypothesis was rejected, because experimental materials displayed color variations that were numerically smaller and statistically different (ΔE_{ab}: $p = 0.044$, ΔE_{00}: $p = 0.014$) than those from OPTB after 10,000 thermocycles. Once fully developed and characterized, materials reported in the present study may hold the promise to increase the service lives of polymer-based restorations and decrease the incidence of secondary caries and the costs of oral health care. Future studies in the field should investigate the long-term (twelve, twenty-four, and thirty-six months) shear bond strength and color stability of experimental materials in challenging conditions [64] that resemble those of the oral cavity to confirm the promising SBS and CS properties of novel experimental materials investigated.

5. Conclusions

The present study has demonstrated that co-doped nanoparticles synthesized via one-step solvothermal reactions can be successfully functionalized into a self-etch and commercially available dental adhesive resin. Experimental materials investigated were observed to display shear bond strength values that were comparable to the parental polymer after twenty-four hours, three months, or six months of water storage. Adhesives containing 30% of NF_TiO_2 were observed to be the most color stable material after 10,000 thermocycles. Experimental materials displayed positive percent changes in SBS values after six months of water storage, which indicates that adhesive interfaces established with experimental materials may become mechanically more stable when compared to the parental polymer. In combination, results reported support the continued development and characterization of the antibacterial nanotechnology investigated here.

Author Contributions: Conceptualization, F.L.E.F. and J.C.M.; data curation, M.K.; formal analysis, A.C.P.-B., M.K. and J.C.M.; funding acquisition, F.L.E.F.; investigation, Q.H. and M.K.; methodology, A.C.P.-B., M.K., M.E.C., F.L.E.F. and J.C.M.; project administration, A.C.P.-B. and J.C.M.; resources, R.D.H., M.E.C., F.L.E.F. and J.C.M.; software, M.E.C. and J.C.M.; supervision, F.L.E.F. and J.C.M.; validation, A.C.P.-B.; writing—original draft, Q.H.; writing—review & editing, A.C.P.-B., M.K., M.E.C., R.D.H., F.L.E.F. and J.C.M. All authors have read and agreed to the published version of the manuscript.

Funding: This research was partially funded by the Oklahoma Center for the Advancement of Science and Technology grant number HR20-121.

Data Availability Statement: Datasets generated and analyzed in the present study are available from the corresponding author on reasonable request.

Conflicts of Interest: The authors declare no conflict of interest. The funders had no role in the design of the study; in the collection, analyses, or interpretation of data; in the writing of the manuscript; or in the decision to publish the results.

References

1. Cho, K.; Rajan, G.; Farrar, P.; Prentice, L.; Prusty, B.G. Dental resin composites: A review on materials to product realizations. *Compos. Part B Eng.* **2022**, *230*, 109495. [CrossRef]
2. Mai, S.; Zhang, Q.; Liao, M.; Ma, X.; Zhong, Y. Recent Advances in Direct Adhesive Restoration Resin-Based Dental Materials With Remineralizing Agents. *Front. Dent. Med.* **2022**, *3*, 22. [CrossRef]
3. Pashley, D.H.; Tay, F.R.; Imazato, S. How to increase the Durability of Resin-Dentin Bonds. *Compend. Contin. Educ. Dent.* **2011**, *32*, 60–64.
4. Burke, F.J.; Wilson, N.H.; Cheung, S.W.; Mjor, I.A. Influence of patient factors on age of restorations at failure and reasons for their placement and replacement. *J. Dent.* **2001**, *29*, 317–324. [CrossRef]
5. Schricker, S.R. 9-Composite resin polymerization and relevant parameters. In *Orthodontic Applications of Biomaterials*; Eliades, T., Brantley, W.A., Eds.; Woodhead Publishing: Amsterdan, The Netherlands, 2017; pp. 153–170. [CrossRef]
6. Gitalis, R.; Bae, J.H.; Preston, M.; Patel, M.; Liu, Z.; Sun, C.; Stewart, C.; Xiao, Y.; Siqueira, W.L.; Glogauer, M.; et al. Human neutrophils compromise the restoration-tooth interface. *Acta Biomater.* **2020**, *117*, 283–293. [CrossRef]
7. Guo, X.; Yu, Y.; Gao, S.; Zhang, Z.; Zhao, H. Biodegradation of Dental Resin-Based Composite—A Potential Factor Affecting the Bonding Effect: A Narrative Review. *Biomedicines* **2022**, *10*, 2313.
8. Bourbia, M.; Ma, D.; Cvitkovitch, D.G.; Santerre, J.P.; Finer, Y. Cariogenic bacteria degrade dental resin composites and adhesives. *J. Dent. Res.* **2013**, *92*, 989–994. [CrossRef]
9. Sarikaya, R.; Song, L.; Yuca, E.; Xie, S.-X.; Boone, K.; Misra, A.; Spencer, P.; Tamerler, C. Bioinspired multifunctional adhesive system for next generation bio-additively designed dental restorations. *J. Mech. Behav. Biomed. Mater.* **2021**, *113*, 104135. [CrossRef]
10. Spencer, P.; Ye, Q.; Misra, A.; Goncalves, S.E.P.; Laurence, J.S. Proteins, Pathogens, and Failure at the Composite-Tooth Interface. *J. Dent. Res.* **2014**, *93*, 1243–1249. [CrossRef]
11. Spencer, P.; Ye, Q.; Park, J.; Topp, E.M.; Misra, A.; Marangos, O.; Wang, Y.; Bohaty, B.S.; Singh, V.; Sene, F.; et al. Adhesive/Dentin interface: The weak link in the composite restoration. *Ann. Biomed. Eng.* **2010**, *38*, 1989–2003. [CrossRef]
12. Yuca, E.; Xie, S.-X.; Song, L.; Boone, K.; Kamathewatta, N.; Woolfolk, S.K.; Elrod, P.; Spencer, P.; Tamerler, C. Reconfigurable Dual Peptide Tethered Polymer System Offers a Synergistic Solution for Next Generation Dental Adhesives. *Int. J. Mol. Sci.* **2021**, *22*, 6552. [CrossRef]
13. Sarikaya, R.; Song, L.; Ye, Q.; Misra, A.; Tamerler, C.; Spencer, P. Evolution of Network Structure and Mechanical Properties in Autonomous-Strengthening Dental Adhesive. *Polymers* **2020**, *12*, 2076. [CrossRef]

14. Xie, S.X.; Song, L.; Yuca, E.; Boone, K.; Sarikaya, R.; VanOosten, S.K.; Misra, A.; Ye, Q.; Spencer, P.; Tamerler, C. Antimicrobial Peptide-Polymer Conjugates for Dentistry. *ACS Appl. Polym. Mater.* **2020**, *2*, 1134–1144. [CrossRef]
15. Rai, M.; Yadav, A.; Gade, A. Silver nanoparticles as a new generation of antimicrobials. *Biotechnol. Adv.* **2009**, *27*, 76–83. [CrossRef]
16. Cai, Y.; Strömme, M.; Zhang, P.; Engqvist, H.; Welch, K. Photocatalysis induces bioactivity of an organic polymer based material. *RSC Adv.* **2014**, *4*, 57715–57723. [CrossRef]
17. Welch, K.; Cai, Y.; Engqvist, H.; Stromme, M. Dental adhesives with bioactive and on-demand bactericidal properties. *Dent. Mater.* **2010**, *26*, 491–499. [CrossRef]
18. Pashley, D.H.; Tay, F.R.; Yiu, C.; Hashimoto, M.; Breschi, L.; Carvalho, R.M.; Ito, S. Collagen degradation by host-derived enzymes during aging. *J. Dent. Res.* **2004**, *83*, 216–221. [CrossRef]
19. Hebling, J.; Pashley, D.H.; Tjaderhane, L.; Tay, F.R. Chlorhexidine arrests subclinical degradation of dentin hybrid layers in vivo. *J. Dent. Res.* **2005**, *84*, 741–746. [CrossRef]
20. Carrilho, M.R.; Carvalho, R.M.; de Goes, M.F.; di Hipolito, V.; Geraldeli, S.; Tay, F.R.; Pashley, D.H.; Tjaderhane, L. Chlorhexidine preserves dentin bond in vitro. *J. Dent. Res.* **2007**, *86*, 90–94. [CrossRef]
21. Makvandi, P.; Jamaledin, R.; Jabbari, M.; Nikfarjam, N.; Borzacchiello, A. Antibacterial quaternary ammonium compounds in dental materials: A systematic review. *Dent. Mater.* **2018**, *34*, 851–867. [CrossRef]
22. Imazato, S.; Ebi, N.; Takahashi, Y.; Kaneko, T.; Ebisu, S.; Russell, R.R.B. Antibacterial activity of bactericide-immobilized filler for resin-based restoratives. *Biomaterials* **2003**, *24*, 3605–3609. [CrossRef]
23. Zane, A.; Zuo, R.; Villamena, F.A.; Rockenbauer, A.; Digeorge Foushee, A.M.; Flores, K.; Dutta, P.K.; Nagy, A. Biocompatibility and antibacterial activity of nitrogen-doped titanium dioxide nanoparticles for use in dental resin formulations. *Int. J. Nanomed.* **2016**, *11*, 6459–6470. [CrossRef]
24. Salehi, P.; Babanouri, N.; Roein-Peikar, M.; Zare, F. Long-term antimicrobial assessment of orthodontic brackets coated with nitrogen-doped titanium dioxide against Streptococcus mutans. *Prog. Orthod.* **2018**, *19*, 35. [CrossRef]
25. Sodagar, A.; Akhoundi, M.S.A.; Bahador, A.; Jalali, Y.F.; Behzadi, Z.; Elhaminejad, F.; Mirhashemi, A.H. Effect of TiO_2 nanoparticles incorporation on antibacterial properties and shear bond strength of dental composite used in Orthodontics. *Dent. Press J. Orthod.* **2017**, *22*, 67–74. [CrossRef]
26. Chambers, C.; Stewart, S.B.; Su, B.; Jenkinson, H.F.; Sandy, J.R.; Ireland, A.J. Silver doped titanium dioxide nanoparticles as antimicrobial additives to dental polymers. *Dent. Mater.* **2017**, *33*, e115–e123. [CrossRef]
27. Esteban Florez, F.L.; Hiers, R.D.; Larson, P.; Johnson, M.; O'Rear, E.; Rondinone, A.J.; Khajotia, S.S. Antibacterial dental adhesive resins containing nitrogen-doped titanium dioxide nanoparticles. *Mater. Sci. Eng. C* **2018**, *93*, 931–943. [CrossRef]
28. Esteban Florez, F.L.; Kraemer, H.; Hiers, R.D.; Sacramento, C.M.; Rondinone, A.J.; Silvério, K.G.; Khajotia, S.S. Sorption, solubility and cytotoxicity of novel antibacterial nanofilled dental adhesive resins. *Sci. Rep.* **2020**, *10*, 13503. [CrossRef]
29. Esteban Florez, F.L.; Trofimov, A.A.; Ievlev, A.; Qian, S.; Rondinone, A.J.; Khajotia, S.S. Advanced characterization of surface-modified nanoparticles and nanofilled antibacterial dental adhesive resins. *Sci. Rep.* **2020**, *10*, 9811. [CrossRef]
30. Melo, M.A.; Cheng, L.; Zhang, K.; Weir, M.D.; Rodrigues, L.K.; Xu, H.H. Novel dental adhesives containing nanoparticles of silver and amorphous calcium phosphate. *Dent. Mater.* **2013**, *29*, 199–210. [CrossRef]
31. Ashraf, M.A.; Peng, W.; Zare, Y.; Rhee, K.Y. Effects of Size and Aggregation/Agglomeration of Nanoparticles on the Interfacial/Interphase Properties and Tensile Strength of Polymer Nanocomposites. *Nanoscale Res. Lett.* **2018**, *13*, 214. [CrossRef]
32. Meerbeek, V. Mechanisms of Resin Adhesion-Dentin and Enamel Bonding. *Aegis Dent. Netw.* **2008**, *2*, 2–8.
33. Preethy, N.A.; Jeevanandan, G.; Govindaraju, L.; Subramanian, E. Comparison of Shear Bond Strength of Three Commercially Available Esthetic Restorative Composite Materials: An In Vitro Study. *Int. J. Clin. Pediatr. Dent.* **2020**, *13*, 635–639. [CrossRef]
34. Nujella, B.P.S.; Choudary, M.T.; Reddy, S.P.; Kumar, M.K.; Gopal, T. Comparison of shear bond strength of aesthetic restorative materials. *Contemp. Clin. Dent.* **2012**, *3*, 22–26. [CrossRef]
35. Sirisha, K.; Rambabu, T.; Shankar, Y.R.; Ravikumar, P. Validity of bond strength tests: A critical review: Part I. *J. Conserv. Dent.* **2014**, *17*, 305–311. [CrossRef]
36. Sirisha, K.; Rambabu, T.; Ravishankar, Y.; Ravikumar, P. Validity of bond strength tests: A critical review-Part II. *J. Conserv. Dent.* **2014**, *17*, 420–426. [CrossRef]
37. Van Meerbeek, B.; Peumans, M.; Poitevin, A.; Mine, A.; Van Ende, A.; Neves, A.; De Munck, J. Relationship between bond-strength tests and clinical outcomes. *Dent. Mater.* **2010**, *26*, e100–e121. [CrossRef]
38. Hidari, T.; Takamizawa, T.; Imai, A.; Hirokane, E.; Ishii, R.; Tsujimoto, A.; Suzuki, T.; Miyazaki, M. Role of the functional monomer 10-methacryloyloxydecyl dihydrogen phosphate in dentin bond durability of universal adhesives in etch-&-rinse mode. *Dent. Mater. J.* **2020**, *39*, 616–623. [CrossRef]
39. Takamizawa, T.; Barkmeier, W.W.; Tsujimoto, A.; Scheidel, D.D.; Watanabe, H.; Erickson, R.L.; Latta, M.A.; Miyazaki, M. Influence of water storage on fatigue strength of self-etch adhesives. *J. Dent.* **2015**, *43*, 1416–1427. [CrossRef]
40. Sano, H.; Yoshikawa, T.; Pereira, P.N.; Kanemura, N.; Morigami, M.; Tagami, J.; Pashley, D.H. Long-term durability of dentin bonds made with a self-etching primer, in vivo. *J. Dent. Res.* **1999**, *78*, 906–911. [CrossRef]
41. Abdalla, A.I. Effect of long-term water aging on microtensile bond strength of self-etch adhesives to dentin. *Am. J. Dent.* **2010**, *23*, 29.
42. Osorio, R.; Pisani-Proenca, J.; Erhardt, M.C.G.; Osorio, E.; Aguilera, F.S.; Tay, F.R.; Toledano, M. Resistance of ten contemporary adhesives to resin–dentine bond degradation. *J. Dent.* **2008**, *36*, 163–169. [CrossRef]

43. Dinh, C.-T.; Nguyen, T.-D.; Kleitz, F.; Do, T.-O. Shape-Controlled Synthesis of Highly Crystalline Titania Nanocrystals. *ACS Nano* **2009**, *3*, 3737–3743. [CrossRef]
44. Huo, Y.; Bian, Z.; Zhang, X.; Jin, Y.; Zhu, J.; Li, H. Highly Active TiO_2-xNx Visible Photocatalyst Prepared by N-Doping in Et3N/EtOH Fluid under Supercritical Conditions. *J. Phys. Chem. C* **2008**, *112*, 6546–6550. [CrossRef]
45. Hiers, R.D.; Huebner, P.; Khajotia, S.S.; Florez, F.L.E. Characterization of Experimental Nanoparticulated Dental Adhesive Resins with Long-Term Antibacterial Properties. *Nanomaterials* **2022**, *12*, 3732. [CrossRef]
46. Esteban Florez, F.L.; Hiers, R.D.; Smart, K.; Kreth, J.; Qi, F.; Merritt, J.; Khajotia, S.S. Real-time assessment of Streptococcus mutans biofilm metabolism on resin composite. *Dent. Mater.* **2016**, *32*, 1263–1269. [CrossRef]
47. Esteban Florez, F.L.; Hiers, R.D.; Zhao, Y.; Merritt, J.; Rondinone, A.J.; Khajotia, S.S. Optimization of a real-time high-throughput assay for assessment of Streptococcus mutans metabolism and screening of antibacterial dental adhesives. *Dent. Mater.* **2020**, *36*, 353–365. [CrossRef]
48. Sharma, G.; Wu, W.; Dalal, E.N. The CIEDE2000 color-difference formula: Implementation notes, supplementary test data, and mathematical observations. *Color Res. Appl.* **2005**, *30*, 21–30. [CrossRef]
49. dos Santos Muniz Mota, G.M.; Kury, M.; Pereira da Silva Braga Tenório, C.; Lucisano Botelho do Amaral, F.; Turssi, C.P.; Cavalli, V. Effects of Artificial Staining and Bleaching Protocols on the Surface Roughness, Color, and Whiteness Changes of an Aged Nanofilled Composite. *Front. Dent. Med.* **2020**, *1*, 610586. [CrossRef]
50. Štruncová, M.; Toma, S.H.; Araki, K.; Bresciani, E.; Rodrigues, F.P.; Medeiros, I.S.; Dutra-Correa, M. Silver nanoparticles added to a commercial adhesive primer: Colour change and resin colour stability with ageing. *Int. J. Adhes. Adhes.* **2020**, *102*, 102694. [CrossRef]
51. Anjum, S.; Ishaque, S.; Fatima, H.; Farooq, W.; Hano, C.; Abbasi, B.H.; Anjum, I. Emerging Applications of Nanotechnology in Healthcare Systems: Grand Challenges and Perspectives. *Pharmaceuticals* **2021**, *14*, 707. [CrossRef]
52. Yu, Z.; Li, Q.; Wang, J.; Yu, Y.; Wang, Y.; Zhou, Q.; Li, P. Reactive Oxygen Species-Related Nanoparticle Toxicity in the Biomedical Field. *Nanoscale Res. Lett.* **2020**, *15*, 115. [CrossRef]
53. Sreenivasalu, P.K.P.; Dora, C.P.; Swami, R.; Jasthi, V.C.; Shiroorkar, P.N.; Nagaraja, S.; Asdaq, S.M.B.; Anwer, M.K. Nanomaterials in Dentistry: Current Applications and Future Scope. *Nanomaterials* **2022**, *12*, 1676. [CrossRef] [PubMed]
54. Shcherbakov, A.B.; Reukov, V.V.; Yakimansky, A.V.; Krasnopeeva, E.L.; Ivanova, O.S.; Popov, A.L.; Ivanov, V.K. CeO2 Nanoparticle-Containing Polymers for Biomedical Applications: A Review. *Polymers* **2021**, *13*, 924. [CrossRef] [PubMed]
55. Montanheiro, T.L.d.A.; Ribas, R.G.; Montagna, L.S.; Menezes, B.R.C.d.; Schatkoski, V.M.; Rodrigues, K.F.; Thim, G.P. A brief review concerning the latest advances in the influence of nanoparticle reinforcement into polymeric-matrix biomaterials. *J. Biomater. Sci. Polym. Ed.* **2020**, *31*, 1869–1893. [CrossRef] [PubMed]
56. Al-Saleh, S.; Alateeq, A.; Alshaya, A.H.; Al-Qahtani, A.S.; Tulbah, H.I.; Binhasan, M.; Shabib, S.; Farooq, I.; Vohra, F.; Abduljabbar, T. Influence of TiO_2 and ZrO_2 Nanoparticles on Adhesive Bond Strength and Viscosity of Dentin Polymer: A Physical and Chemical Evaluation. *Polymers* **2021**, *13*, 3794. [CrossRef] [PubMed]
57. Sun, J.; Forster, A.M.; Johnson, P.M.; Eidelman, N.; Quinn, G.; Schumacher, G.; Zhang, X.; Wu, W.-L. Improving performance of dental resins by adding titanium dioxide nanoparticles. *Dent. Mater.* **2011**, *27*, 972–982. [CrossRef]
58. Sensi, L.G.; Lopes, G.C.; Monteiro, S., Jr.; Baratieri, L.N.; Vieira, L.C. Dentin bond strength of self-etching primers/adhesives. *Oper. Dent.* **2005**, *30*, 63–68.
59. Šimunović, L.; Blagec, T.; Vrankić, A.; Meštrović, S. Color Stability of Orthodontic Brackets and Adhesives in Potentially Staining Beverages—In Vitro Study. *Dent. J.* **2022**, *10*, 115. [CrossRef]
60. Porojan, L.; Toma, F.R.; Uțu, I.-D.; Vasiliu, R.D. Optical Behavior and Surface Analysis of Dental Resin Matrix Ceramics Related to Thermocycling and Finishing. *Appl. Sci.* **2022**, *12*, 4346. [CrossRef]
61. Reddy, A.; Kumar, D.; Shivanna, V. Thermal cyclic changes on water sorption and solubility of composite restoratives—An in-vitro study. *J. Conserv. Dent.* **2006**, *9*, 63–71. [CrossRef]
62. Gale, M.S.; Darvell, B.W. Thermal cycling procedures for laboratory testing of dental restorations. *J. Dent.* **1999**, *27*, 89–99. [CrossRef] [PubMed]
63. El-Rashidy, A.A.; Abdelraouf, R.M.; Habib, N.A. Effect of two artificial aging protocols on color and gloss of single-shade versus multi-shade resin composites. *BMC Oral Health* **2022**, *22*, 321. [CrossRef] [PubMed]
64. Peris, A.R.; Mitsui, F.H.; Lobo, M.M.; Bedran-russo, A.K.; Marchi, G.M. Adhesive systems and secondary caries formation: Assessment of dentin bond strength, caries lesions depth and fluoride release. *Dent. Mater.* **2007**, *23*, 308–316. [CrossRef] [PubMed]

nanomaterials

MDPI

Article

Strategy for Conjugating Oligopeptides to Mesoporous Silica Nanoparticles Using Diazirine-Based Heterobifunctional Linkers

Md Arif Khan [1], Ramy W. Ghanim [1], Maelyn R. Kiser [1], Mahsa Moradipour [1], Dennis T. Rogers [2], John M. Littleton [2], Luke H. Bradley [3], Bert C. Lynn [4], Stephen E. Rankin [1,*] and Barbara L. Knutson [1,*]

1 Department of Chemical and Materials Engineering, University of Kentucky, Lexington, KY 40506-0046, USA; makhan226@uky.edu (M.A.K.); Ramy.Ghanim@uky.edu (R.W.G.); maelyn.kiser@uky.edu (M.R.K.); mahsa.moradipour93@gmail.com (M.M.)

2 Naprogenix Inc., UK-ASTeCC, Lexington, KY 40506-0286, USA; dennistrogers@gmail.com (D.T.R.); john.littleton@uky.edu (J.M.L.)

3 Department of Neuroscience, University of Kentucky, Lexington, KY 40536-0298, USA; lhbradley@uky.edu

4 Department of Chemistry, University of Kentucky, Lexington, KY 40506-0055, USA; bclynn2@uky.edu

* Correspondence: stephen.rankin@uky.edu (S.E.R.); bknut2@uky.edu (B.L.K.); Tel.: +1-859-257-9799 (S.E.R.); +1-859-257-5715 (B.L.K.)

Abstract: Successful strategies for the attachment of oligopeptides to mesoporous silica with pores large enough to load biomolecules should utilize the high surface area of pores to provide an accessible, protective environment. A two-step oligopeptide functionalization strategy is examined here using diazirine-based heterobifunctional linkers. Mesoporous silica nanoparticles (MSNPs) with average pore diameter of ~8 nm and surface area of ~730 m^2/g were synthesized and amine-functionalized. Tetrapeptides Gly-Gly-Gly-Gly (GGGG) and Arg-Ser-Ser-Val (RSSV), and a peptide comprised of four copies of RSSV (4RSSV), were covalently attached via their N-terminus to the amine groups on the particle surface by a heterobifunctional linker, sulfo-succinimidyl 6-(4,4′-azipentanamido)hexanoate (sulfo-NHS-LC-diazirine, or SNLD). SNLD consists of an amine-reactive NHS ester group and UV-activable diazirine group, providing precise control over the sequence of attachment steps. Attachment efficiency of RSSV was measured using fluorescein isothiocyanate (FITC)-tagged RSSV (RSSV-FITC). TGA analysis shows similar efficiency (0.29, 0.31 and 0.26 mol peptide/mol amine, respectively) for 4G, RSSV and 4RSSV, suggesting a generalizable method of peptide conjugation. The technique developed here for the conjugation of peptides to MSNPs provides for their attachment in pores and can be translated to selective peptide-based separation and concentration of therapeutics from aqueous process and waste streams.

Keywords: mesoporous silica; nanoparticle; conjugation; oligopeptide; heterobifunctional linker; diazirine

1. Introduction

Synthetic organic functional groups that can mimic the biological specificity of host–guest interactions have been used for analysis, sensing and isolation of different biomolecules, especially in affinity column chromatography [1–4]. Recent progress in supramolecular chemistry has resulted in tailor-made organic functionalities with high selectivity and specificity toward an array of biomolecules and therapeutic ligands, which can be used for their selective separation [5,6]. Synthetic peptides, in particular, have tremendous molecular recognition and selective binding capabilities and a large volume of peptide libraries with different binding properties has been developed during the last two decades [7–9]. Oligopeptide mimics of biological binding sites (of longer proteins) on solid supports are durable, reusable and cost-effective media for affinity separations [7].

Mesoporous silica materials are an ideal support for high affinity functional groups due to their high surface area, tunable pore size and ease of surface modification [10–12]. Use of mesoporous silica functionalized with affinity binding sites is widespread in chromatography [13,14].The most common biomolecule linking strategy is to functionalize the silica surface with amine groups and then to covalently link the amine moieties with the N-terminus of peptide/protein molecules. Bifunctional linkers [15–17] for amine–amine conjugation can be mainly divided into two types: homo-bifunctional and hetero-bifunctional. Homo-bifunctional linkers such as bis(sulfosuccinimidyl)suberate (BS^3) have amine-reactive NHS ester groups at both ends. While developed to crosslink two proteins or peptides in solution, they can hypothetically react with amine-functionalized particles at one end and with the amine terminal group of a peptide at the other to covalently attach the peptide to the surface [15,18,19]. During peptide conjugation to amine-functionalized porous particles, competitive rapid hydrolysis of the ester groups of homo-bifunctional linkers poses a significant problem during attachment, and proteins may be physically adsorbed rather than attaching covalently [20–22]. Controlling the rate of reaction of the linker with both the peptide and the surface is also challenging. Because of this, homo-bifunctional linkers lack specificity and precise control of attachment orientation.

A hetero-bifunctional linker containing an amine-reactive end and a UV-activable end overcomes the limitations of homo-bifunctional linkers for peptide conjugation. The activation of the UV active group by a light source provides precise control over the attachment mechanism and location. One of the most common hetero-bifunctional linkers for peptide/protein conjugation is the phenyl azide-based sulfosuccinimidyl 6-(4-azido-2-nitrophenyl-amino) hexanoate (sulfo-SANPAH) [20,23,24]. The UV-activable end of sulfo-SANPAH is activated at a wavelength < 320 nm (as low as 260 nm) [15,25], which can cause denaturation of proteins. In addition, the large size of the aromatic photoactive group in sulfo-SANPAH can create steric hindrance [26]. On the other hand, the diazirine group is activated at higher wavelength (340–370 nm) and, thus, does not cause denaturation of proteins and peptides. As a result, diazirine has been used extensively as a photo-labeling agent for proteins [25,27,28]. Diazirine-based photoreactive linkers have better stability compared to commonly used aromatic azide photo-linkers [29,30]. The linkers can also be designed to provide optimal peptide orientation for ligand attachment from the bound surface using spacers. For example, succinimidyl 6-(4,4′-azipentanamido) hexanoate (NHS-LC-diazirine) provides a sufficient spacing arm for the biomolecule to avoid effects of surface interactions on its properties [31]. The NHS-LC-diazirine linker has been shown to be stable under ambient lighting conditions [31,32]. The derivative sulfo-NHS-LC-diazirine (SNLD) contains a charged sulfate group that enhances its aqueous solubility and as a result can be used in physiological media [33].

Covalently linking peptides on the inner surface of the pores of mesoporous silica nanoparticles using hetero-bifunctional linkers has the potential to utilize their high pore volumes, but has limited applicability to mesoporous silica with insufficient pore size (<4 nm) synthesized by traditional templating methods [34,35]. Large pores are required for peptides and linkers to infiltrate the interior surface without pore blockage, as discussed in recent reviews of protein loading in mesoporous silica-based materials [36,37]. Having pores slightly larger than proteins has been shown to allow them to fully access the large pore volume of MSNPs while providing protection from factors such as temperature, solvent, pH and proteolytic hydrolysis [38–40]. Immobilizing peptides and proteins inside of mesoporous silica allows their use in diverse applications including therapeutic drug delivery, enzymatic catalysis in harsh environments and for affinity separations [41–43]. The hydrodynamic radius of proteins and polypeptides varies considerably depending on folding, but typical sizes are less than 8 nm for denatured proteins with up to 160 residues in their sequence, and considerably more residues for proteins in their native, folded state [44]. Surfactant templating with cationic surfactants such as alkyltrimethylammonium salts typically produces pore sizes of ~3 nm in diameter [45–47]. Only recently has surfactant templated synthesis of MSNPs with <170 nm particle diameter and pore sizes of 4.5 nm to 8 nm been reported by using pore expanding agents [48,49]. These particles facilitate the conjugation of long peptide sequences and proteins along with sufficient spacer arms to prevent strong surface effects. A study of peptide conjugation to the inner surface of silica microbubbles (cavity size: 0.55–0.65 mm) using NHS-diazirine has been reported [50]. However, the pore size of a microbubble is thousands of times larger than MSNPs, so there is a need to investigate the use of NHS-diazirine linkers in pores comparable in size to peptides and proteins. Based on previous studies [20,50,51], we hypothesize that effective attachment strategies for oligopeptides within expanded mesopores (>4 nm diameter) can be developed using NHS-diazirine linkers to the peptide N-terminus.

This work examines strategies to conjugate functional oligopeptides to large-pore amine-functionalized MSNPs (~8 nm diameter pores) using the diazirine-based hetero-bifunctional linker sulfo-NHS-LC-diazirine (SNLD) with a spacer of 1.25 nm (SNLD extended length [52]). The model system of four-amino-acid peptide RSSV (Arg-Ser-Ser-Val) was selected for this application from a combinatorial peptide library based on its ability to bind β-estradiol (equilibrium constant, $K = 6 \times 10^4$ M^{-1}) via column chromatography with good selectivity versus other steroids [53]. Conjugation of the peptide 4RSSV (Arg-Ser-Ser-Val-Arg-Ser-Ser-Val-Arg-Ser-Ser-Val-Arg-Ser-Ser-Val), a four-repeat peptide of the original RSSV 4-mer, and 4G (Gly-Gly-Gly-Gly) to the nanoparticles is investigated to test the versatility of the conjugation strategy for peptides of varying length and amino acid sequence. Two conjugation strategies are proposed based on the sequence of attaching the linker to the particle or peptide (Figure 1). For Type-1 attachment, the linker is first attached to the surface amine group using the NHS-ester and then the peptide amine group is attached to the linker using UV activation of diazirine. For Type-2 attachment, the linker is first attached to the peptide using the NHS-ester followed by attachment to the surface using UV activation. The anticipated advantage of Type-1 attachment is that unbound linkers can be removed by washing after the first step and before attachment to peptides. However, activated diazirine is indiscriminate toward N-H or O-H moieties and can result in attachment to either terminus of the peptide. Activation of the diazirine group under UV light has also been shown to bind the carboxyl moieties of proteins to some degree [54]. Another potential problem with Type-1 attachment in a porous system is the possible attachment of the diazirine moiety with another amine group inside the pores. Specific attachment of the peptide N-terminus to the linker during the first step is the main advantage of Type-2 attachment. In addition, the peptide-attached linker can diffuse into the pore prior to the photoactivated conjugation to the surface, promoting reactions within the pores and not just at the surface. Attachment of diazirine to O-H moieties of particles provide no disadvantages (in fact, it is probably advantageous in keeping some of amine groups unattached and positively charged) compared to Type-1 attachment. However, during the second step of Type-2 attachment, the activated diazirine of the peptide-attached linker can react with the C-terminus of another peptide. These solution-based complexes would be removed during washing but would result in inefficient use of peptide and linker. Considering these pros and cons, the attachment efficiency is examined for both of these methods using fluorescence spectroscopy of fluorescein isothiocyanate (FITC)-labeled peptide and thermogravimetric (TGA) analysis.

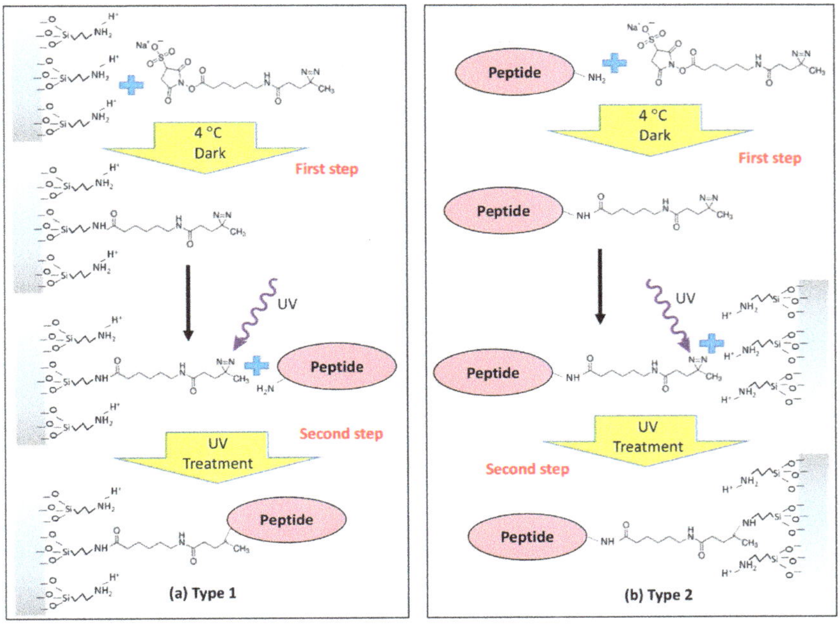

Figure 1. Schematic diagram of the peptide attachment strategies using hetero-bifunctional cross-linker Sulfo-NHS-LC-Diazirine (SNLD): (**a**) Type-1 attachment of the linker to the particle amine group first using the NHS group and then attaching to the peptide amine group using the UV-reactive diazirine group and (**b**) Type-2 attachment of the linker to the peptide amine group first using the NHS group and then attaching to the particle amine group using the UV-reactive diazirine group.

2. Materials and Methods

2.1. Chemicals and Reagents

Tetrapropyl orthosilicate (TPOS, 95%), phosphate-buffered saline (PBS) tablets, (3-aminopropyl)triethoxysilane (APTES, 99%) and N,N-dimethylformamide (DMF, molecular biology grade) were purchased from Sigma-Aldrich (St. Louis, MO, USA). Cetyltrimethylammonium bromide (CTAB, 99.8%) was purchased from MP Biomedicals (Solon, OH, USA); NaOH pellets (≥97%) from EMD Millipore; acetone (≥99.5%) from BDH analytical; 1,3,5-triisopropylbenzene (TIPB, >95%), triethanolamine (TEA, >98%) and fluorescamine from Alfa-Aesar (Tewksbury, MA, USA); and ultrapure deionized ultra-filtrated (DIUF) water, ethanol (200 proof) and 12 N HCl (ACS grade) from Fisher Scientific (Pittsburgh, PA, USA). Heterobifunctional linker sulfo-NHS-LC-diazirine (SNLD, Pierce^{TM}) was purchased from Thermo Fisher Scientific (Pittsburgh, PA, USA) and used for peptide conjugation.

Oligopeptide synthesis. Peptide 4G (GGGG) was obtained from Sigma-Aldrich (St. Louis, MO, USA); RSSV and 4RSSV were synthesized by Genscript (Piscataway, NJ, USA); and RSSV-FITC was synthesized by Lifetein (Hillsborough, NJ, USA) using solid-phase peptide synthesis and purified to >95% yield using reverse phase high-pressure liquid chromatography (RP-HPLC). RSSV-FITC was designed (sequence Arg-Ser-Lys(FITC)-Val) by replacing the serine at position 3 with a lysine to incorporate the fluorescein isothiocyanate (FITC) fluorescent label while maintaining the peptide arginine positive charge for peptide function and the single free amine (N-terminus) for attachment. All peptide sequences were confirmed to be of the correct molecular mass by LC-MSMS analysis following purification. Lyophilized peptide was resuspended in PBS buffer (pH 7.4) to a final concentration of 1.7 mg/mL 4G, 3 mg/mL RSSV or 12 mg/mL 4RSSV before conjugation to particles.

2.2. Mesoporous Silica Nanoparticles (MSNPs) Synthesis

MSNPs were synthesized following the method described by Yamada et al. [49], where TIPB was used to swell the CTAB micelles, the pore forming agent, during surfactant-templated synthesis. Initially, 0.56 mL of TEA and 3.0 g of CTAB were added to 360 mL of DIUF water. The solution was stirred at 80 °C for 2 h for complete mixing and emulsion formation, and 16 mL of TIPB was added under vigorous mixing. After 30 min, a complete colloidal state (oil-in-water) was obtained and 4.77 mL of TPOS was added with constant stirring. The solution was stirred vigorously for 12 h at 80 °C to obtain white solid particles. The particles were then separated by repeated centrifugation and washing, and the surfactant was removed by acidic ethanol (2 M HCl in ethanol) washing before drying at 84 °C overnight.

2.3. Amine Functionalization and Quantification

Amine-functionalized MSNPs (MSNPAs) were obtained by condensing APTES on the particle surface using modified literature procedures [10,55–57]. Two-hundred milligrams of MSNPs were uniformly dispersed in 25 mL of dry ethanol by sonication for 15 min. An amount of 0.5 mL of APTES was added dropwise under constant stirring in a nitrogen-filled glove bag, and the solution was kept stirring in a closed vessel for 24 h at room temperature. Particles were centrifuged with repeated ethanol washing and cured at 84 °C for 24 h. After curing, particles were stirred in excess ethanol for 24 h to remove any remaining loosely bound amine species. The functionalized particles were again washed 3 times with anhydrous ethanol and dried at 84 °C.

Amine quantification. The amount of amine groups on the functionalized particle surface was determined by a previously reported fluorescamine assay after particle dissolution [58,59]. A total of 30 mg of particles was dissolved over an 8 h period in 30 mL of 0.02 M NaOH at room temperature under vigorous stirring. One-hundred microliters of this solution and 1.0 mL of 1.0 mM fluorescamine in acetone were mixed with 2.0 mL of PBS solution at pH 7.4. The emitted fluorescence intensity of this solution was measured at 480 nm after excitation at 366 nm using an Agilent (Santa Clara, CA, USA) Varian Cary Eclipse fluorescent spectrophotometer. The amount of amine on the particle surface was determined by a calibration curve prepared using known amounts of APTES and non-functionalized MSNPs.

2.4. Peptide Attachment to MSNPAs

For Type-1 attachment, 10 mg of MSNPAs were sonicated in 1 mL of PBS solution (pH 7.4) for 15 min to make a uniform dispersion and mixed with 3 mg of SNLD in 100 µL DMF at 4 °C. The mixture was stirred at 4 °C overnight followed by centrifugation of the particles. Particles were washed with fresh PBS solution three times to remove excess, unbound linker and then dispersed in 2 mL of PBS solution containing 3 mg of RSSV peptide at room temperature with vortex mixing. The solution was then stirred overnight to allow adsorption of peptide on the particle surface and finally treated with UV light (Thorlabs, Newton, NJ, USA, model M00284926, λ = 365 nm, 1.2 A) with continuous stirring for 60 min, which was selected based on a series of attachment experiments using RSSV-FITC with different UV treatment times (0, 10, 30, 45, 60, 90 and 120 min) (see Section 2.6 for quantification). After UV treatment, peptide-conjugated particles (MSNPA-RSSV) were separated by centrifugation and washed 5 times with fresh PBS solution and dried overnight in vacuum at room temperature.

For the Type-2 attachment, 3 mg of SNLD in 100 µL DMF was mixed with 3 mg of RSSV peptide in 1 mL PBS solution, and the mixture was allowed to stir at 4 °C overnight for the completion of linking with the amine terminus of peptide. Ten milligrams of MSNPAs was dispersed uniformly in 1 mL PBS with sonication and added to the peptide-linker solution. The combined mixture was allowed to stir overnight for adsorption of peptide–linker conjugation on the particle surface and then UV treated for 60 min with continuous stirring to allow for the attachment to the particle surface. Finally, peptide-attached particles were separated by centrifugation, washed thoroughly with fresh PBS solution and dried overnight in vacuum at room temperature. 4G and 4RSSV attachment using the Type-2 attachment method is similar, but 1.7 mg of 4G or 12 mg of 4RSSV was used to keep the molar ratio of peptide:amine approximately the same as for RSSV.

2.5. Material Characterization

A FEI (Hillsboro, OR, USA) Helios Nanolab 660 Focused Ion Beam/Scanning Electron Microscope (SEM) was used to examine the particle morphology. Particles were dispersed onto a 15 mm aluminum stub using double-sided carbon tape, excess materials were blown off with dry N_2 and the samples were stored in a desiccator for 24 h. Prior to SEM analysis, the particles were coated with conductive Au-Pd alloy using an Emscope (Hercules, CA, USA) SC400 sputtering system. Average and standard deviation of particle diameters were calculated using 20 random particles using ImageJ Software, version 1.53. The internal pore structure of the particles was characterized using a FEI (Hillsboro, OR, USA) Talos F200X Transmission Electron Microscopy (TEM). Particles were dispersed in ethanol within a sonication bath for 5 min. Using a pipette, a small volume of the suspension was drop cast onto a lacey carbon-coated copper TEM grid, which was allowed to dry for a minimum of 5 min. Afterward, the samples were stored in a TEM storage box overnight before imaging. Surface characterization was performed using nitrogen adsorption conducted at -196 °C with a Micromeritics (Norcross, GA, USA) TriStar 3000 gas sorption instrument. Samples were degassed at 135 °C for 4 h under flowing N_2 gas before analysis. The specific surface area, average pore diameter and pore size distribution were estimated using the Brunauer–Emmett–Teller (BET) isotherm and by Barrett–Joyner–Halenda (BJH) method, respectively. To confirm the covalent linkage of peptides, Fourier transform infrared (FTIR) spectroscopy was conducted using a Thermo Fisher Scientific (Pittsburgh, PA) Nicolet Nexus 470 spectrometer with a deuterated triglycine sulfate (DTGS) detector. A total of 0.5 g of anhydrous KBr and particles (0.5–1.0 wt%) was crushed with a mortar and pestle, and some of this powder was pressed into a pellet for transmission analysis. Dynamic light scattering (DLS) was used to measure the zeta potential of the particles with an Anton-Paar (Ashland, VA, USA) Lightsizer 500 instrument. Initially, a 1 mg/mL uniform particle suspensions was made in DIUF water with sonication and diluted to around 0.1 mg/mL concentration before measurement. The pH values of the solutions were adjusted by adding a very small amount of either 0.1 N HCl or 0.1 N NaOH solution in water as required to obtain the desired pH, which was checked before every measurement with a benchtop pH meter (Accumet Research AR25 dual channel pH meter from Fisher Scientific, Pittsburgh, PA, USA). For measurements, the solutions were carefully placed in an Omega Cuvette consisting of an inverted omega-shaped capillary tube without any air bubbles.

2.6. Quantification of Peptide Attachment

Fluorescein isothiocyanate (FITC)-labeled RSSV peptide (RSSV-FITC) was used to quantify RSSV attachment efficiency to the particle amine groups by solution depletion with fluorescence spectroscopy. During quantification of peptide attachment, the amount of particles and linkers and the solution volume were the same as during RSSV attachment. The amount of RSSV-FITC was adjusted to 5.5 mg (instead of 3 mg RSSV) to keep the molar ratio of peptide:amine the same. Fluorescence intensity of the solution after attachment was measured at an emission wavelength of 520 nm (peak fluorescence) after excitation at 495 nm (peak absorbance) and compared to a calibration curve prepared with known amounts of RSSV-FITC. Type-1 attachment was quantified using the calibration curve of RSSV-FITC (only peptide) fluorescence intensity, whereas Type-2 attachment was quantified using the calibration curve of SNLD-RSSV-FITC (peptide conjugated with linker) after correcting to account for photo-bleaching (intensity reduction) during the UV treatment period.

Thermogravimetric analysis (TGA) was performed to quantify the amount of organic groups (peptides) conjugated to the particle surface with a TA-SDT-Q600 simultaneous

TGA/DSC instrument (TA Instruments, New Castle, DE, USA). Particle samples were dried at 50 °C under vacuum overnight before performing TGA analysis from 25 °C to 500 °C with a ramp rate of 10 °C/min and under constant dry air flow of 100 mL/min. Functional group contents were analyzed by thermal decomposition and combustion of organics for MSNPA, MSNPA-4G, MSNPA-RSSV and MSNPA-4RSSV compared to bare MSNP. Specifically, mass losses from 150 °C to 500 °C were used to determine the amount of amine grafting (before conjugation) and peptide grafting (after conjugation).

3. Results and Discussion

MSNPs with large pores (~8 nm average pore size) were synthesized by the method of Yamada et al. [49], where CTAB surfactant template was used to create pores and TIPB served to expand the micelle pore templates. The conditions used here were intended to reproduce the particles reported by Yamada et al. with a TIPB:CTAB molar ratio of 8, which were shown to have an open, accessible pore morphology. After synthesis, the surfactants were removed from the pores by acidic washing. Particles were then functionalized with amine groups by the condensation of the organosilane precursor, APTES, on the particle and pore surface. Spherical particles with average diameter of 146 ± 27 nm and radially oriented, 8 nm pores were obtained after template extraction, as seen in the TEM and SEM image presented in Figure 2. This particle size is consistent with the size reported by Yamada et al. [49] and suitable for cell uptake [60].

Surface characterization (surface area, pore volume and average pore size) was performed using nitrogen adsorption before and after amine functionalization (Figure S1 in electronic supplementary information (ESI) and Table 1). Nitrogen sorption of particles showed Type-IV isotherms (Figure S1a), consistent with the presence of uniformly sized mesopores and textural porosity giving an uptick in adsorption at high relative pressure. X-ray diffraction patterns did not show any peaks (data not shown), as the radially oriented pores (Figure 2a) do not have large domains of uniformly oriented mesopores. The average pore diameter, as determined by the BJH method, was reduced with functionalization (7.9 nm to 7.6 nm) (Figure S1b). The pore size distribution of the MSNP base material matches that reported by Yamada et al. for particles prepared with 8:1 TIPB:CTAB ratio [49]. The width of the pore size distribution is consistent with the radial mesopore structure evident in Figure 2b. The surface area and pore volume were also reduced after functionalization, consistent with amine grafting inside the mesopores. The large pore size of the particles (relative to CTAB-templating alone) allows the amine functional groups in the pores to be accessible for covalent attachment to the linkers and peptides. The amount of amines on the particle surface was estimated by chemical analysis to be 1.53 mmol amine/g silica, corresponding to 64% of a monolayer coverage on the particle surface considering the projected area per aminopropyl group on the surface (0.5 nm^2/aminopropyl group) [58,61].

Oligopeptides RSSV, 4G and 4RSSV were conjugated to amine functional groups on the particle using hetero-bifunctional linker SNLD, which provides a combination of amine-reactive chemistry with the photochemistry of diazirine groups for UV activation. The silica pore walls should not hinder UV light reaching the interior of the particles; silica cavities with pore walls of much greater thickness (2 μm) have been functionalized using a similar UV treatment process [50]. UV-activated conjugation is relatively rapid and efficient compared to NHS conjugation [62] and was determined to be optimized with 60 min of treatment using RSSV-FITC attachment (data not shown).

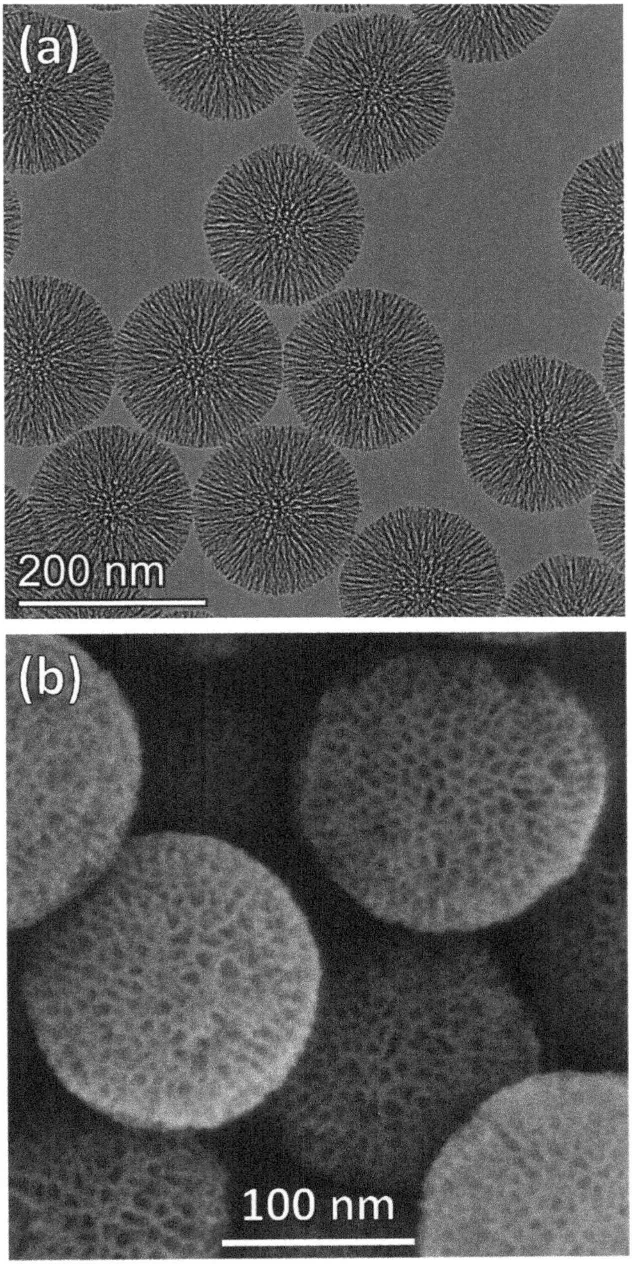

Figure 2. Representative (**a**) transmission electron micrograph and (**b**) scanning electron micrograph of bare MSNPs showing spherical particles with average particle diameter 146 ± 27 nm and radially oriented mesopores with an average pore diameter of 8 nm.

Table 1. Surface properties of MSNPs from nitrogen adsorption before and after amine functionalization.

Particle Type	BET Surface Area (m^2/g)	Total Pore Volume (cm^3/g)	Average Pore Diameter (nm) [a]
MSNP	729	2.32	7.9 ± 2.2
MSNPA	469	1.50	7.6 ± 1.9

[a] The range is determined from the full width at half maximum (FWHM) of the BJH pore size distribution.

For RSSV, two different peptide conjugation strategies were used with a bifunctional linker (Figure 1). For Type-1 attachment, SNLD was first conjugated to amine groups on the particles by using amine-reactive NHS moieties, and then the peptide N-terminus was attached to the linker using the diazirine end via UV treatment. In the Type-2 attachment sequence, the peptide N-terminus was first attached to the linker NHS end before attaching the diazirine end to particle amine groups by UV treatment. FITC-conjugated RSSV (RSSV-FITC) was used to evaluate these two oligiopeptide attachment strategies. Fluorophore-tagged peptides/proteins are widely used to calculate the attachment efficiency using homo- or hetero-bifunctional linkers [63]. The UV–Vis absorbance spectra of RSSV-FITC were measured (Figure S2 of ESI), where the absorbance peak and intensities do not change with UV treatment up to 120 min of treatment. Fluorescence intensities were used to measure RSSV-FITC attachment by solution depletion, while accounting for photobleaching of the fluorescent moiety (using a control UV treatment in the absence of any particles) during the functionalization process. UV illumination caused 25% and 35% reduction in intensity for RSSV-FITC and SNLD-RSSV-FITC after 60 min of treatment for Type-1 (Figures S3 and S4) and Type-2 attachment (Figures S5 and S6), respectively.

Based on the calculation of solution depletion after peptide conjugation, the attachment efficiency of Type-2 conjugation was found to be 0.43 mol peptide/mol amine, whereas for Type-1 conjugation, it was 0.24 mol peptide/mol amine. Better efficiency of Type-2 attachment is consistent with literature describing antibody conjugation to polyamine yarns using NHS-LC-diazirine [64]. The improved attachment efficiency may be due to the high concentration of diazirine groups accessible to amine groups in the pores during Type-2 attachment. It is also possible for the diazirine group to bind with another amine group or unintended peptide moieties during Type-1 attachment, which along with lower attachment efficiency, makes Type-1 a less attractive option. During Type-2 attachment, activated diazirine can attach to the carboxyl moiety of another peptide. However, if activated diazirine binds to the C-terminus of another peptide, the complex will be removed during particle washing steps. Thus, Type-2 attachment is preferable to preserve peptide functionality. UV–Vis absorbance and fluorescence intensity of the particles and the supernatant after Type-2 conjugation using RSSV-FITC with UV treatment are provided in Figure 3, compared to identical systems in the absence of UV treatment (no diazirine activation and subsequent covalent bond formation). The absorbance and fluorescence intensity of the particles increased only when they underwent UV treatment, whereas absorbance and fluorescence intensity decreased in the supernatant, consistent with RSSV-FITC attachment to the particles. Note that the particle external surface represents less than 3% of the total surface area (considering spherical particles), and as a result, peptide attachment should be only 0.03 mol peptide/mol amine if peptides were only able to attach to the outer surface amines of the particles. Considering the high relative amount of peptide attachment, the peptides are conjugated primarily to the amines inside the pores, as hypothesized.

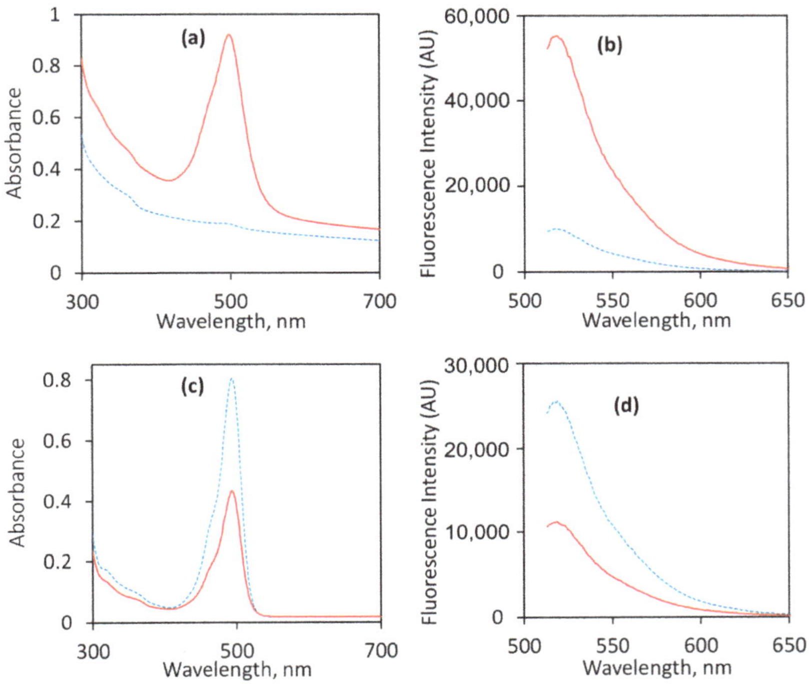

Figure 3. UV–Vis absorbance (left) and fluorescence intensity (right) of (**a**,**b**) particles re-suspended in solution and (**c**,**d**) supernatant after fluorescein isothiocyanate (FITC)-labeled RSSV (Arg-Ser-Ser-Val tetrapeptide) attachment to the particles using Type-2 conjugation. Solid red lines and dashed blue lines represent results with or without UV treatment, respectively. Both UV absorbance and fluorescence intensity of the particles increases after UV treatment, whereas supernatant intensity decreases, suggesting successful binding of peptide.

To show covalent linkage (not merely physical adsorption), FTIR spectra of the particles after peptide attachment were measured. The FTIR spectra of MSNPs (bare, amine-functionalized, RSSV-functionalized and 4RSSV-functionalized prepared by Type-2 attachment) are compared in Figure 4, along with the spectra of fresh linker and peptides. The FTIR spectra of the bare MPSNs do not contain a peak due to –CH_2 stretching (2800–3000 cm^{-1}), suggesting complete removal of the organic template following particle synthesis. For bare MSNPs and MSNPAs, the most prominent peaks are bands corresponding to Si-O-Si and Si-OH vibration, located at 1080 and 960 cm^{-1}, respectively [38]. Primary amine peaks are not visible, but amine functionalization was quantified by chemical analysis, as described previously. The linker, SNLD, has symmetric and asymmetric (-C=O from the ester) stretching vibrations at 1788 and 1736 cm^{-1}, respectively, whereas peaks at 1223 and 1051 cm^{-1} can be assigned as asymmetric C-N-C and N-C-O stretching vibrations, respectively [21]. There are also two peaks correspond to diazirine (N=N stretching at 1643 and N-H amide bond stretching at 1540 cm^{-1}) [65–68]. After peptide attachment, the intensity of –CH_2 stretching vibrations (2800–3000 cm^{-1}) increases, which indicates the presence of linking spacer between particle surface and peptide, whereas peaks corresponding to NHS ester and diazirine groups disappear, suggesting their conversion during the attachment process. Both MSNPA-RSSV and MSNPA-4RSSV show increased intensity corresponding to the arginine side chain stretching vibration, which confirms peptide attachment to the particle surface. Specifically, the spectra of the RSSV oligopeptides have characteristic peaks from the arginine side chain ($CN_3H_5^+$) (asymmetric and symmetric stretching vibrations of 1673 and 1586 cm^{-1}, respectively), -CH_3 bending vibration at 1460 cm^{-1} from valine and C-OH bending vibration from serine side chain (1181 cm^{-1}) [69]. On the other hand, characteristic peaks of the inner groups from 4G (CH_2 bending at 1435 cm^{-1}, COO- symmetric and asymmetric stretching at 1788 and 1736 cm^{-1}, and C=O stretching at 1637 cm^{-1} [70]) remain, while N-H symmetric and asymmetric stretching peaks (at 3311 and 3276 cm^{-1} [71]) disappear, suggesting covalent linkage through the N-terminus.

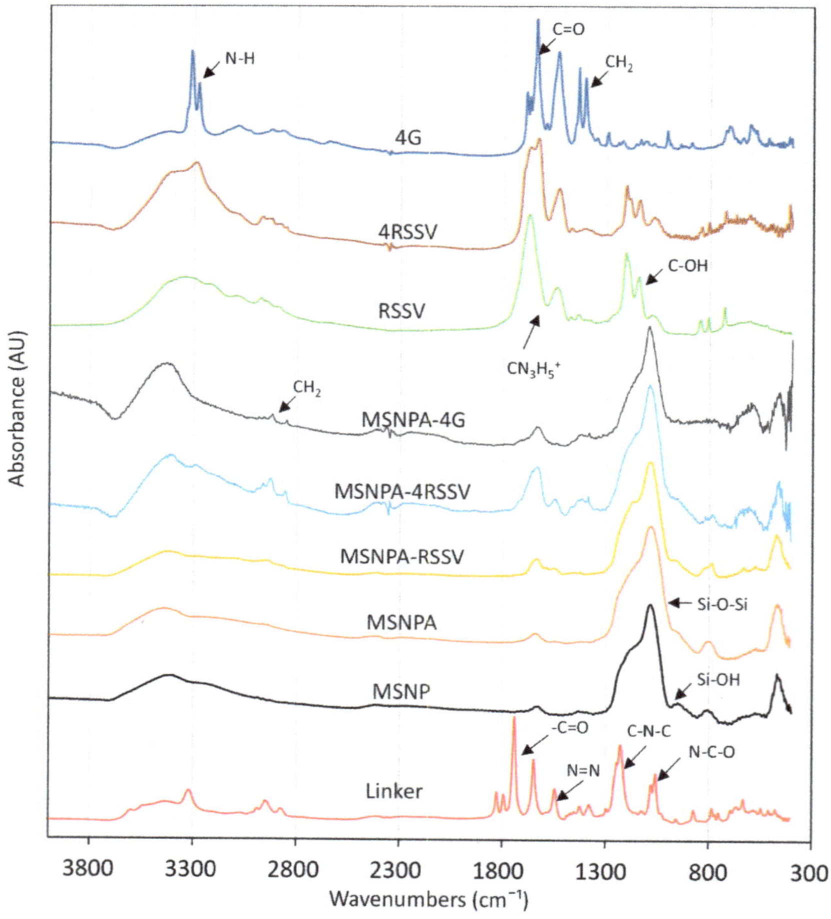

Figure 4. FTIR spectra of peptide functionalized particles relative to bare MSNPs and MSNPAs, as well as fresh linker (sulfo-NHS-LC-diazirine) and peptides with peaks corresponding to major functional groups labeled. In the figure, 4RSSV is a peptide made of 4 sequential RSSV units and 4G is a tetraglycine.

The quantity of non-fluorescent peptides (RSSV, 4RSSV and 4G) conjugated to the MSPAs was measured directly by TGA to demonstrate that the Type-2 attachment strategy is generalizable. Specifically, the mass loss by the peptide (RSSV, 4G and 4RSSV)–linker-conjugated MSNPAs (synthesized using Type-2 attachment) was compared to that of MSNPs and MSNPAs in the range 150 °C to 500 °C, which corresponds to the thermal degradation and combustion of the organic groups (Figure 5). For bare MSNPs, 2.6% mass is lost from 150 to 500 °C (0.0278 mg/mg silica; representing impurities such as residual template and further temperature-induced condensation of silica) and is subtracted before calculating the organic content of the other particles based on mass loss in this temperature range.

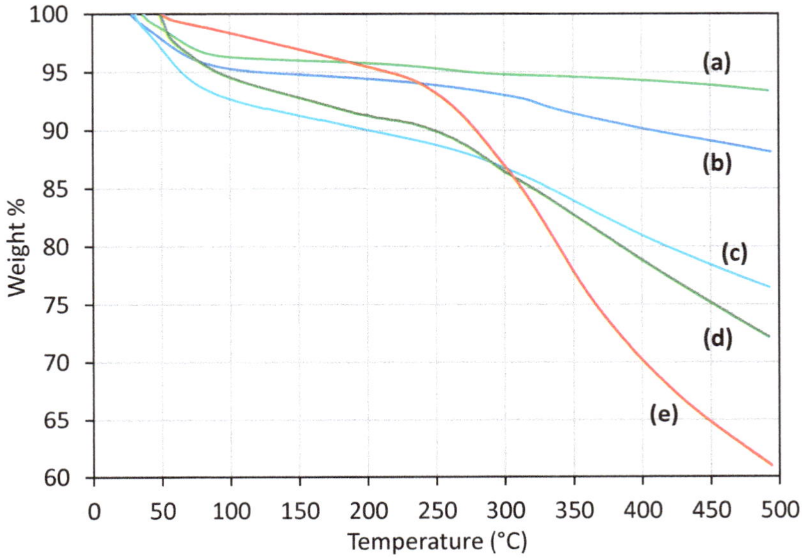

Figure 5. Thermogravimetric analysis (TGA) profiles of particles showing relative mass loss with temperature increase for (**a**) MSNP, (**b**) MSNPA, (**c**) MSNPA-4G, (**d**) MSNPA-RSSV and (**e**) MSNPA-4RSSV.

The amount of aminopropyl groups on MSNPAs was found to be 0.494 mg/mg silica (1.04 mmol amine/g silica). Therefore, TGA analysis underestimates the amount of amine by a factor of 0.67 compared to that of chemical analysis (see above). The reason for this underestimation may be the presence of carbon residue on the particle surface (which was visually observed on the particles). After subtraction of the weight loss of organics for MSNPAs, an additional 0.132, 0.208 and 0.516 mg/(mg silica) weight loss (due to removal of peptides and linkers) was observed for MSNPA-4G, MSNPA-RSSV and MSNPA-4RSSV, respectively. This corresponds to 0.300 mmol 4G/g silica, 0.324 mmol RSSV/g silica and 0.267 mmol 4RSSV/g silica. The molar attachment efficiency with amine is similar for the three peptides: 0.288, 0.311 and 0.257 mol peptide/mol amine for 4G, RSSV and 4RSSV, respectively, based on the masses of amines calculated using TGA. It is usually difficult to estimate protein length due to the presence of different secondary structures, but for oligopeptides, an average length of 3.6 Å can be assumed per amino acid [72]. Thus, a contour length of 1.4 nm, 1.4 nm and 5.8 nm can be approximated for 4G, RSSV and 4RSSV, respectively, in addition to 1.25 nm of spacer length from the linker [52]. Therefore, the method of peptide attachment to MSNPAs is robust to amino acid sequences and oligopeptide length. Furthermore, note that TGA analysis underestimated RSSV attachment by a factor of 0.72 compared to fluorescence spectroscopy measurements, which is close to aminopropyl underestimation for MSNPAs, again due to carbon residue on the particle surface. Thus, chemical methods are more accurate in estimation of peptide attachment, but TGA can provide a reasonable estimation of relative attachment.

All of the particles after peptide conjugation (MSNPA-4G, MSNPA-RSSV and MSNPA-4RSSV) remain positively charged at physiological pH (Figure S7), consistent with the presence of unreacted surface amine groups. MSNPA-4G has a slightly lower zeta potential compared to bare MSNPAs, but MSNPA-RSSV and MSNPA-4RSSV have comparable charge to MSNPAs. Possible charge reduction after reaction of surface amine groups due to peptide conjugation may be offset by the charge of the positive moieties of arginine side chains for MSNPA-RSSV and MSNPA-4RSSV. Positive charge is important for the colloidal stability of the particles during suspension and re-suspension. Due to high positive charge, these particles should be appropriate for intracellular penetration where positive charge has been shown to be beneficial [73].

A challenge of covalent immobilization of peptides on surfaces is the possible loss of the desired selective ligand binding functionality after attachment [7]. Selective binding of biomolecules with peptide immobilized to an affinity column surface was demonstrated [7,53], and even utilized for the fractionation of some biomolecules [74–76]. For biomolecule binding with covalently linked peptides on particle surfaces, although hydrophobic interactions are the main driving force for binding, hydrogen bonding with hydroxyl and aminopropyl groups may also be present [77,78], which makes the binding more complicated. A large amount of non-specific binding of biomolecules from aqueous solutions is also expected. Here, controlled covalent linkage of peptide to mesoporous silica nanoparticles provides a better opportunity to effectively use the pores, in contrast to noncovalent functionalization [34,79], which has been used for RNA delivery or theranostics. Functionalization inside the pores is also superior compared to only external surface functionalization by peptide for drug delivery [10] or cellular receptor binding [80]. High β-estradiol binding capacity of tetrapeptide RSSV [53] and its repeat 4-mer should also provide an opportunity for the removal of these types of compounds from polluted water sources, which are well-recognized endocrine disrupting compounds [81]. However, selective biomolecule binding and separation is beyond the scope of current investigation and will be pursued in future studies.

4. Conclusions

Functional oligopeptides 4G (Gly-Gly-Gly-Gly), RSSV (Arg-Ser-Ser-Val) and 4RSSV were attached to large pore (7.9 nm diameter), amine-functionalized MSNPS using a hetero-bifunctional linker, sulfo-NHS-LC-diazirine (SNLD), which contains an amine-reactive NHS ester group and UV-activable diazirine group. Hetero-bifunctional peptide linkers containing a diazirine group provide precise control of the mechanism and orientation during attachment with a high activation wavelength (365 nm), which is more benign to proteins and peptides compared to other linkers that are activated at lower wavelength. Two different conjugation schemes were compared to attach oligopeptide RSSV to the MSNPs based on the order of addition (Type-1: functionalize particle with linker and then attach the peptide; Type-2: attach the linker to the peptide and then functionalize the particle with peptide–linker conjugate). The efficiency of peptide attachment was measured by fluorescence spectroscopy using FITC-labeled peptides (RSSV-FITC). Higher attachment efficiency per mol amine groups was found for Type-2 attachment (0.43 mol RSSV/mol amine) compared with Type-1 attachment (0.24 mol RSSV/mol amine). Type-2 attachment efficiencies of 4G, RSSV and 4RSSV on particles, as determined by TGA analysis, were similar. This demonstrates that the attachment strategy is generalizable and can be used to attach a range of sizes of oligopeptide to MSNPA.

The functional oligopeptide conjugation to "large pore" MSNPs appropriate for biomolecule loading was demonstrated using a versatile and robust hetero-bifunctional linking strategy, which provides precise control of binding moieties of peptide molecules. High-capacity platforms for selective separation of biomolecules with therapeutic value can be designed by selecting oligopeptides that mimic the specific binding sites of biomolecules. For example, MSNPAs can be used to selectively isolate different small molecular therapeutics from living plants, a technique recently demonstrated by using engineered silica nanoparticles [73]. Further, high-capacity adsorbents for the removal of specific therapeutics in polluted water sources can be developed based on functionalizing silica particles with selective members of peptide libraries.

Supplementary Materials: The following supporting information can be downloaded at: https://www.mdpi.com/article/10.3390/nano12040608/s1, Figure S1: (a) Nitrogen sorption isotherms and (b) pore size distributions of non-functionalized and amine functionalized MSNPs; Figure S2: Absorbance spectra of RSSV-FITC before and after UV exposure; Figure S3: Calibration curve for the concentration of RSSV-FITC in solution; Figure S4: Fluorescence intensity of RSSV-FITC solution during UV photobleaching; Figure S5: Calibration curve for fluorescence intensities of SNLD-RSSV-FITC before UV treatment; Figure S6: Fluorescence intensity of SNLD-RSSV-FITC solution during UV photo-bleaching; and Figure S7: Zeta potentials vs pH for MSNPA-RSSV, MSNPA-4RSSV, MSNPA-4G, and MSNPA.

Author Contributions: Conceptualization, J.M.L., B.L.K. and S.E.R.; methodology, M.A.K., B.L.K., S.E.R., L.H.B., D.T.R. and B.C.L.; formal analysis, M.A.K.; investigation, M.A.K., M.R.K., R.W.G. and M.M.; resources, L.H.B.; writing—original draft preparation, M.A.K.; writing—review and editing, S.E.R. and B.L.K.; visualization, M.A.K.; project administration, B.L.K. and S.E.R.; funding acquisition, J.M.L., B.L.K., L.H.B. and S.E.R. All authors have read and agreed to the published version of the manuscript.

Funding: This research was funded by the United States National Institutes of Health (NIH), grant numbers R41AT008312 and 2R44AT008312-02; and Kentucky Science and Engineering Foundation, grant number KSEF-2929-RDE-016. This work was performed in part at the University of Kentucky Electron Microscopy Center, a member of the National Nanotechnology Coordinated Infrastructure (NNCI), which is supported by the National Science Foundation (NNCI-2025075).

Data Availability Statement: The data presented in this study are available on request from the corresponding authors.

Conflicts of Interest: J.M.L. and D.T.R. are employees of Naprogenix Inc. and J.M.L. owns stock in the company. The funders of this work had no role in the design of the study; in the collection, analyses or interpretation of data; in the writing of the manuscript; or in the decision to publish the results.

References

1. Yang, H.; Yuan, B.; Zhang, X.; Scherman, O.A. Supramolecular chemistry at interfaces: Host–guest interactions for fabricating multifunctional biointerfaces. *Acc. Chem. Res.* **2014**, *47*, 2106–2115. [CrossRef] [PubMed]
2. Houk, K.N.; Leach, A.G.; Kim, S.P.; Zhang, X. Binding Affinities of Host–Guest, Protein–Ligand, and Protein–Transition-State Complexes. *Angew. Chem. Int. Ed.* **2003**, *42*, 4872–4897. [CrossRef]
3. Uhlenheuer, D.A.; Petkau, K.; Brunsveld, L. Combining supramolecular chemistry with biology. *Chem. Soc. Rev.* **2010**, *39*, 2817–2826. [CrossRef] [PubMed]
4. Hage, D.S. Analysis of biological interactions by affinity chromatography: Clinical and pharmaceutical applications. *Clin. Chem.* **2017**, *63*, 1083–1093. [CrossRef] [PubMed]
5. Frączyk, J.; Walczak, M.; Kamiński, Z.J. Cellulose template assembled synthetic peptides as molecular receptors. *Curr. Protein Peptide Sci.* **2016**, *17*, 117–126. [CrossRef]
6. Menegatti, S. Peptoid Affinity Ligands. U.S. Patent 10,065,988 B2, 4 September 2018.
7. Tozzi, C.; Anfossi, L.; Giraudi, G. Affinity chromatography techniques based on the immobilisation of peptides exhibiting specific binding activity. *J. Chromatogr. B* **2003**, *797*, 289–304. [CrossRef]
8. Tothill, I.E. Peptides as molecular receptors. In *Recognition Receptors in Biosensors*; Springer: Berlin/Heidelberg, Germany, 2010; pp. 249–274.
9. Noppe, W.; Plieva, F.M.; Galaev, I.Y.; Vanhoorelbeke, K.; Mattiasson, B.; Deckmyn, H. Immobilised peptide displaying phages as affinity ligands: Purification of lactoferrin from defatted milk. *J. Chromatogr. A* **2006**, *1101*, 79–85. [CrossRef]
10. Pan, L.; He, Q.; Liu, J.; Chen, Y.; Ma, M.; Zhang, L.; Shi, J. Nuclear-Targeted Drug Delivery of TAT Peptide-Conjugated Monodisperse Mesoporous Silica Nanoparticles. *J. Am. Chem. Soc.* **2012**, *134*, 5722–5725. [CrossRef]
11. Song, N.; Yang, Y.-W. Molecular and supramolecular switches on mesoporous silica nanoparticles. *Chem. Soc. Rev.* **2015**, *44*, 3474–3504. [CrossRef]
12. Moradipour, M.; Chase, E.K.; Khan, M.A.; Asare, S.O.; Lynn, B.C.; Rankin, S.E.; Knutson, B.L. Interaction of lignin-derived dimer and eugenol-functionalized silica nanoparticles with supported lipid bilayers. *Colloids Surf. B Biointerfaces* **2020**, *191*, 111028. [CrossRef]
13. Ma, Y.; Qi, L.; Ma, J.; Wu, Y.; Liu, O.; Cheng, H. Large-pore mesoporous silica spheres: Synthesis and application in HPLC. *Colloids Surf. A* **2003**, *229*, 1–8. [CrossRef]
14. Zhao, J.; Gao, F.; Fu, Y.; Jin, W.; Yang, P.; Zhao, D. Biomolecule separation using large pore mesoporous SBA-15 as a substrate in high performance liquid chromatography. *Chem. Commun.* **2002**, *2*, 752–753. [CrossRef] [PubMed]

15. Hermanson, G.T. *Bioconjugate Techniques*; Academic Press: London, UK, 2013.
16. Farkaš, P.; Bystrický, S. Chemical conjugation of biomacromolecules: A mini-review. *Chem. Pap.* **2010**, *64*, 683–695. [CrossRef]
17. Sundoro, B.M. Bifunctional Linker. U.S. Patent 4,680,338, 14 July 1987.
18. Shi, J.-M.; Pei, J.; Liu, E.-Q.; Zhang, L. Bis (sulfosuccinimidyl) suberate (BS^3) crosslinking analysis of the behavior of amyloid-β peptide in solution and in phospholipid membranes. *PLoS ONE* **2017**, *12*, e0173871. [CrossRef] [PubMed]
19. Xia, B.; Xiao, S.-J.; Guo, D.-J.; Wang, J.; Chao, J.; Liu, H.-B.; Pei, J.; Chen, Y.-Q.; Tang, Y.-C.; Liu, J.-N. Biofunctionalisation of porous silicon (PS) surfaces by using homobifunctional cross-linkers. *J. Mater. Chem.* **2006**, *16*, 570–578. [CrossRef]
20. Lee, J.P.; Kassianidou, E.; MacDonald, J.I.; Francis, M.B.; Kumar, S. N-terminal specific conjugation of extracellular matrix proteins to 2-pyridinecarboxaldehyde functionalized polyacrylamide hydrogels. *Biomaterials* **2016**, *102*, 268–276. [CrossRef] [PubMed]
21. Lim, C.Y.; Owens, N.A.; Wampler, R.D.; Ying, Y.; Granger, J.H.; Porter, M.D.; Takahashi, M.; Shimazu, K. Succinimidyl Ester Surface Chemistry: Implications of the Competition between Aminolysis and Hydrolysis on Covalent Protein Immobilization. *Langmuir* **2014**, *30*, 12868–12878. [CrossRef]
22. Kalkhof, S.; Sinz, A. Chances and pitfalls of chemical cross-linking with amine-reactive N-hydroxysuccinimide esters. *Anal. Bioanal. Chem.* **2008**, *392*, 305–312. [CrossRef]
23. Karakeçili, A.G.; Satriano, C.; Gümüşderelioğlu, M.; Marletta, G. Enhancement of fibroblastic proliferation on chitosan surfaces by immobilized epidermal growth factor. *Acta Biomater.* **2008**, *4*, 989–996. [CrossRef]
24. Li, B.; Chen, J.; Wang, J.H.C. RGD peptide-conjugated poly (dimethylsiloxane) promotes adhesion, proliferation, and collagen secretion of human fibroblasts. *J. Biomed. Mater. Res. A* **2006**, *79*, 989–998. [CrossRef]
25. Oluwabusola, E. Development of Diazirine-Based Crosslinking Agents for Covalently Linking Protein. Master's Thesis, University of Salford, Manchester, UK, 2015.
26. Shigdel, U.K.; Zhang, J.; He, C. Diazirine-Based DNA Photo-Cross-Linking Probes for the Study of Protein–DNA Interactions. *Angew. Chem. Int. Ed.* **2008**, *47*, 90–93. [CrossRef] [PubMed]
27. Chim, L.K.-L. Immobilizing Gradients of Neurotrophic Factors for Directed Peripheral Nervous System Cell Migration Guidance. Master's Thesis, Johns Hopkins University, Baltimore, MD, USA, 2016.
28. Das, J. Aliphatic diazirines as photoaffinity probes for proteins: Recent developments. *Chem. Rev.* **2011**, *111*, 4405–4417. [CrossRef] [PubMed]
29. Suchanek, M.; Radzikowska, A.; Thiele, C. Photo-leucine and photo-methionine allow identification of protein-protein interactions in living cells. *Nat. Meth.* **2005**, *2*, 261. [CrossRef] [PubMed]
30. Tanaka, Y.; Kohler, J.J. Photoactivatable crosslinking sugars for capturing glycoprotein interactions. *J. Am. Chem. Soc.* **2008**, *130*, 3278–3279. [CrossRef]
31. Liashkovich, I.; Rosso, G.; Rangl, M.; Ebner, A.; Hafezi, W.; Kühn, J.; Schön, P.; Hinterdorfer, P.; Shahin, V. Photopicking: In Situ Approach for Site-Specific Attachment of Single Multiprotein Nanoparticles to Atomic Force Microscopy Tips. *Adv. Func. Mater.* **2017**, *27*, 1604506. [CrossRef]
32. Moussus, M.; der Loughian, C.; Fuard, D.; Courçon, M.; Gulino-Debrac, D.; Delanoë-Ayari, H.; Nicolas, A. Intracellular stresses in patterned cell assemblies. *Soft Matter* **2014**, *10*, 2414–2423. [CrossRef]
33. Lozito, T.P.; Tuan, R.S. Mesenchymal stem cells inhibit both endogenous and exogenous MMPs via secreted TIMPs. *J. Cell. Physiol.* **2011**, *226*, 385–396. [CrossRef]
34. Lu, J.; Shen, H.-H.; Wu, Z.; Wang, B.; Zhao, D.; He, L. Self-assembly of bi-functional peptides on large-pore mesoporous silica nanoparticles for miRNA binding and delivery. *J. Mater. Chem. B* **2015**, *3*, 7653–7657. [CrossRef]
35. Terracciano, R.; Casadonte, F.; Pasqua, L.; Candeloro, P.; Di Fabrizio, E.; Urbani, A.; Savino, R. Enhancing plasma peptide MALDI-TOF-MS profiling by mesoporous silica assisted crystallization. *Talanta* **2010**, *80*, 1532–1538. [CrossRef]
36. Liu, H.-J.; Xu, P. Smart Mesoporous Silica Nanoparticles for Protein Delivery. *Nanomaterials* **2019**, *9*, 511. [CrossRef]
37. Xu, C.; Lei, C.; Yu, C. Mesoporous Silica Nanoparticles for Protein Protection and Delivery. *Front. Chem.* **2019**, *7*, 290. [CrossRef] [PubMed]
38. Schlipf, D.M.; Zhou, S.; Khan, M.A.; Rankin, S.E.; Knutson, B.L. Effects of Pore Size and Tethering on the Diffusivity of Lipids Confined in Mesoporous Silica. *Adv. Mater. Interfaces* **2017**, *4*, 1601103. [CrossRef]
39. Yushkova, E.D.; Nazarova, E.A.; Matyuhina, A.V.; Noskova, A.O.; Shavronskaya, D.O.; Vinogradov, V.V.; Skvortsova, N.N.; Krivoshapkina, E.F. Application of Immobilized Enzymes in Food Industry. *J. Agric. Food Chem.* **2019**, *67*, 11553–11567. [CrossRef] [PubMed]
40. Bolivar, J.M.; Nidetzky, B. The Microenvironment in Immobilized Enzymes: Methods of Characterization and Its Role in Determining Enzyme Performance. *Molecules* **2019**, *24*, 3460. [CrossRef]
41. Aggarwal, S.; Chakravarty, A.; Ikram, S. A comprehensive review on incredible renewable carriers as promising platforms for enzyme immobilization & thereof strategies. *Int. J. Biol. Macromol.* **2021**, *167*, 962–986. [CrossRef]
42. Castillo, R.R.; Lozano, D.; Vallet-Regí, M. Mesoporous Silica Nanoparticles as Carriers for Therapeutic Biomolecules. *Pharmaceutics* **2020**, *12*, 432. [CrossRef]
43. Kumar, P.; Tambe, P.; Paknikar, K.M.; Gajbhiye, V. Mesoporous silica nanoparticles as cutting-edge theranostics: Advancement from merely a carrier to tailor-made smart delivery platform. *J. Control. Release* **2018**, *287*, 35–57. [CrossRef]
44. Wilkins, D.K.; Grimshaw, S.B.; Receveur, V.; Dobson, C.M.; Jones, J.A.; Smith, L.J. Hydrodynamic Radii of Native and Denatured Proteins Measured by Pulse Field Gradient NMR Techniques. *Biochemistry* **1999**, *38*, 16424–16431. [CrossRef]

45. Khan, M.A.; Wallace, W.T.; Islam, S.Z.; Nagpure, S.; Strzalka, J.; Littleton, J.M.; Rankin, S.E.; Knutson, B.L. Adsorption and Recovery of Polyphenolic Flavonoids Using TiO2-Functionalized Mesoporous Silica Nanoparticles. *ACS Appl. Mater. Interfaces* **2017**, *9*, 32114–32125. [CrossRef]
46. Kim, T.-W.; Chung, P.-W.; Lin, V.S.Y. Facile Synthesis of Monodisperse Spherical MCM-48 Mesoporous Silica Nanoparticles with Controlled Particle Size. *Chem. Mater.* **2010**, *22*, 5093–5104. [CrossRef]
47. Tan, B.; Rankin, S.E. Interfacial Alignment Mechanism of Forming Spherical Silica with Radially Oriented Nanopores. *J. Phys. Chem. B* **2004**, *108*, 20122–20129. [CrossRef]
48. Gu, J.; Huang, K.; Zhu, X.; Li, Y.; Wei, J.; Zhao, W.; Liu, C.; Shi, J. Sub-150 nm mesoporous silica nanoparticles with tunable pore sizes and well-ordered mesostructure for protein encapsulation. *J. Colloid Interface Sci.* **2013**, *407*, 236–242. [CrossRef] [PubMed]
49. Yamada, H.; Ujiie, H.; Urata, C.; Yamamoto, E.; Yamauchi, Y.; Kuroda, K. A multifunctional role of trialkylbenzenes for the preparation of aqueous colloidal mesostructured/mesoporous silica nanoparticles with controlled pore size, particle diameter, and morphology. *Nanoscale* **2015**, *7*, 19557–19567. [CrossRef] [PubMed]
50. Berneschi, S.; Baldini, F.; Cosci, A.; Farnesi, D.; Conti, G.N.; Tombelli, S.; Trono, C.; Pelli, S.; Giannetti, A. Fluorescence biosensing in selectively photo–activated microbubble resonators. *Sens. Actuators B* **2017**, *242*, 1057–1064. [CrossRef]
51. Lozito, T.P.; Jackson, W.M.; Nesti, L.J.; Tuan, R.S. Human mesenchymal stem cells generate a distinct pericellular zone of MMP activities via binding of MMPs and secretion of high levels of TIMPs. *Matrix Biol.* **2014**, *34*, 132–143. [CrossRef]
52. ThermoFisher Scientific. *Bioconjugation Technical Handbook*; Thermo Fisher Scientific Inc.: Pittsburgh, PA, USA, 2018.
53. Tozzi, C.; Anfossi, L.; Giraudi, G.; Giovannoli, C.; Baggiani, C.; Vanni, A. Chromatographic characterisation of an estrogen-binding affinity column containing tetrapeptides selected by a combinatorial-binding approach. *J. Chromatogr. A* **2002**, *966*, 71–79. [CrossRef]
54. Iacobucci, C.; Götze, M.; Piotrowski, C.; Arlt, C.; Rehkamp, A.; Ihling, C.; Hage, C.; Sinz, A. Carboxyl-photo-reactive MS-cleavable cross-linkers: Unveiling a hidden aspect of diazirine-based reagents. *Anal. Chem.* **2018**, *90*, 2805–2809. [CrossRef]
55. Ezzeddine, Z.; Batonneau-Gener, I.; Pouilloux, Y.; Hamad, H.; Saad, Z.; Kazpard, V. Divalent heavy metals adsorption onto different types of EDTA-modified mesoporous materials: Effectiveness and complexation rate. *Microporous Mesoporous Mater.* **2015**, *212*, 125–136. [CrossRef]
56. Na, H.-K.; Kim, M.-H.; Park, K.; Ryoo, S.-R.; Lee, K.E.; Jeon, H.; Ryoo, R.; Hyeon, C.; Min, D.-H. Efficient Functional Delivery of siRNA using Mesoporous Silica Nanoparticles with Ultralarge Pores. *Small* **2012**, *8*, 1752–1761. [CrossRef]
57. Lee, J.E.; Lee, N.; Kim, H.; Kim, J.; Choi, S.H.; Kim, J.H.; Kim, T.; Song, I.C.; Park, S.P.; Moon, W.K. Uniform mesoporous dye-doped silica nanoparticles decorated with multiple magnetite nanocrystals for simultaneous enhanced magnetic resonance imaging, fluorescence imaging, and drug delivery. *J. Am. Chem. Soc.* **2009**, *132*, 552–557. [CrossRef]
58. Schlipf, D.M.; Rankin, S.E.; Knutson, B.L. Selective external surface functionalization of large-pore silica materials capable of protein loading. *Microporous Mesoporous Mater.* **2017**, *244*, 199–207. [CrossRef]
59. Ritter, H.; Nieminen, M.; Karppinen, M.; Brühwiler, D. A comparative study of the functionalization of mesoporous silica MCM-41 by deposition of 3-aminopropyltrimethoxysilane from toluene and from the vapor phase. *Microporous Mesoporous Mater.* **2009**, *121*, 79–83. [CrossRef]
60. Mu, Q.; Jiang, G.; Chen, L.; Zhou, H.; Fourches, D.; Tropsha, A.; Yan, B. Chemical Basis of Interactions Between Engineered Nanoparticles and Biological Systems. *Chem. Rev.* **2014**, *114*, 7740–7781. [CrossRef]
61. Khan, M.A. Nanoharvesting and Delivery of Bioactive Materials Using Engineered Silica Nanoparticles. Ph.D. Thesis, University of Kentucky, Lexington, KE, USA, 2019.
62. Van Itallie, C.M.; Mitic, L.L.; Anderson, J.M. Claudin-2 forms homodimers and is a component of a high molecular weight protein complex. *J. Biol. Chem.* **2011**, *286*, 3442–3450. [CrossRef] [PubMed]
63. Casanova, D.; Giaume, D.; Moreau, M.; Martin, J.-L.; Gacoin, T.; Boilot, J.-P.; Alexandrou, A. Counting the number of proteins coupled to single nanoparticles. *J. Am. Chem. Soc.* **2007**, *129*, 12592–12593. [CrossRef]
64. Waddell, C. The Development of a Rapid Fiber-Based Immunoassay as a Point-of-Care or In-Home Diagnostic Test. Master's Thesis, Clemson University, Clemson, SC, USA, 2013.
65. Ettinger, R. Infrared Spectrum of Diazirine. *J. Chem. Phys.* **1964**, *40*, 1693–1699. [CrossRef]
66. Gambi, A.; Winnewisser, M.; Christiansen, J.J. The infrared spectrum of diazirine: H2C<(N=N). Rovibrational analysis of the ν3 fundamental. *J. Mol. Spectrosc.* **1983**, *98*, 413–424. [CrossRef]
67. Martucci, N.M.; Migliaccio, N.; Ruggiero, I.; Rea, I.; Terracciano, M.; De Stefano, L.; Arcari, P.; Rendina, I.; Lamberti, A. Bioengineered Surfaces for Real-Time Label-Free Detection of Cancer Cells. In *Lab-on-a-Chip Fabrication and Application;* IntechOpen: London, UK, 2016; p. 179.
68. Mitchell, R.; Merritt, J. The infrared spectra of 3,3-dimethyldiazirine and 3,3-dimethyl-d6-diazirine. *J. Mol. Spectrosc.* **1968**, *27*, 197–209. [CrossRef]
69. Barth, A. The infrared absorption of amino acid side chains. *Prog. Biophys. Mol. Biol.* **2000**, *74*, 141–173. [CrossRef]
70. Kolev, T.; Spiteller, M.; Koleva, B. Spectroscopic and structural elucidation of amino acid derivatives and small peptides: Experimental and theoretical tools. *Amino Acids* **2010**, *38*, 45–50. [CrossRef]
71. Zhang, G.; Xu, B.; Han, F.; Zhou, Y.; Liu, H.; Li, Y.; Cui, L.; Tan, T.; Wang, N. Green synthesis, composition analysis and surface active properties of sodium cocoyl glycinate. *Am. J. Anal. Chem.* **2013**, *4*, 445. [CrossRef]

72. Ainavarapu, S.R.K.; Brujić, J.; Huang, H.H.; Wiita, A.P.; Lu, H.; Li, L.; Walther, K.A.; Carrion-Vazquez, M.; Li, H.; Fernandez, J.M. Contour Length and Refolding Rate of a Small Protein Controlled by Engineered Disulfide Bonds. *Biophys. J.* **2007**, *92*, 225–233. [CrossRef] [PubMed]
73. Khan, M.A.; Wallace, W.T.; Sambi, J.; Rogers, D.T.; Littleton, J.M.; Rankin, S.E.; Knutson, B.L. Nanoharvesting of bioactive materials from living plant cultures using engineered silica nanoparticles. *Mater. Sci. Eng. C* **2020**, *106*, 110190. [CrossRef] [PubMed]
74. Ejima, D.; Yumioka, R.; Tsumoto, K.; Arakawa, T. Effective elution of antibodies by arginine and arginine derivatives in affinity column chromatography. *Anal. Biochem.* **2005**, *345*, 250–257. [CrossRef] [PubMed]
75. Endo, T. Fractionation of glycoprotein-derived oligosaccharides by affinity chromatography using immobilized lectin columns. *J. Chromatogr. A* **1996**, *720*, 251–261. [CrossRef]
76. Liau, C.Y.; Chang, T.M.; Pan, J.P.; Chen, W.L.; Mao, S.J. Purification of human plasma haptoglobin by hemoglobin-affinity column chromatography. *J. Chromatogr. B* **2003**, *790*, 209–216. [CrossRef]
77. Hu, Y.; Cai, K.; Luo, Z.; Jandt, K.D. Layer-by-layer assembly of β-estradiol loaded mesoporous silica nanoparticles on titanium substrates and its implication for bone homeostasis. *Adv. Mater.* **2010**, *22*, 4146–4150. [CrossRef]
78. Steffi, C.; Wang, D.; Kong, C.H.; Wang, Z.; Lim, P.N.; Shi, Z.; San Thian, E.; Wang, W. Estradiol-loaded poly (ε-caprolactone)/silk fibroin electrospun microfibers decrease osteoclast activity and retain osteoblast function. *ACS Appl. Mater. Interfaces* **2018**, *10*, 9988–9998. [CrossRef]
79. Sardan, M.; Yildirim, A.; Mumcuoglu, D.; Tekinay, A.B.; Guler, M.O. Noncovalent functionalization of mesoporous silica nanoparticles with amphiphilic peptides. *J. Mater. Chem. B* **2014**, *2*, 2168–2174. [CrossRef]
80. Sweeney, S.K.; Luo, Y.; O'Donnell, M.A.; Assouline, J.G. Peptide-mediated targeting mesoporous silica nanoparticles: A novel tool for fighting bladder cancer. *J. Biomed. Nanotechnol.* **2017**, *13*, 232–242. [CrossRef]
81. Lahcen, A.A.; Baleg, A.A.; Baker, P.; Iwuoha, E.; Amine, A. Synthesis and electrochemical characterization of nanostructured magnetic molecularly imprinted polymers for 17-β-estradiol determination. *Sens. Actuators B* **2017**, *241*, 698–705. [CrossRef]

nanomaterials

MDPI

Article

PEGylated Gold Nanoparticles Grafted with N-Acetyl-L-Cysteine for Polymer Modification

Dominik Fajstavr [1], Adéla Karasová [1], Alena Michalcová [2], Pavel Ulbrich [3], Nikola Slepičková Kasálková [1], Jakub Siegel [1], Václav Švorčík [1] and Petr Slepička [1,*]

[1] Department of Solid State Engineering, The University of Chemistry and Technology, 166 28 Prague, Czech Republic; dominik.fajstavr@vscht.cz (D.F.); adelakarasova@gmail.com (A.K.); nikola.kasalkova@vscht.cz (N.S.K.); jakub.siegel@vscht.cz (J.S.); vaclav.svorcik@vscht.cz (V.Š.)

[2] The Department of Metals and Corrosion Engineering, The University of Chemistry and Technology, 166 28 Prague, Czech Republic; alena.michalcova@vscht.cz

[3] Department of Biochemistry and Microbiology, University of Chemistry and Technology Prague, 166 28 Prague, Czech Republic; pavel.ulbrich@vscht.cz

* Correspondence: petr.slepicka@vscht.cz

Abstract: The subjects of this work were the enhancement and determination of the stability and other properties of gold nanoparticles (AuNPs) in an aqueous solution, gold nanoparticle immobilization, and further surface grafting on polyethylene naphthalate (PEN). Gold nanoparticles in PEG with a subsequent water solution addition were prepared using cathode sputtering; for the subsequent surface activation, two different solutions were used: (i) sodium citrate dihydrate (TCD) and (ii) N-acetyl-L-cysteine (NALC). The aim of this work was to study the effect of the concentration of these solutions on AuNPs stability, and further, the effect of the concentration of gold nanoparticles and their morphology, and to describe the aging process of solutions, namely, the optical properties of samples over 28 days. Stabilized AuNPs were prepared in an N-acetyl-L-cysteine (NALC) system and subsequently immobilized with NALC. The surface chemistry modification of AuNPs was confirmed using HRTEM/EDS. Gold nanoparticles were successfully immobilized with NALC. Grafting of the modified PEN from a solution of colloidal gold stabilized in the PEG–H_2O–NALC system led to the polymer surface functionalization.

Keywords: nanoparticles; noble metal; sputtering; grafting; polymer; plasma modification

1. Introduction

Metallic colloidal nanoparticles can be prepared via top-down methods, which are usually physical methods, using bulk materials in a controlled process to form nanostructures/nanoparticles. These include molecular epitaxy, vacuum evaporation, or cathode sputtering [1–3]. In contrast, bottom-up methods, mainly wet chemical preparations of nanoparticles, are based on molecular aggregation, nucleation, and controlled growth of nanoclusters using donor ligands, polymers, and surfactants, which also serve as stabilizers. These include the chemical reduction of metal salts [4–6], electrochemical processes, or the controlled decomposition of metastable organometallic compounds [7].

For the first time, an innovative study was conducted to describe a versatile process for preparing magnetic colloidal solutions via sputtering on liquid surfaces [8]. It was shown for the first time that metallic colloidal NPs can be obtained by sputtering on silicone oil. However, silicone oil is not a good stabilizing agent and the obtained colloids were not stable. To avoid agglomeration, another liquid medium must be used. An ionic liquid (IL) seems to be the ideal type of liquid for this method of NP preparation. The sputtering of nanoparticles on ILs is a research area that began in 2006 with a groundbreaking study by Torimoto and co-authors [9]. Gold nanoparticles (AuNPs) with a diameter of about 5.5 nm were obtained by sputtering from an Au target in an ionic

(Ar+) medium directly onto the surface of 1-ethyl-3-methylimidazolium tetrafluoroborate. Nanoparticles (silver nanoparticles) were sputtered into a liquid medium (silicone oil) in 1996 using radiofrequency magnetron deposition [10]. However, ILs are not the only liquid substrates used for sputtering metal nanoparticles. For example, vegetable oils, glycerol, or liquid polyethylene glycol are widely used for this method [11–15].

Electrical and optical properties depend on the structure of nanoparticles. Spherical AuNPs exhibit a variety of colors (e.g., brown, orange, red, and purple) in aqueous solution as their core size increases from 1–100 nm, and generally, AuNPs exhibit an absorption peak between 500–550 nm. AuNPs are considered to be one of the most suitable carrier systems due to their increased biocompatibility, stability, low toxicity, and oxidation resistance. Thus, colloidal gold (which may also be attached to a polymer surface) is useful in various medical disciplines, including biosensors [16,17] and biological detection, catalysis, bioelectronics, drug delivery and macromolecular carriers, and bioimaging [18]. PEG is a very weakly immunogenic substance, making it suitable for the development of PEG–protein conjugates as drugs [19,20]. Polyethylene glycol (PEG) modification, or PEGylation, is a common method of functionalizing gold nanoparticles. AuNPs are prepared simply via cathodic sputtering into liquid PEG as a capture medium without additional chemical reactions or added stabilizers. The stability and functional integrity of PEGylated AuNPs is dependent on many factors, including the size of the AuNPs used, the molecular weight of PEG, the attached functional groups, and the ligand used for PEGylation [21]. Nanoparticles can be coated with a layer of PEG alone or PEG in conjunction with other molecules, such as biotin, peptides, or oligonucleotides, to assist in the implementation of AuNPs into target cells. Due to their ability to bind cell membranes, these functionalized AuNPs can serve as good drug carriers. PEGylated AuNPs functionalized with biomolecules, such as lectin, lactose, and biotin, were synthesized in [22,23]. PEGylated AuNPs are useful in the cellular and intracellular targeting of biological materials. Hetero-bifunctional PEGylated AuNPs were synthesized in which AuNPs were functionalized with a thiol group at one end and coumarin, a fluorescent dye, on the other. These functionalized AuNPs could reach cells that could be easily monitored due to the attached dye [24].

Another effective way to increase the specificity and efficiency of nanoparticle systems is via functionalization with amino acids and peptides. AuNPs conjugated to amino acids, such as lysine, polylysine, and glycine, bind DNA with greater efficiency to gene transport without toxicity. The primary ammonium groups of these amino acids contributed to a higher binding ability of cationic groups to DNA [25]. Amino-acid-functionalized gold colloids provide a backbone for efficient DNA binding. The amino-acid-based NPs responded to intracellular levels of glutathione and provided a tool for controlled release and concomitant DNA expression [26]. In his work [27], Russier-Antoine described a simple synthesis that produces large chiral supramolecular nanocomplexes of gold and cysteine. Gold–cysteine polymers of NP induced remarkable two-photon nonlinear spectra and so-called hyper-polarizability. Thus, they appear to be good candidates for nonlinear optical microscopy. The high arginine peptide (CLANNR8) was conjugated to AuNPs for transport into a cell membrane [28]. Some research groups have developed methods for functionalizing gold nanoparticles and other nanoparticles using oligonucleotides either alone or with some modifications. DNA-conjugated nanostructures can be synthesized in a controlled manner, either by attaching a specific number of single-stranded DNA molecules via thiol caps or by saturating the surface of gold nanoparticles with single-stranded DNA molecules [29]. Kinetic and thermodynamic studies of DNA hybridized to AuNP have shown that DNA first adheres to nanoparticles and then slowly diffuses to its surface [30]. Aptamer–AuNP conjugation was recently used to target prostate cancer cells [31]. This was accomplished by attaching AuNP to an oligonucleotide complementary to the anti-PSMA (prostate-specific membrane antigen) sequence, which facilitates the attachment of PSMA–AuNP to the anti-PSMA antibody. In summary, the most commonly used functionalization methods involve application of the following functional groups: PEG [32–35], amine group [36], carboxyl group [37], peptide [38,39], or DNA [40,41]. For applications in tissue engineering for the support of cell growth or as antibacterial properties, the as-sputtered nanoparticles in water solution or grafted nanoparticles on polymer substrates were used recently [42–45].

AuNPs were prepared via cathode sputtering into PEG, followed by transfer into solutions of sodium citrate dihydrate or N-acetyl-L-cysteine (LNAC). The effects of the concentrations of these solutions on the stability of AuNPs were studied. In addition, the concentration of nanoparticles, their morphology, and the change in optical properties of colloidal solutions during aging were monitored. Sample analysis was performed using UV-Vis spectroscopy, atomic absorption spectrometry (AAS), transmission electron microscopy (TEM), and high-resolution TEM (HRTEM). The described AuNP/NALC/H_2O/PEG solutions were then used to graft the surface of polyethylene naphthalate. To our best knowledge, the combination of PEN plasma treatment for its activation combined with subsequent grafting in a colloidal gold solution for gold grafting has never been published. It was shown to successfully functionalize the polymer surface with AuNPs.

2. Materials and Methods

2.1. Materials

Colloidal gold was obtained via cathodic sputtering from a Au target (Safina s.r.o., purity 99.9999%, Vestec, Czech republic) into a liquid medium. PEG-600 (polyethylene glycol with an average molecular mass of 600 Da, Sigma-Aldrich, St. Louis, MO, USA) was selected as the medium. PEG is a hygroscopic colorless viscous liquid at room temperature, is fully water-soluble, and is miscible with water in all proportions. PEG is a commonly used cathode sputtering medium because it has a sufficiently low saturated vapor pressure at deposition pressures, stabilizes nanoparticles without the need for additional compounds, and is a biocompatible material with low toxicity [1].

Sputtering was carried out at 20 °C in a Sputter Coater SCD 050 (Baltec, Balzers, Liechtenstein) at an argon pressure of approximately 8 Pa, a current of 30 mA, and an electrode gap of 50 mm. Immediately after the deposition of the AuNPs into PEG-600, the mixture was diluted to an aqueous solution in a PEG:water ratio of 1:9. Either an aqueous solution of sodium citrate dihydrate (TCD—a white crystalline powder, M_w = 294.10 g·mol^{-1}, Sigma Aldrich,) or the amino acid solution N-acetyl-L-cysteine (NALC—white crystalline solid, M_w = 163.19 g·mol^{-1}, Sigma Aldrich) was used to modify the nanoparticles. The stability of the solutions and the size of the nanoparticles were monitored such as the dependence of this factors on the amount of TCD or amino acid in the solution. TCD solutions were prepared at four concentrations as follows: (a) 17.0 mmol·L^{-1} (by mixing 0.5 g TCD + 100 mL H_2O), (b) 34.0 mmol·L^{-1} (1.0 g TCD), (c) 51.0 mmol·L^{-1} (1.5 g TCD), and (d) 68.0 mmol·L^{-1} (2.0 g TCD). The sample preparation procedure was the same as for the TCD samples, but the deposition times (100, 300, 600, 900 s) and concentrations of the NALC solutions varied: (a) 30.6 mmol·L^{-1} (by mixing 0.5 g NALC + 100 mL H_2O), (b) 61.3 mmol·L^{-1} (1.0 g NALC), and (c) 91.9 mmol·L^{-1} (1.5 g NALC). A demonstration of the experiment is shown in Figure 1.

The last part of this work was focused on the study and modification of the surface of polyethylene naphthalate (PEN). Biaxially oriented PEN (density 1.36 g·cm^{-3}, thickness 50 μm, T_m ~ 250–290 °C, T_g ~ 120 °C, supplied by Goodfellow Ltd., Cambridge Ltd., Huntington, UK) was used.

The plasma modification method was chosen for the PEN surface activation. Balzers SCD 050 from BAL-TEC was used and the etching mode was applied. The modification conditions were as follows: temperature 20 °C, pressure 8 Pa, and modification power of 8 W. Modification was carried out for the following deposition times: 200, 400, 600, 800, and 1000 s.

For the next experiment, a solution of AuNPs in PEG/H_2O/NALC at an NALC/H_2O ratio of 1.5 g/100 mL was prepared. PEN samples that were modified by plasma exposure for 200 and 400 s were immersed in a colloidal gold solution for 24 h. After 24 h, the PEN samples were removed from the solution, washed with distilled water, and dried at room temperature. The PEN samples thus prepared were subjected to AFM (Bruker Corp., Billerica, MA, USA) and XPS (Scienta Omicron GmbH, Taunusstein, Germany) analysis. Changes to the PEN morphology after grafting to the Au nanoparticles or the amino acid NALC were investigated in combination with a detailed analysis of the chemistry of the grafted surface.

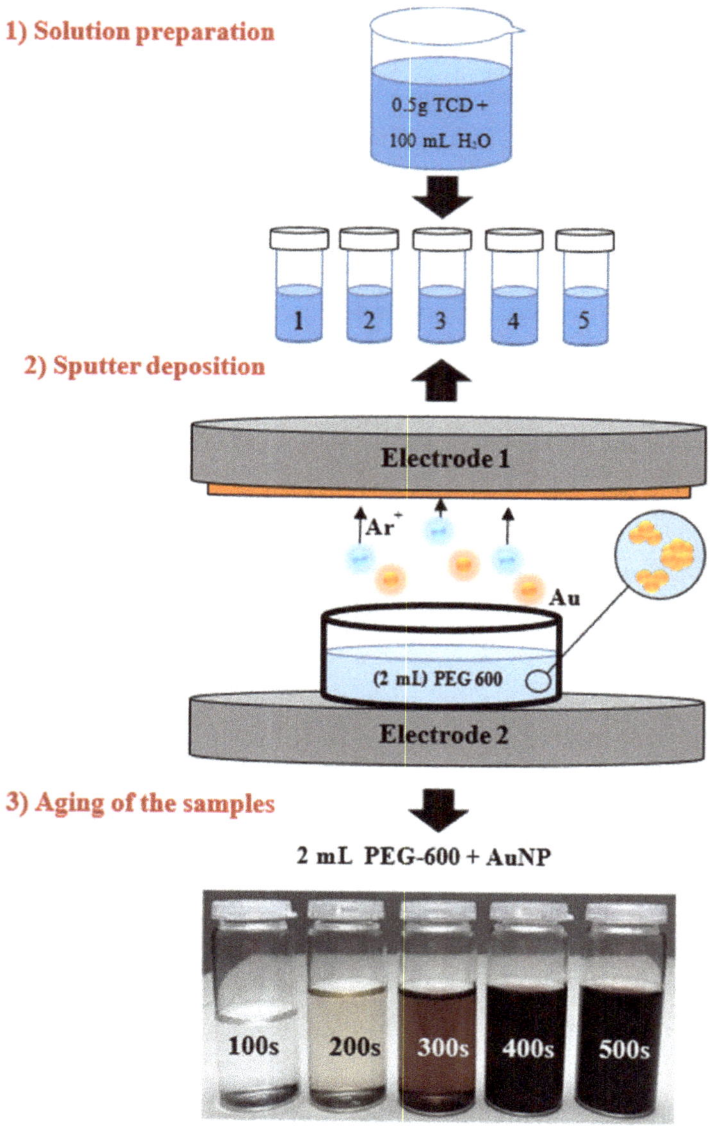

Figure 1. Principle of the preparation of a series of AuNPs solutions for one concentration of TCD (c = 17.0 mmol·L^{-1}): (**1**) preparation of the TCD solution and subsequent dosing of 18 mL into glass sample containers; (**2**) deposition of AuNPs into 2 mL PEG-600 at selected times (100–500 s), followed by the addition of TCD; (**3**) after aging of the products.

2.2. Characterization Techniques

Samples were weighed using the Metler Toledo UMX2 automatic microbalance (Mettler-Toledo, Columbus, GA, USA). UV-Vis spectroscopy was used to characterize the optical properties of the samples. The absorbance of AuNPs solutions was measured in a 1 cm glass cuvette using a Perkin-Elmer spectrophotometer Lambda 25 (Waltham, MA, USA). Spectra were recorded at room temperature in the range of 250 to 800 nm. Gold nanoparticles with a spherical shape exhibited only one peak at 520 nm at this interface. This optical property changes with the AuNPs shape.

The determination of the gold concentration in the solutions was performed on an Agilent 280FS AAS spectroscope (Santa Clara, CA, USA) with a flame atomization technique. The determination was carried out by flame atomization (acetylene–air) at a wavelength of 242.8 nm. Atomic absorption spectrometry (AAS, Agilent, Santa Clara, CA, USA) is a spectrometric analytical method for the determination of trace and significant concentrations of individual elements in the analyzed solution. It utilizes the absorption of radiation by free atoms of the monitored element. For the generation of free atoms, a flame is most often used in the AAS.

Transmission electron microscopy (TEM) was used to monitor and study the size and shape of nanoparticles in solutions. TEM samples were analyzed on a JEOL JEM-1010 transmission electron microscope. The images were taken with an SIS Megaview III digital camera (Soft Imaging Systems, 80 kV acceleration voltage, Tokyo, Japan), and analysis was performed with AnalySIS 2.0 software (Tokyo, Japan). High-resolution transmission electron microscopy (HRTEM) was used to characterize the solutions in more detail. An EFTEM Jeol 2200 FS (Tokyo, Japan) with a dot resolution of 0.23 nm was used. A transmission electron microscope equipped with an energy filter enabled operation at accelerating voltages up to 200 kV. HRTEM analysis was used in conjunction with electron dispersion spectroscopy (EDS, Joel Ltd., Tokyo, Japan). The resolution of the EDS analysis was 1–2.4 nm. The chemical compositions of the samples were measured using EDS. Some solutions were subjected to EDS element mapping and point analysis.

Atomic force microscope (AFM Dimension ICON Bruker Corp., Billerica, MA, USA) in QNM mode was used to study the surface morphology of the PEN layer in an ambient atmosphere. The silicon tip was mounted on a SCANASYST-AIR nitride cantilever with a spring constant of 0.4 $N{\cdot}m^{-1}$. Data were processed using NanoScope Analysis software (Billerica, MA, USA).

The quantitative composition of the PEN surface was obtained using X-ray photoelectron spectroscopy (XPS, Scienta Omicron GmbH, Taunusstein, Germany). An ESCAProbeP spectrometer manufactured by Omicron Nanotechnology Ltd. (East Grinstead, UK) was used for the measurements. The sample area analyzed was 3 × 4 mm2. The surface of the sample was excited using X-rays from a 1488.7 eV monochromatic source. The relative representation of elements was evaluated from the peak intensities in the XPS spectra using CasaXPS software (2.3.1, Wilmslow, UK).

3. Results

3.1. Stability Studies of AuNP Solutions in Sodium Citrate Dihydrate

Evidence of the presence of gold nanoparticles was obtained from the UV-Vis spectra. The peak height provides information on the concentration of the AuNPs in the solution. According to the peak width, it is possible to conclude information about the stabilization of homogeneous solutions. Shifts to the longer wavelengths are typical of increasing NP sizes. According to the shapes of the peaks, these were all spherical nanoparticles.

Colloidal solutions of prepared AuNPs that were stabilized with distilled water with TCD showed absorption peaks of UV-Vis spectra. In Figure 2A, we can see that the absorbance was around 0.8 for all samples, except for the sample with the highest TCD concentration (2 g/100 mL).

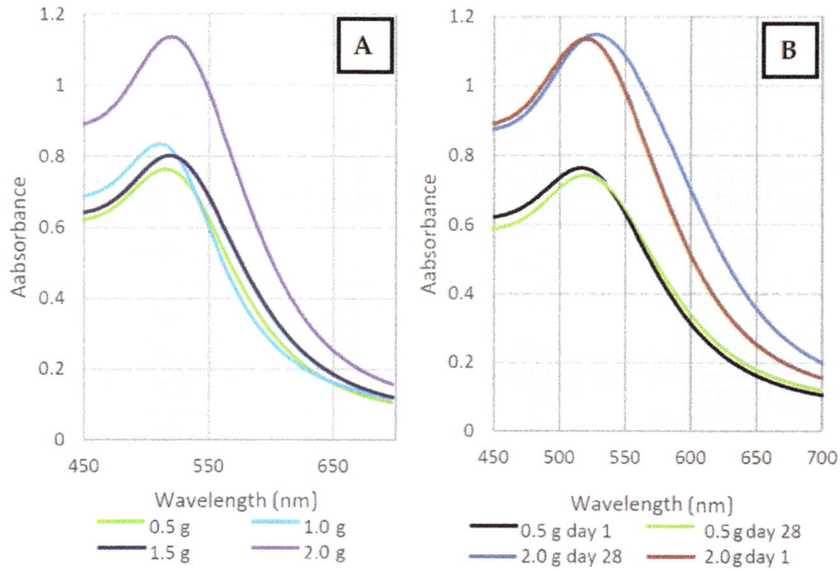

Figure 2. (**A**) Changes in the absorbances of AuNPs in PEG/TCD/H_2O with different TCD/H_2O ratios (0.5 g/100 mL, 1.0 g/100 mL, 1.5 g/100 mL, and 2.0 g/100 mL) prepared with a deposition time of 300 s; (**B**) changes in the AuNPs' absorbances in PEG/TCD/H_2O with different TCD/H_2O ratios (0.5 g/100 mL and 2.0 g/100 mL) prepared with a deposition time of 300 s during the aging process.

It is apparent from both graphs (Figure 2A,B) that the changes in the TCD/H_2O ratio had no significant effect on the distribution of the AuNPs sizes. In Figure 2B, we can observe the change in size, shift, and broadening of the absorption peaks, indicating a slight aggregation and growth of the nanoparticles during aging. The high absorbance value of AuNPs in the TCD/H_2O solution at a ratio of 2 g/100 mL may have been caused by the higher concentration of TCD in the solution. Peak positions were in the range of 516–532 nm. The peak shifts during the 28-day aging are shown in the following graphs in Figure 3.

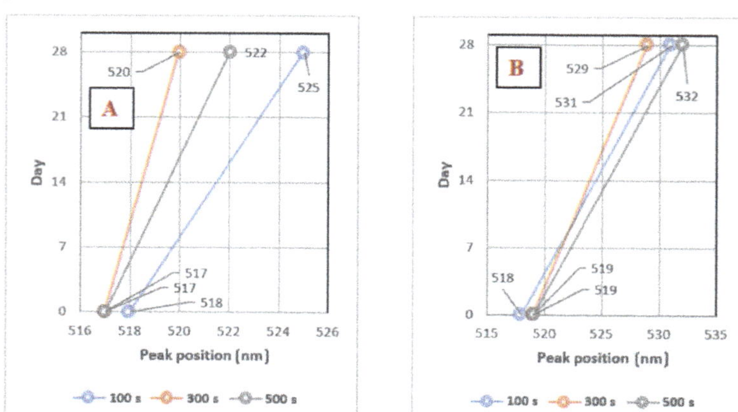

Figure 3. (**A**) AuNPs absorption peak positions in PEG/TCD/H_2O with a TCD/H_2O ratio of 0.5 g/100 mL prepared with deposition times of 100, 300, and 500 s; (**B**) AuNPs' absorption peak positions in PEG/TCD/H_2O with a TCD/H_2O ratio of 1.5 g/100 mL prepared with deposition times of 100, 300, and 500 s.

Solutions prepared with a TCD/H_2O ratio of 0.5 g/100 mL demonstrated high stability, with the absorption peak shifted by only 4–7 nm over 28 days. For a solution with a higher TCD/H_2O ratio (1.5 g/100 mL), the absorption peaks shifted between 10–13 nm. For this reason, both solutions could be considered stable, but higher concentrations of TCD had a greater influence on the optical properties of the solutions during the aging process. In this work, many different combinations of solutions (TCD/H_2O) were made; therefore, only two samples were chosen for further comparison of the AuNPs stability in $TCD/H_2O/PEG$ solutions, which differed the most in both the deposition time and TCD/H_2O ratio. The comparisons of the UV-Vis solution spectra are shown in Figure 4A,B.

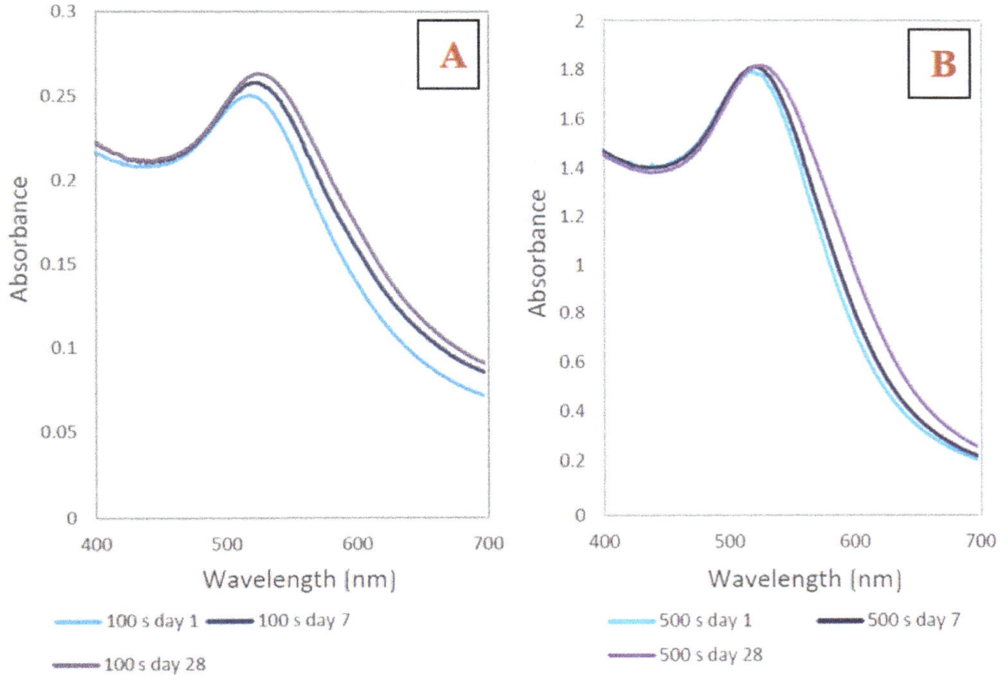

Figure 4. (**A**) Absorbance of AuNPs in $PEG/TCD/H_2O$ with a TCD/H_2O ratio of 0.5 g/100 mL prepared with a deposition time of 100 s during the aging process; (**B**) absorbance of AuNPs in $PEG/TCD/H_2O$ with a TCD/H_2O ratio of 2 g/100 mL prepared with a deposition time of 500 s during the aging process.

The UV-Vis spectra peak maximum for both samples shown in Figure 4A,B ranged between 513 and 532 nm. We managed to obtain stable AuNPs in all solutions. However, these two solutions differed significantly in terms of their absorption maxima. A shorter deposition time (100 s, Figure 4A) resulted in a lower concentration of AuNPs in the solution. Longer deposition times (500 s, Figure 4B) led to a higher content of nanoparticles in solution and an increase in absorbance.

Selected samples were studied using TEM analysis. In Figure 5, we can see that the prepared AuNPs had a spherical shape. From the images of Figure 5C,D, it can be seen that longer deposition times led to more concentrated solutions and larger particles; AuNP agglomerates of the 'chain' type were up to 200 nm in size. A solution with a higher concentration of TCD formed a denser AuNPs network due to the affinity of AuNPs for the TCD that the nanoparticles enveloped.

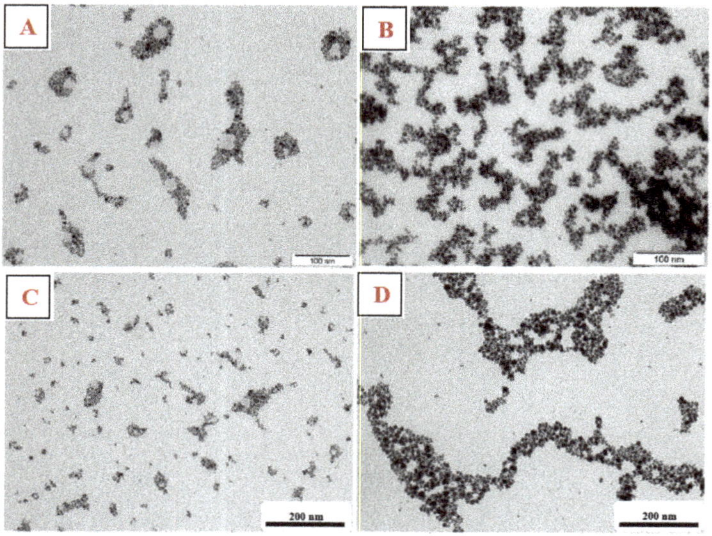

Figure 5. TEM AuNP images in PEG/TCD/H_2O prepared with a deposition time of 300 s. (**A**) TCD/H_2O (0.5 g/100 mL) solution and (**B**) TCD/H_2O (1.5 g/100 mL) solution. TEM AuNP images in PEG/TCD/H_2O at a TCD/H_2O ratio of 0.5 g/100 mL prepared with deposition times of 100 (**C**) and 500 s (**D**).

3.2. Study of AuNP Stability in N-Acetyl-L-Cysteine

The prepared AuNPs solutions stabilized in NALC were analyzed immediately after preparation and the UV-Vis spectra of the prepared AuNPs solutions were obtained. The UV-Vis spectra of the solutions with NALC with a deposition time of 100 and 300 s showed a broad band with no absorption maximum. The presence of colloidal gold in these solutions was not confirmed using this method. For solutions with higher deposition times (600 and 900 s), the positions of the absorption maxima ranged between 459 and 515 nm, but the peaks had a very flat profile. The solution with the highest concentration of NALC showed the most obvious absorption maxima (1.5 g NALC/100 mL H_2O), where the absorption maxima ranged from 513 to 515 nm (for 600 and 900 s), but the curve profiles were still very wide. With higher deposition times, the absorbance of the solutions also increased.

The solutions described in Figure 6 showed slightly different absorbances during the aging process. The positions of the absorption peaks shifted by 0–5 nm over 28 days, indicating their good stability. By day 2, the peak position was red-shifted, indicating possible aggregation of the particles. After the 21st day, the values started to fall again, and AuNP could already be degraded. The peak positions for NALC/H_2O 0.5 g/100 mL solutions ranged from 498 to 503 nm; for the solutions with 1.0 g/100 mL, the positions ranged from 490 to 496 nm; for the solutions with 1.5 g/100 mL, the positions ranged from 513 to 519 nm (Figure 6).

From the TEM images (Figure 7), we can see that we managed to create spherical AuNPs independent of the concentration of NALC/H_2O in the solutions. However, solutions with a higher proportion of NALC (1.0 and 1.5 g/100 mL H_2O) showed significant agglomerations of gold nanoparticles. NALC probably wrapped individual AuNPs and together formed a core–shell structure of Au–NALC. This increased their affinity for other non-enveloped AuNPs and clustering could occur. After washing these samples with distilled water, the clusters were no longer so apparent; there was probably a partial 'washing' of the coating. This may have also been affected by the existence of weak van der Waals interactions between the particles such that washing the particles may have affected the interactions between them.

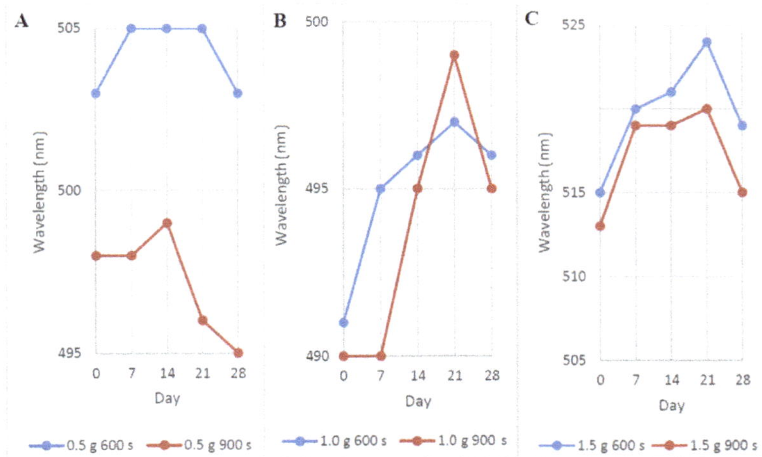

Figure 6. AuNP absorption peak positions in PEG/NALC/H_2O prepared with a deposition time of 600 or 900 s in NALC/H_2O: (**A**) 0.5 g/100 mL, (**B**) 1.0 g/100 mL, and (**C**) 1.5 g/100 mL.

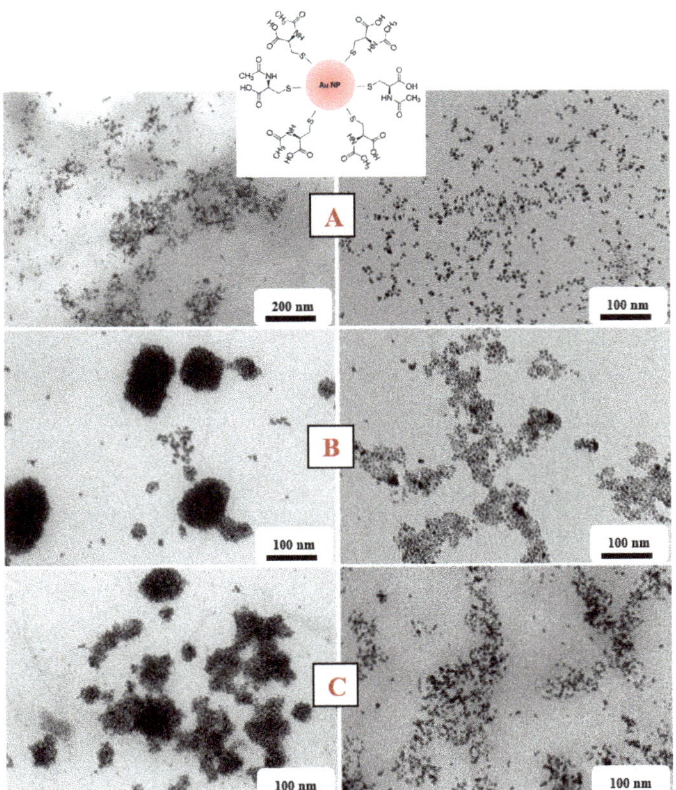

Figure 7. TEM AuNP images in PEG/NALC/H_2O prepared with a deposition time of 900 s. (**A**) NALC/H_2O solution (0.5 g/100 mL), (**B**) TCD/H_2O solution (1.0 g/100 mL), and (**C**) TCD/H_2O solution (1.5 g/100 mL). Right column after washing with distilled water. A scheme of a grafted Au nanoparticle with NALC is also introduced.

We managed to obtain spherical gold nanoparticles with sizes in the range of 6–10 nm (Figure 8). The concentration of the amino acid in the solution did not affect the size of the generated AuNPs, only their aggregation. EDS analysis confirmed the presence of Au and S in the solutions (Figure 9).

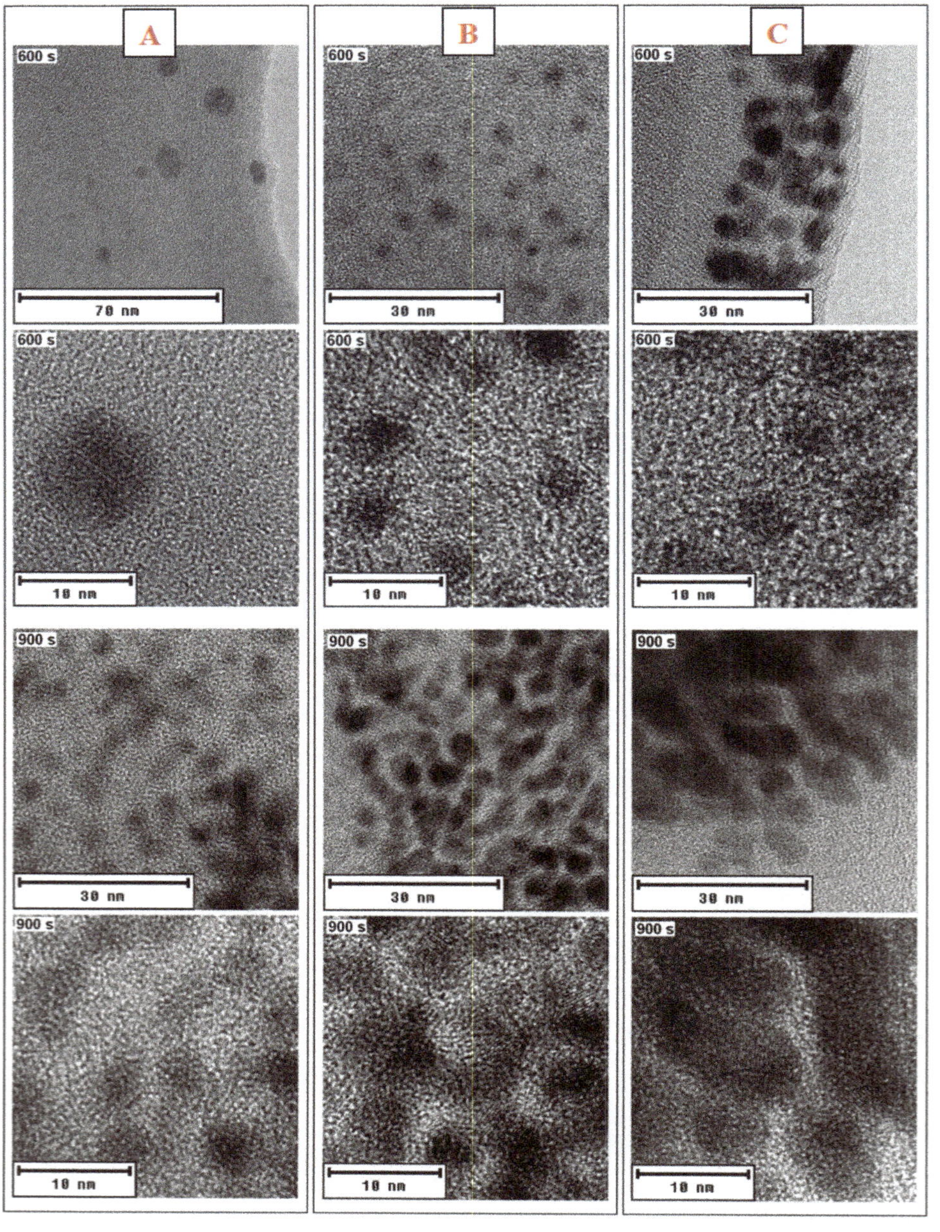

Figure 8. HRTEM AuNP images in PEG/NALC/H_2O prepared with a deposition time of 600 or 900 s: (**A**) NALC/H_2O solution (0.5 g/100 mL), (**B**) TCD/H_2O solution (1.0 g/100 mL), and (**C**) TCD/H_2O solution (1.5 g/100 mL).

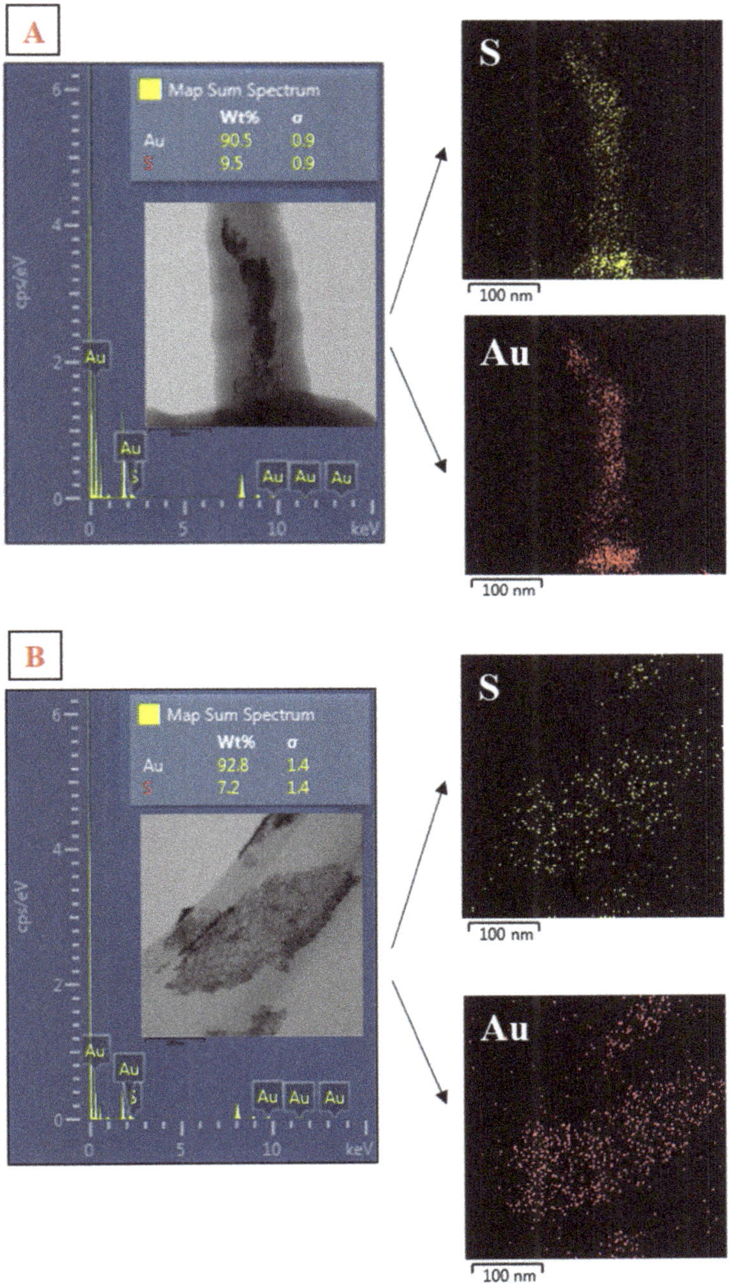

Figure 9. (**A**) Sample analysis of AuNP-900 s/NALC/H_2O at an NALC/H_2O ratio of 0.5 g/100 mL; EDS spectrum—Au and S in the sample. HRTEM image of the sample, EDS mapping of Au and S content in solution. (**B**) Analysis of AuNP-600 s/NALC/H_2O sample at an NALC/H_2O ratio of 0.5 g/100 mL; EDS spectrum—Au and S in the sample. HRTEM image of the sample, EDS mapping of Au and S content in solution.

3.3. Modification of PEN Using AuNP/PEG/NALC/H_2O

PEN was modified using plasma with exposure times of 200–1000 s. Ablation times of 200 and 400 s were chosen for grafting PEN foils since these plasma modification times are sufficient to activate the surface of the material based on our previous experiments. After the PEN activation, samples were exposed to the AuNP/PEG/NALC/H_2O solution for 24 h, followed by changes in the PEN morphology and chemistry.

Changes in the chemical composition of the surface layer were studied using the XPS method (Figure 10). XPS analysis showed the presence of sulfur (S) and gold on the PEN surface for all studied samples. The values of the individual atomic concentrations of sulfur and gold are demonstrated in Table 1 below. Since the original PEN did not contain gold or sulfur, the results show that the immersion of PEN samples in AuNP/NALC solutions bound to AuNP–NALC in the polymer surface layer.

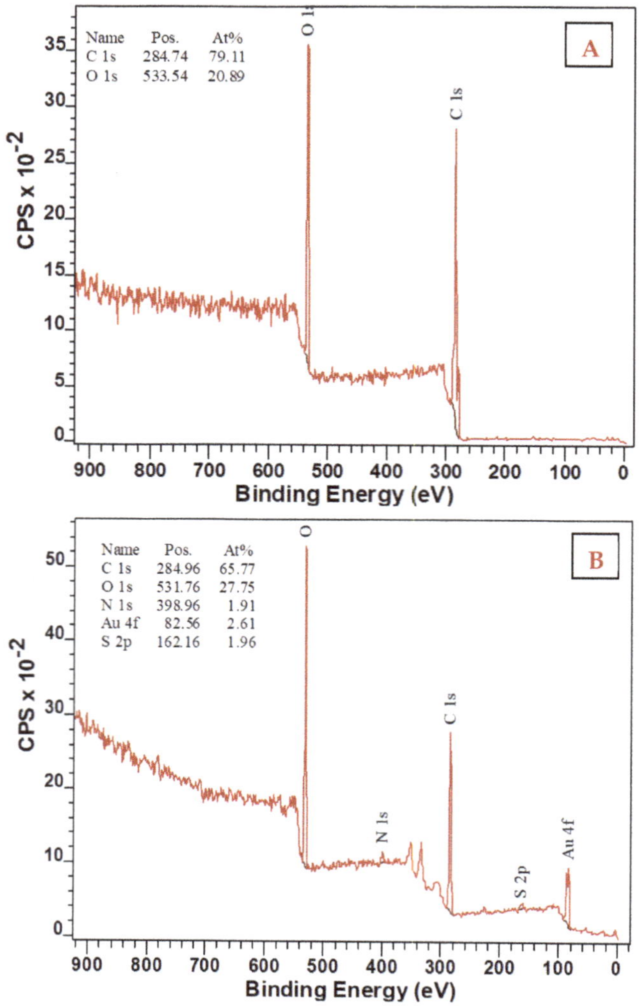

Figure 10. XPS analysis: (**A**) non-modified PEN and (**B**) PEN plasma modified for 400 s after 24 h immersion in AuNP/PEG/NALC/H_2O solution (NALC/H_2O ratio-1.5 g/100 mL, AuNP deposition time-600 s).

Table 1. Atomic concentrations of elements (S, Au) determined using the XPS method in pristine PEN, PEN modified with plasma 8 W for 200 and 400 s, and PEN modified with plasma 200 and 400 s after 24 h of immersion in AuNP/NALC/H_2O solution (NALC/H_2O ratio—1.5 g/100 mL), deposition time 600 and 900 s.

Sample	Element Concentration (at%)	
	S	Au
Pristin PEN; PEN modif. 200 s; PEN modif. 400 s	-	-
PEN modif. 200 s, AuNP 600 s	1.26	0.12
PEN modif. 200 s, AuNP 900 s	1.26	2.32
PEN modif. 400 s, AuNP 600 s	1.96	2.61
PEN modif. 400 s, AuNP 900 s	0.83	0.59

The concentrations of bound gold and sulfur were the highest for the 400 s plasma-modified sample that was immersed in the AuNP solution prepared with a deposition time of 600 s. In the previous part of this work, it was discussed that a high concentration of NALC in solution may cause AuNPs amino acid 'wrapping' or 'chaining,' which was also probably the reason for lower gold concentration values on the PEN surface.

The binding of elements to the polymer surface was also reflected in the change in its surface morphology, which was studied using AFM. In Figures 11 and 12, there are 2D images of the surface of the modified PEN before and after the Au/amino acid grafting. At the same time, the surface roughness values R_a and effective area S are presented. It is apparent from the AFM results that the surface morphology changed due to grafting.

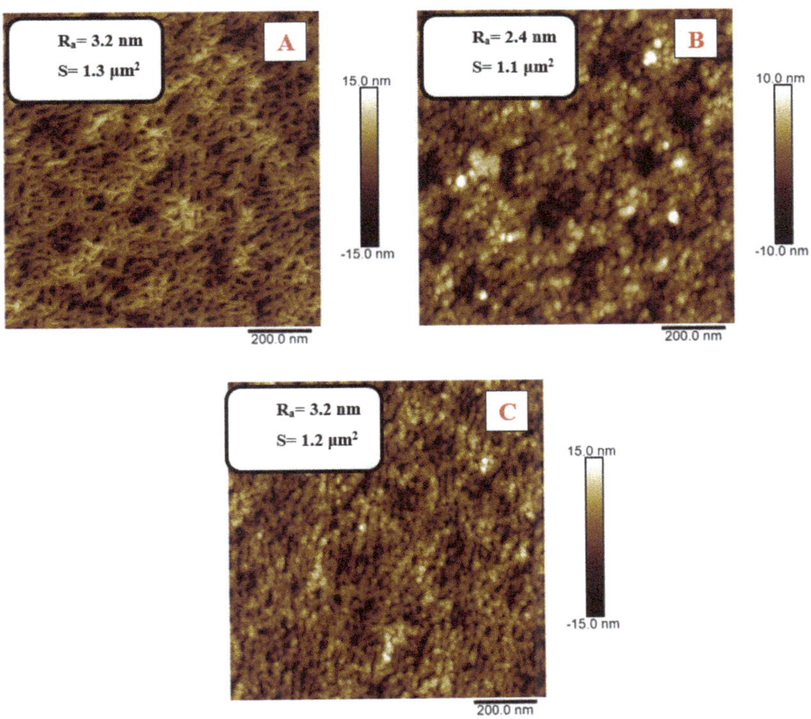

Figure 11. Two-dimensional AFM images of PEN that was modified using plasma for 200 s at 8 W power: (**A**) non-grafted sample; (**B**) grafted in AuNP/NALC (1.5 g)/H_2O (100 mL)/PEG solution, AuNP deposition time 600 s; (**C**) grafted in AuNP/NALC (1.5 g)/H_2O (100 mL)/PEG solution, AuNP deposition time 900 s.

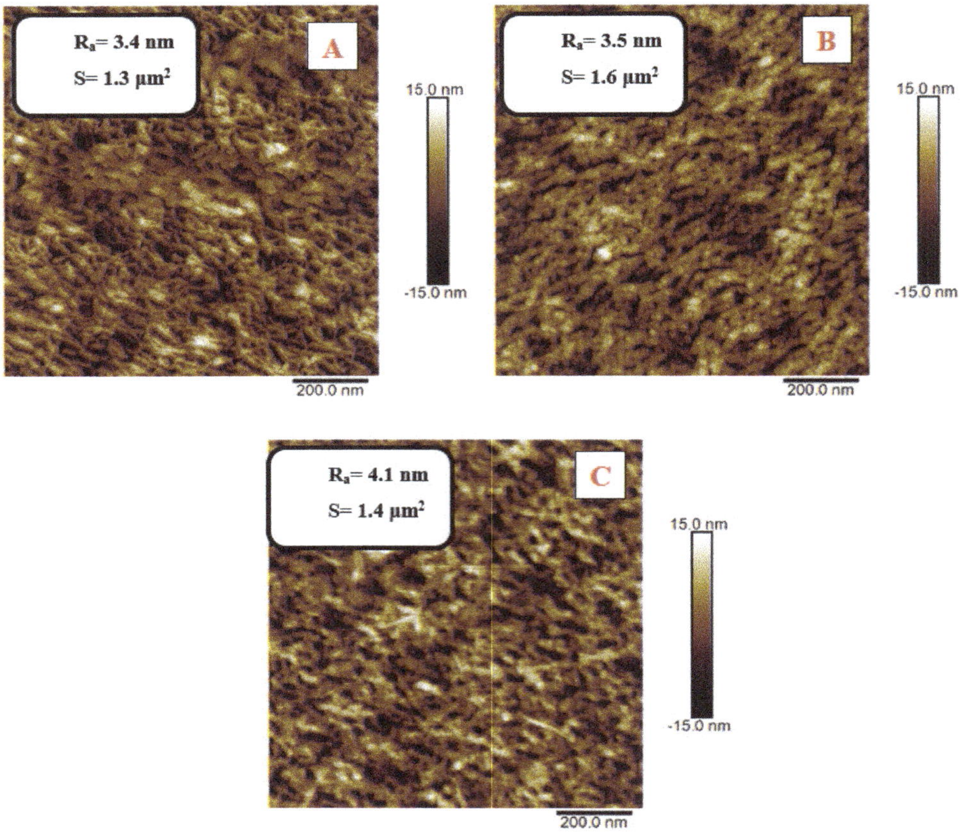

Figure 12. Two-dimensional AFM images of PEN that was modified using plasma for 400 s at 8 W power: (**A**) non-grafted sample; (**B**) grafted in AuNP/NALC (1.5 g)/H_2O (100 mL)/PEG solution, AuNP deposition time 600 s; (**C**) grafted in AuNP/NALC (1.5 g)/H_2O (100 mL)/PEG solution, AuNP deposition time 900 s.

In Figure 11, we can see that after the plasma modification for 200 s, the surface of the PEN had a lamellar structure; after grafting, a globular structure appeared on the surface of the PEN with a significant increase in the surface segmentation. The globular structure results from the interaction of PEN with the solution, but mainly from the binding of Au, which was confirmed by the abovementioned XPS analysis. Furthermore, a decrease in the effective area S after grafting in solution with AuNP deposited at 600 s and a subsequent increase in S after grafting in solution with AuNP deposited at 900 s are shown.

As can be seen in Figure 12, after the 400 s plasma modification, the PEN surface had a more structured lamellar structure; after grafting, globular structures appeared on the activated PEN surface. The change in effective area S had the same trend as for the 200 s modified PEN. The grafting in the AuNP-600 s solution led to a decrease in S, while the grafting in the AuNP-900 s solution increased the effective area.

In both cases (200 and 400 s), a significant change in the surface roughness is visible in the images compared to the AuNP-600 s and AuNP-900 s. This change in R_a was due to the formation of AuNP aggregates due to the higher concentration of AuNP in the solution.

4. Conclusions

Gold nanoparticles were prepared via sputter deposition into polyethylene glycol. AuNPs were stabilized in the PEG–H_2O–TCD system. Spherical AuNPs of sizes up to 10 nm were formed. Higher sodium citrate concentrations in solutions were found to have a significant effect on the optical properties of the colloidal solutions. In the PEG–H_2O–TCD system, AuNPs began to cluster into visible structures in as early as 14 days. This additional aggregation was probably due to the coating and agglomeration of unreacted citrate particles.

AuNPs were also stabilized in the PEG–H_2O–NALC system. We managed to prepare spherical grafted Au/NALC nanoparticles with sizes up to 10 nm. Samples with higher AuNP deposition times showed higher stability. However, the AuNP size increased during the aging of the solutions. Higher concentrations of NALC in solutions led to the aggregation of nanoparticles and even 'chaining.' Otherwise, the concentration of NALC in the solutions did not affect the size of the formed nanoparticles.

Grafting of the modified PEN from a solution of colloidal gold stabilized in the PEG–H_2O–NALC system led to the polymer surface functionalization. XPS and AFM analyses confirmed the presence of Au in the PEN surface layer. The grafting of gold nanoparticles onto the polymer had a significant effect on the surface morphology of PEN when the original linear structure changed to a globular structure.

Author Contributions: Conceptualization, P.S.; methodology, J.S. and N.S.K.; validation, P.S. and V.Š.; investigation, D.F., A.K., P.U. and P.S.; resources, P.S.; data curation, N.S.K. and A.M.; writing—original draft preparation, D.F. and P.S.; supervision, P.S.; writing—review and editing P.S. and D.F. All authors have read and agreed to the published version of the manuscript.

Funding: This research was funded by the Czech Science Foundation, grant number 19-02482S and by OP VVV Project NANOTECH ITI II. No. CZ.02.1.01.

Institutional Review Board Statement: Not applicable.

Informed Consent Statement: Not applicable.

Data Availability Statement: Data are contained within the article.

Conflicts of Interest: The authors declare no conflict of interest.

References

1. Jamkhande, P.G.; Ghule, N.W.; HaqueBamer, A.; Kalaskar, M.G. Metal nanoparticles synthesis: An overview on methods of preparation, advantages and disadvantages, and applications. *J. Drug Deliv. Sci. Technol.* **2019**, *53*, 101174. [CrossRef]
2. Matsuyama, K.; Tsubaki, T.; Kato, T.; Okuyama, T.; Muto, H. Preparation of catalytically active Au nanoparticles by sputter deposition and their encapsulation in metal-organic framework of Cu3(BTC)2. *Mater. Lett.* **2020**, *261*, 127124. [CrossRef]
3. Slepička, P.; Slepičková Kasálková, N.; Siegel, J.; Kolská, Z.; Švorčík, V. Methods of Gold and Silver Nanoparticles Preparation. *Materials* **2020**, *13*, 1. [CrossRef] [PubMed]
4. Harada, M.; Yamamoto, M.; Sakata, M. Temperature dependence on the size control of palladium nanoparticles by chemical reduction in nonionic surfactant/ionic liquid hybrid systems. *J. Mol. Liquids* **2020**, *311*, 113255. [CrossRef]
5. De Souza, C.D.; Ribeiro Nogueira, B.; Rostelato, M.E.C.M. Review of the methodologies used in the synthesis gold nanoparticles by chemical reduction. *J. Alloys Compound.* **2019**, *798*, 714–740. [CrossRef]
6. Machida, H.; Sugahara, T.; Hirasawa, I. Preparation of dispersed metal nanoparticles in the aqueous solution of metal carboxylate and the tetra-n-butylammonium carboxylate. *J. Crystal Growth* **2019**, *514*, 14–20. [CrossRef]
7. Zhou, J.; Ralston, J.; Sedev, R.; Beattie, D.A. Functionalized gold nanoparticles: Synthesis, structure and colloid stability. *J. Colloid Interface Sci.* **2008**, *331*, 251–262. [CrossRef]
8. Wagener, M.; Günther, B. Sputtering on liquids—A versatile process for the production of magnetic suspensions? *J. Magn. Magn. Mater.* **1999**, *201*, 41–44. [CrossRef]
9. Torimoto, T.; Okazaki, K. Sputter deposition onto ionic liquids: Simple and clean synthesis of highly dispersed ultrafine metal nanoparticles. *Appl. Phys. Lett.* **2006**, *89*, 24311. [CrossRef]
10. Ye, G.X.; Zhang, Q.R.; Feng, C.M.; Ge, H.L.; Jiao, Z.K. Structural and electrical properties of a metallic rough-thin-film system deposited on liquid substrates. *Phys. Rev. B Condens. Matter* **1996**, *54*, 14754–14757.
11. Slepička, P.; Elashnikov, R.; Ulbrich, P.; Staszek, M.; Kolská, Z.; Švorčík, V. Stabilization of sputtered gold and silver nanoparticles in PEG colloid solutions. *J. Nanopart. Res.* **2015**, *17*, 11–26. [CrossRef]

12. Slepička, P.; Přibyl, M.; Fajstavr, D.; Ulbrich, P.; Siegel, J.; Řezníčková, A.; Švorčík, V. Grafting of platinum nanostructures on biopolymer at elevated temperature. *Colloids Surf. A* **2018**, *546*, 316–325. [CrossRef]
13. Reznickova, A.; Slepicka, P.; Slavikova, N.; Staszek, M.; Svorcik, V. Preparation, aging and temperature stability of PEGylated gold nanoparticles. *Colloids Surf. A* **2017**, *523*, 91–97. [CrossRef]
14. Parveen, R.; Ullah, S.; Sgarbi, R.; Tremiliosi-Filho, G. One-pot ligand-free synthesis of gold nanoparticles: The role of glycerol as reducing-cum-stabilizing agent. *Colloids Surf. A* **2019**, *565*, 162–171. [CrossRef]
15. Leopold, N.; Chiş, V.; Mircescu, N.E.; Marişca, O.T.; Buja, O.M.; Leopold, L.F.; Socaciu, C.; Braicu, C.; Irimie, A.; Berindan-Neagoe, I. One step synthesis of SERS active colloidal gold nanoparticles by reduction with polyethylene glycol. *Colloids Surf. A* **2013**, *436*, 133–138. [CrossRef]
16. Zhao, J.; Wang, L.; Fu, D.; Zhao, D.; Wang, Y.; Yuan, Q.; Zhu, Y.; Yang, J.; Yang, F. Gold nanoparticles amplified microcantilever biosensor for detecting protein biomarkers with high sensitivity. *Sens. Actuators A* **2021**, *321*, 112563. [CrossRef]
17. Hua, Z.; Yu, T.; Liu, D.; Xianyu, Y. Recent advances in gold nanoparticles-based biosensors for food safety detection. *Biosensors Bioelectron.* **2021**, *179*, 113076. [CrossRef] [PubMed]
18. Suárez-García, S.; Solórzano, R.; Novio, F.; Alibés, R.; Busqué, F.; Ruiz-Molina, D. Coordination polymers nanoparticles for bioimaging. *Coord. Chem. Rev.* **2021**, *432*, 213716. [CrossRef]
19. Wang, Y.; Langley, R.J.; Tamshen, K.; Harms, J.; Middleditch, M.J.; Maynard, H.D.; Jamieson, S.M.F.; Perry, J.K. Enhanced Bioactivity of a Human GHR Antagonist Generated by Solid-Phase Site-Specific PEGylation. *Biomacromolecules* **2021**, *22*, 299–308. [CrossRef]
20. Matsuhira, T.; Sakai, H. Entropy-Driven Supramolecular Ring-Opening Polymerization of a Cyclic Hemoglobin Monomer for Constructing a Hemoglobin–PEG Alternating Polymer with Structural Regularity. *Biomacromolecules* **2021**, *22*, 1944–1954. [CrossRef]
21. Qian, W.; Murakami, M.; Ichikawa, Y.; Che, Y. Highly efficient and controllable PEGylation of gold nanoparticles prepared by femtosecond laser ablation in water. *J. Phys. Chem. C* **2011**, *115*, 23293–23298.
22. Takae, S.; Akiyama, Y.; Otsuka, H.; Nakamura, T.; Nagasaki, Y.; Kataoka, K. Ligand density effect on biorecognition by PEGylated gold nanoparticles: Regulated interaction of RCA120 lectin with lactose installed to the distal end of tethered PEG strands on gold surface. *Biomacromolecules* **2005**, *6*, 818–824. [CrossRef] [PubMed]
23. Shimmin, R.G.; Schoch, A.B.; Braun, P.V. Polymer size and concentration of effects on the size of gold nanoparticles cappped by polymeric thiols. *Langmuir* **2007**, *20*, 5613–5620. [CrossRef]
24. Shenoy, D.; Fu, W.; Li, J.; Crasto, C.; Jones, G.; DiMarzio, C.; Amiji, M. Surface functionalization of gold nanoparticles using hetero-bifunctional poly (ethylene glycol) spacer for intracellular tracking and delivery. *Int. J. Nanomed.* **2006**, *1*, 51–57. [CrossRef]
25. Valkenier, H.; Malytskyi, V.; Blond, P.; Retout, M.; Mattiuzzi, A.; Goole, J.; Raussens, V.; Jabin, I.; Bruylants, G. Orcid Controlled Functionalization of Gold Nanoparticles with Mixtures of Calix[4]arenes Revealed by Infrared Spectroscopy. *Langmuir* **2017**, *33*, 8253–8259. [CrossRef] [PubMed]
26. Fu, W.; Shenoy, D.; Li, J.; Crasto, C.; Jones, G.; Dimarzio, C.; Sridhar, S.; Amiji, M. Biomedical applications of gold nanoparticles functionalized using hetero-bifunctional poly(ethylene glycol) spacer. *MRS Online Proc. Libr.* **2004**, *1*, 199–204. [CrossRef]
27. Wangoo, N.; Bhasin, K.; Mehta, S.; Suri, C. Synthesis and capping of water-dispersed gold nanoparticles by an amino acid: Bioconjugation and binding studies. *J. Colloids. Interface Sci.* **2008**, *323*, 247–254. [CrossRef] [PubMed]
28. Russier-Antoine, I.; Bertorelle, F.; Kulesza, A.; Soleilhac, A.; Bensalah-Ledoux, A.; Guy, S.; Dugourd, P.; Brevet, P.; Antoine, R. Chiral supramolecular gold-cysteine nanoparticles: Chiroptical and nonlinear optical properties. *Prog. Natur. Sci. Mater. Int.* **2016**, *26*, 455–460. [CrossRef]
29. Sun, L. Functional Gold Nanoparticle−Peptide Complexes as Cell-Targeting Agents. *Langmuir* **2008**, *24*, 10293–10297. [CrossRef] [PubMed]
30. Chen, C.; Wang, W.; Ge, J.; Zhao, X.S. Kinetics and thermodynamics of DNA hybridization on gold nanoparticles. *Nucl. Acid Res.* **2009**, *37*, 3756–3765. [CrossRef]
31. Chang, T.L.; Tsai, C.Y.; Sun, C.C.; Uppala, R.; Chen, C.C.; Lin, C.H.; Chen, P.H. Electrical detection of DNA using gold and magnetic nanoparticles and bio bar-code DNA between nanogap electrodes. *Microelectron. Eng.* **2006**, *83*, 1630–1633. [CrossRef]
32. Lipka, J.; Semmler-Behnke, M.; Sperling, R.A.; Wenk, A.; Takenaka, S.; Schleh, C.; Kissel, T.; Parak, W.J.; Kreyling, W.G. Biodistribution of PEG-modified gold nanoparticles following intratracheal instillation and intravenous injection. *Biomaterials* **2010**, *31*, 6574–6581. [CrossRef] [PubMed]
33. Cho, W.S.; Cho, M.; Jeong, J.; Choi, M.; Han, B.S.; Shin, H.S.; Hong, J.; Chung, B.H.; Jeong, J.; Cho, M.H. Size-dependent tissue kinetics of PEG-coated gold nanoparticles. *Toxicol. Appl. Pharmacol.* **2010**, *245*, 116–123. [CrossRef]
34. Kolska, Z.; Valha, P.; Slepička, P.; Švorčík, V. Refractometric study of systems water-poly (ethylene glycol) for preparation and characterization of Au nanoparticles dispersion. *Arabian J. Chem.* **2019**, *12*, 5019–5027. [CrossRef]
35. Ishii, T.; Otsuka, H.; Kataoka, K.; Nagasaki, Y. Preparation of functionally PEGylated gold nanoparticles with narrow distribution through autoreduction of auric cation by alpha-biotinyl-PEG-block-[poly(2-*N*, *N*-dimethylamino)ethyl methacrylate)]. *Langmuir* **2004**, *20*, 561–564. [CrossRef]
36. Lee, S.H.; Bae, K.H.; Kim, S.H.; Lee, K.R.; Park, T.G. Amine functionalized gold nanoparticles as non-cytotoxic and efficient intracellular siRNA delivery carriers. *Int. J. Pharm.* **2008**, *364*, 94–101. [CrossRef] [PubMed]

37. Aziz, M.A.; Kim, J.P.; Oyama, M. Preparation of monodispersed carboxylate-functionalized gold nanoparticles using pamoic acid as a reducingand capping reagent. *Gold Bull.* **2014**, *47*, 127–132. [CrossRef]
38. Chen, P.; Selegård, R.; Ailic, D.; Liedberg, B. Peptide functionalized gold nanoparticles for colorimetric detection of matrilysin (MMP-7) activity. *Nanoscale* **2013**, *5*, 8973–8976. [CrossRef]
39. Bastis, N.G.; Sanchez-Tillo, E.; Pujals, S.; Farrera, C.; Kogan, M.J.; Giralt, E.; Celada, A.; Iloberas, J.; Puntes, V. Peptides conjugated to gold nanoparticles induce macrophage activation. *Mol. Immunol.* **2009**, *46*, 743–748. [CrossRef]
40. Javier, D.J.; Nitin, N.; Levy, M.; Ellington, A.; Richards-Kortum, R. Aptamer-targeted gold nanoparticles as molecular specific contrast agents for refelectance imaging. *Bioconjugate Chem.* **2008**, *19*, 1309–1312. [CrossRef]
41. Kim, J.H.; Jang, H.H.; Ryou, S.M.; Kim, S.; Bae, J.; Lee, K.; Han, M.S. A functionalized gold nanoparticles-assisted universal carrier for antisense DNA. *Chem. Commun.* **2010**, *46*, 4151–4153. [CrossRef]
42. Neděla, O.; Slepička, P.; Švorčík, V. Surface Modification of Polymer Substrates for Biomedical Applications. *Materials* **2017**, *10*, 1115. [CrossRef]
43. Slepicka, P.; Slepickova Kasalkova, N.; Siegel, J.; Kolska, Z.; Bacakova, L.; Svorcik, V. Nano-structured and functionalized surfaces for cytocompatibility improvement and bactericidal action. *Biotechnol. Adv.* **2015**, *33*, 1120–1129. [CrossRef] [PubMed]
44. Slepicka, P.; Siegel, J.; Lyutakov, O.; Slepickova Kasalkova, N.; Kolska, Z.; Bacakova, L.; Svorcik, V. Polymer nanostructures for bioapplications induced by laser treatment. *Biotechnol. Adv.* **2018**, *36*, 839–855.
45. Slepička, P.; Malá, Z.; Rimpelová, S.; Švorčík, V. Antibacterial properties of modified biodegradable PHB non-woven fabric. *Mater. Sci. Eng. C* **2016**, *65*, 364–368. [CrossRef]

MDPI AG
Grosspeteranlage 5
4052 Basel
Switzerland
Tel.: +41 61 683 77 34

Nanomaterials Editorial Office
E-mail: nanomaterials@mdpi.com
www.mdpi.com/journal/nanomaterials

www.ingramcontent.com/pod-product-compliance
Lightning Source LLC
Chambersburg PA
CBHW061546210926
51850CB00009B/2011
* 9 7 8 3 7 2 5 8 5 3 8 9 2 *